Oil Recovery and Extraction

Oil Recovery and Extraction

Edited by **Andy Margo**

SYRAWOOD
PUBLISHING HOUSE

New York

Published by Syrawood Publishing House,
750 Third Avenue, 9th Floor,
New York, NY 10017, USA
www.syrawoodpublishinghouse.com

Oil Recovery and Extraction
Edited by Andy Margo

International Standard Book Number: 978-1-68286-103-5 (Hardback)

Printed in the United States of America.

Contents

Preface

Every book is initially just a concept; it takes months of research and hard work to give it the final shape in which the readers receive it. In its early stages, this book also went through rigorous reviewing. The notable contributions made by experts from across the globe were first molded into patterned chapters and then arranged in a sensibly sequential manner to bring out the best results.

In contemporary times, oil is one of the most crucial non-renewable sources of energy. Exploration and extraction of oil is a complex process. The task of managing oil resources is challenging, thus it becomes important to discover and implement innovative technologies. This book covers all the important areas within petroleum science and crude oil exploration such as reservoir simulations, subsurface analysis and drilling technology. It outlines various advanced techniques of extraction and the cost involved. The aim of this book is to serve as a great source of information for students, geoscientists, researchers and engineers engaged in the petroleum industry.

It has been my immense pleasure to be a part of this project and to contribute my years of learning in such a meaningful form. I would like to take this opportunity to thank all the people who have been associated with the completion of this book at any step.

Editor

Experimental study on the effects of kerosene-doped gasoline on gasoline-powered engine performance characteristics

O. Obodeh* and N. C. Akhere

Mechanical Engineering Department, Ambrose Alli University, Ekpoma, Edo State, Nigeria.

This investigation was carried out to study the engine-out emissions from a four-stroke, four-cylinder, water-cooled spark ignition (SI) engine with kerosene blended gasoline with different proportions of kerosene ranging from 0 - 50% by volume in step of 10%. Gaseous exhaust emissions were measured with the aid of pocket gas [TM]-portable gas analyzer. The experimental results showed that the engine-out emissions increase with increase concentration of kerosene in the blend. The analyses gave increase ranging from 21.7 - 53% for carbon monoxide (CO), 23.4 - 57.1% for unburnt hydrocarbon (HC) and 2.4 - 8.2% for particulate matter (PM). The experiment also showed increase in specific fuel consumption (SFC) for all load conditions ranging form 34 - 36%. Measures needed to reduce fuel adulteration were suggested. The measures encompass effective monitoring mechanism and enforcement of heavy penalty on sale of adulterated fuels.

Key words: Kerosene-gasoline blend, gasoline engine, engine-out emissions.

INTRODUCTION

Where different products of comparable qualities have different prices or consumers have no efficient tools to distinguish similar products of different qualities, unscrupulous operators will always try to exploit the situation for illegal profits. Illegal practices in the retail business is a global phenomenon, and fuel adulteration is one of the major abuses along with under-dispensing products to customers, mislabeling the octane number of gasoline, labeling leaded gasoline as unleaded and forging customs declarations or smuggling fuel to avoid or reduce excise duty payments (NNPC, 2007). These practices lead to losses in several areas, which include damaging engines and worsening air quality (Ale, 2003; Fonseca et al., 2007; Kamil et al., 2008). Evading fuel taxes reduces government revenue. Under-dispensing supplies to consumers lead to consumer losses. Doping gasoline with solvents and other chemicals can leave harmful deposits in engines (Biswas and Ray, 2001).

In Nigeria, the adulteration of gasoline is normally indulged primary due to the significant price differential between products. Adulteration is defined as the introduction of foreign substance into gasoline illegally or unauthorized with the result that the product does not conform to the requirements and specifications of the product (NNPC, 2007). The foreign substances are also called adulterants which when introduced alter and degrade the quality of the base transport fuels. Gasoline is a major transport fuel in Nigeria. Adulteration of the fuel at the point of sale and during transportation has become an acute problem in the country (Igbafe and Ogbe, 2005). Transport fuels (gasoline and diesel) are often adulterated with other cheaper products or byproduct or waste hydrocarbon stream for monetary gains. For example, gasoline is widely adulterated with kerosene. With large number of adulterants available in the market, both indigenous and imported, the magnitude of the problem of fuel adulterations has grown into alarming proportions in the past few years.

The poor in Nigeria depend on kerosene for their coo-king energy needs. Kerosene is subsidized to address

---

*Corresponding author. E-mail: engobodeh@yahoo.com.

the energy accessibility issues of the poor especially the cooking energy needs (Igbafe and Ogbe, 2005). Currently, gasoline is sold in major cities for ₦65 per liter while kerosene is sold in Mega-Filling stations at ₦50 per liter. The price differentiate has resulted in illegal diversion of kerosene meant for cooking energy needs of the poor towards adulteration of transport fuels.

Stringent norms are being advocated all over the world to reduce emissions from transport systems (Gupta, 2001; Mohan et al., 2006). However no strategy for emission reduction can be successful without addressing the issue of fuel adulteration (Mohan et al., 2006). The objective of this study is to analyze and identify the different exhaust emissions from the combustion of kerosene-doped gasoline and compare the emissions with that of pure gasoline operation.

MATERIALS AND METHODS

A four-stroke, four-cylinder, water-cooled, spark-ignition engine of brake power 60 kW at rated speed of 4600 rpm was used. The engine specifications are as shown in Table 1.

Fuels (gasoline and kerosene) were bought from a major oil marketer (Texaco Filling Station, Ekpoma). The chemical compositions of fuels used are presented in Table 2.

For the mixture preparation of gasoline and kerosene, six sets of sample mixture were prepared in 100: 00, 90:10, 80: 20, 70: 30, 60: 40, 50: 50 ratios.

Tests of engine performance on pure gasoline (100: 00) were conducted as a basis for comparison. The engine was run on "no load" condition and its speed adjusted to 4500 rpm by adjusting the fuel control valve. The test engine was run to attain uniform speed and then it was gradually loaded. The experiments were conducted at five load conditions of 0 (no load), 25, 50, 75 and 100% of the rated load. For each load condition, the engine was run for at least 20 min. The experiments were repeated with other samples namely: 90:10, 80:20, 70:30, 60: 40, and 50: 50 ratios.

The speed and load of the engine were controlled independently by the fuel control system and dynamometer. Air flow rate was measured using a laminar flow element and fuel flow rate was measured using a positive displacement meter. DIGICON model DT-240P non-contact tachometer with the range from 5 - 10,000 rpm and resolution of 1 rpm was used to measure the engine speed. Gaseous exhaust emissions were measured with the aid of pocket gas [TM]-portable gas analyzer. During the experiments, the average ambient temperature and atmospheric pressure were recorded as 30 °C and 756 mmHg respectively.

RESULTS AND DISCUSSION

The variation of carbon monoxide (CO) emissions with different proportions of kerosene by volume in the kerosene-gasoline blend at different load conditions is shown in Figure 1.

As the concentration of kerosene in gasoline increases, the value of CO increases, for 10% adulteration it was 21.7% and for 50% mix, the value was 53%. This is due to incomplete combustion of fuel owing to higher density and viscosity with poor volatile property of kerosene when compared to pure gasoline (Usha et al., 2003; Igbafe and Ogbe, 2005).

Figure 2 illustrates the percentage change of hydrocarbon (HC) emissions due to adulteration of gasoline with kerosene.

On adulteration with kerosene, HC increases significantly. On 10% adulteration, the percentage increase was 23.4% while that for 50% adulteration was 57.1%. This is due to increase in quenching effect with poor volatility of kerosene when compared to pure gasoline. Kerosene, being a mixture of low volatility, high molecular weight hydrocarbons ($C_{10}H_{22}$ to $C_{16}H_{34}$) than gasoline (C_5H_{12} to C_9H_{34}) is more prone to emit more HC and CO in the exhaust of spark ignition (SI) engine due to less effective combustion (Muralikrisha et al., 2006).

Consequences of incomplete combustion of fuel are increased absorption of heavier HC components in engine oil film which escapes the combustion process. Increased portion of heavier HC components remain in liquid phase and may escapes the combustion process. This contributes to photochemical smog (Perola et al., 2003; Ghose et al., 2004).

The particulate matter (PM) emissions increase with increase in percentage of kerosene in the blend as shown in Figure 3.

On 10% adulteration, the percentage increase was 2.4% while that for 50% adulteration was 8.2%. The in-cylinder liquid fuel (droplets or pool), if ignited by the flame, produces PM via formation, growth and pyrolysis of polycyclic aromatic hydrocarbons (PAHs). PAHs are also adsorbed on PM (Perola et al., 2003). PAH increases with increase in the naphthene content of the fuel (Table 2). In addition, the use of fuels with higher density results in higher emissions of PM and smoke (Heywood, 1988). This is due to the fact that PM emission is dependent of molecular weight and volatility of fuel (Heywood, 1988). The effects of blending gasoline with kerosene are increased density, decreased volatility and reduced octane rating (Fonseca et al., 2007). When gasoline was doped with kerosene in higher concentration, it was difficult to start the engine and there was also possibility of knock (knock is noise generated when auto-ignition of a portion of end gas takes place ahead of propagating flame). Since the octane number of kerosene is lower than that of gasoline, kerosene-adulterated gasoline will cause knocking of the engine (Fonseca et al., 2007). This was noticed when the engine was run with the adulterated fuel in the ratio of 50% kerosene and 50% gasoline. There is also possibility of carbon deposits on the spark plug, piston head and valves (Heywood, 1988).

The variation of specific fuel consumption (SFC) as a function of different proportions of kerosene by volume in the kerosene-gasoline blend at different load conditions is shown in Figure 4.

Increase in SFC for all load conditions ranges from 34 - 36% compared to pure gasoline operation. Increase in fuel consumption rate is attributed to lower heating value of kerosene as compared to pure gasoline (Biswas and Ray, 2001; Kamil et al., 2008). In other words, more fuel

Experimental study on the effects of kerosene-doped gasoline on gasoline-powered engine...

3

Table 1. Test engine specifications.

Parameter	Value
Make and model	2.0 SLX Nissan Gasoline
Year of manufacture	1988
Type	4-Stroke, in-line
Number of cylinder	4
Bore	88 mm
Stroke	82 mm
Displacement	1994 mm^3
Compression ratio	8.2:1
Air induction	Naturally aspirated, water cooled
Valves per cylinder	4
Number of plugs	4
Maximum power	60 kW at 4600 rpm
Maximum torque	144 Nm at 3000 rpm
Maximum speed	5000 rpm

Table 2. Chemical composition of fuel used.

Fuel	Paraffins (%vol.)	Naphthenes (%vol.)	Olefins (%vol.)	Aromatics (%vol.)
Gasoline	42	4	17	35
Kerosene	55	28	0	16

Source: Fonseca et al., 2007.

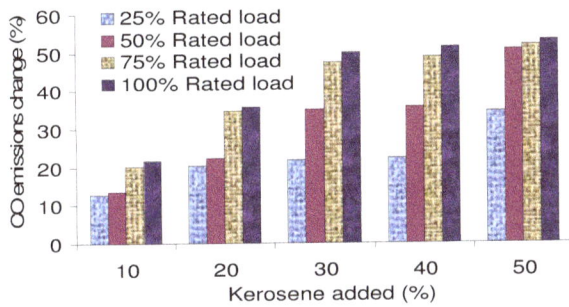

Figure 1. Variation of co-emission as a function of kerosene added.

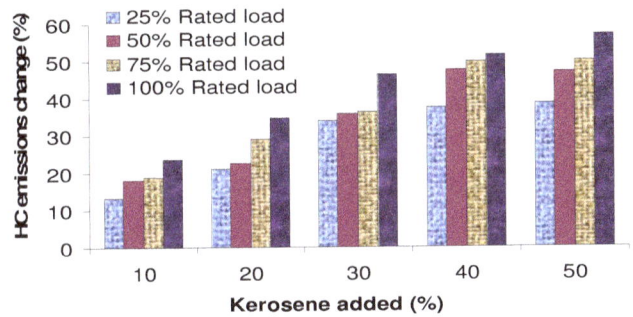

Figure 2. Variation of HC as a function of kerosene added.

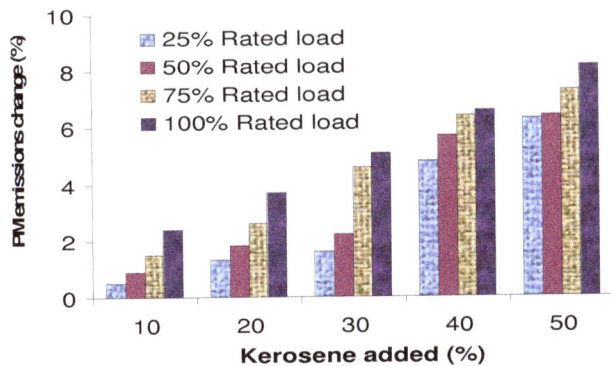

Figure 3. Variation of PM emission as a function of Kerosene added.

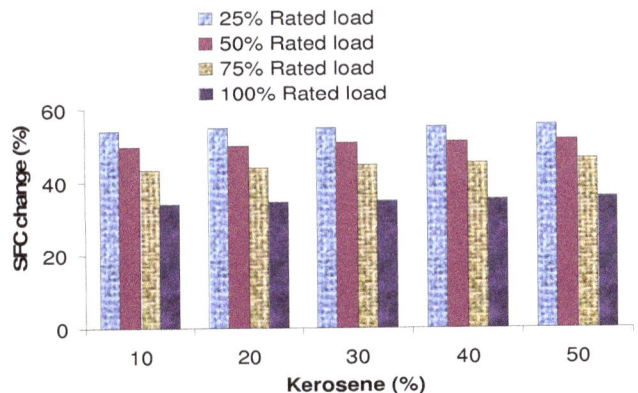

Figure 4. Variation of SFC as a function of kerosene added.

is needed in order to produce the same amount of energy. The consistent increase in the SFC with increase amount of kerosene in the fuel blends is due to more fuel supplied to the engine in order to maintain constant brake mean effective pressure.

Conclusion

An experimental investigation has been carried out to evaluate the effect of kerosene-doped gasoline on the engine-out emissions and performance of gasoline-powered engine. The experimental results showed that the engine-out emissions increase with the increase concentration of kerosene in the blend. The analyses gave increase ranging form 21.7 - 53% for CO, 23.4 - 57.1% for HC and 2.4 - 8.25 for PM. Increase in SFC for all load conditions ranges from 34 - 36% compared to pure gasoline operation. On 50% adulteration, it was a bit difficult to start the engine.

The experiment also showed that blending kerosene with gasoline increased knocking tendency drastically. This is due to the accumulation of the heavier fractions in the cylinders of the engine (Heywood, 1988).

To curb fuel adulteration, oil companies should carryout filter paper test, density checks, blue dyeing of kerosene. Oil companies and government agencies should carryout surprise and regular inspections of retail outlets with mobile laboratories. Heavy penalty on sale of adulterated fuels should be enforced in order to discourage fuel adulteration.

REFERENCES

Ale BB (2003). Fuel Adulteration and Tailpipe Emissions, J. Inst. Eng. 3(1): 12 - 16.

Biswas D, Ray R (2001). Evaluation of Adulterated Petrol-Fuels, Indian Chem. Eng. J. 43(4): 314 - 317.

Fonseca MM, Yoshida MI, Fortes ICP, Pasa VMD (2007). Thermogravimetric Study of Kerosene-Doped Gasoline, J. Therm. Anal. Calorim. 87(2): 499 - 503.

Ghose MK, Paul R, Benerjee SK (2004). Assessment of the Impact of Vehicle Pollution on Urban Air Quality, J. Environ. Sci. Eng. 46(1): 33 - 40.

Gupta A (2001). Fuel Adulteration- Complexities and Options to Combat, Proceedings of 4th International Petroleum Conference and Exhibition, New Delhi, October 10-12: 162-165.

Heywood JB (1988). Internal Combustion Engine Fundamentals, McGraw-Hill Book Co., New York pp. 915-916.

Igbafe AI, Ogbe MP (2005). Ambient Air Monitoring for Carbon monoxide from Engine Emission in Benin City, Nigeria, Afr. J. Sci. Technol. 1(2): 208 - 212.

Kamil M, Sardar N, Ansari MY (2008). Experimental Study on Adulterated Gasoline and Diesel Fuels, Indian Chem. Eng. J. 89(1): 23 - 28.

Mohan D, Agrawal AK, Singh RS (2006). Standardization for Automotive Exhaust Pollution: Some Issues in Indian Perspective, J. Inst. Eng. 86: 39 - 43.

Muralikrisha MVS, Kishor K, Venkata RD (2006). Studies on Exhaust Emissions of Catalytic Coated Spark Ignition Engine with Adulterated Gasoline, J. Environ. Sci. Eng. 48(2): 97 - 102.

NNPC (2008). Warri Refining and Petrochemical Co. LTD, Technical Report 4: 74 - 76.

Perola VC, Zacarias D, Pires AF, Pool CS, Carvlho RF (2003). Measurements of Polycyclic Aromatic Hydrocarbons in Airborne Particles from the Metropolitan Area of Sao Paulo City, Brazil, Atmospheric Environ. 37(21): 3009 - 3018.

Usha MT, Srinivas T, Ramakrishna KA (2003). Study on Automobile Exhaust Pollution with Regard to Carbon monoxide Emissions, Nat., Environ. Pollut. Technol. 2(4): 473 - 474.

A comparative study of recycling of used lubrication Oils using distillation, acid and activated charcoal with clay methods

Udonne J. D.

Department of Chemical and Polymer Engineering, Lagos State University, Lagos, Nigeria.
E-mail: udonne.joseph@gmail.com.

Lubricating oils are viscous liquids used for lubricating moving part of engines and machines. Since lubricating oils are obtained from petroleum – a finite product, and with dwindling production from world oil reserves, the need arises more than ever, to recycle used lubricating oils. Accordingly, this research effort focuses on comparative study of four methods of recycling of used lubrication oils: acid/clay treatment, distillation/clay, acid treatment and activated charcoal/clay treatment methods. Test carried out on the recycled lubrication oil include: flash point, pour point, specific gravity, metal contents, viscosity and sulphur contents. The results from the tests showed that, viscosity increased from 25.5 for used lube oil to 86.2 for distillation, 89.10 for acid/clay treatment and 80.5 is for activated/clay treatment. This is compared with 92.8 cs for fresh lube oil. Other results from the different tests showed varied degrees of improvement with the best results obtained using the acid/clay treatment.

Key words: Lubrication, refining, environmental pollution, viscosity.

INTRODUCTION

Lubricating oils from petroleum consists essentially of complex mixtures of hydrocarbon molecules. They are mostly composed of isoalkanes having slightly longer branches and the monocycloalkanes and monoaromatics which have several short branches on the ring (Cutler, 2009). These hydrocarbon molecules generally range from low viscosity oils having molecular weights as low as 250, up to very viscous lubricants with molecular weight as high as 1000 (Concawe, 1985). The carbon atoms range from 20 to 34. Lubricating oils are viscous liquid and are used for lubricating moving parts of engines and machines. Grease, which is a semi-solid, also belongs to this group. There are three major classes of lubricating oils, namely: lubricating greases, automotive oils and industrial lubricating oil.

When lubricating oils are used in service, they help to protect rubbing surfaces and promote easier motion of connected parts. In the process, they serve as a medium to remove high build up of temperature on the moving surfaces. Further build up of temperature degrade the lubricating oils, thus leading to reduction in properties such as: viscosity, specific gravity, etc. Dirts and metal parts worn out from the surfaces are deposited into the lubricating oils. With increased time of usage, the lubricating oil loses its lubricating properties as a result of over-reduction of desired properties, and thus must be evacuated and a fresh one replaced. With the large amount of engine oils used, the disposal of lubricating oils has now become a major problem. Many nations are now addressing the problem of environmental pollution posed by waste or used lubricating oils in their countries (Cooke, 1982). In USA, for example, about 2 billion gallons of oils are generated annually (Coyler, 2000). This has led industries and governments for find satisfactory solutions that will reduce the contribution of used lubrication oil to pollution and also recover these valuable hydrocarbon resource (Whisman et al., 1978).

In disposing used oil, many people use it as a dust cure; that is, for dust prevention (Bennet et al., 1960). This method of disposal is in many ways unsatisfactory as the lead-bearing dust and run-off, constitute air and water pollution. Another method by which used oil is being disposed is by incineration. This method represents another poor use of such a valuable product, and the attendant emission of probably carcinogenous products, contribute to environmental pollution (Georgel and La

Tour, 1977). Recycling of used lubricants is now attracting more attention than before. This is partly because of the fear of dwindling of world oil reserves and more as a result of the environment concern which it posses. The following three distinctive reasons explain the interest in the re-cycling of waste lubricating oils (Iarc, 1984):

1. The need to conserve crude reserves.
2. Minimizing unemployment through the building/construction of used lubricating oil recycling plant.
3. The elimination of environment pollution source of used lubricant.

The recycling of used lubricants has been practiced to various degrees since the 1930s and particularly during the Second World War when the scarcity of adequate supplies of crude oil during the conflict encouraged the reuse of all types of materials including lubricants (Asseff, 1961). Environmental considerations regarding the conservation of resources have maintained interest in the concept of recycling up to the present day. The reclamation of spent crankcase oils is now a subject of pressing national interest in some countries. On the other hand, pollution by used lubrication oils is recognized now to account for greater pollution than all oil spills at sea and off-shore put together. Some countries have petroleum storage and spent oils represent a precious commodity, which must not be wasted. It was because of this, the conservation of petroleum resources have been declared national policy for several countries and the benefit of wise resource management are obvious (Mortier and Orszulik, 1994). Recent impetus on waste recovery leads to renewed interest in re-distilling to convert this used lubricating oil into useful original lubricating oils.

Re-distilling or re-fining is the use of distilling or refining processes on used lubrication oil to produce high quality base stock for lubricants or other petroleum products. The use of this method has increased tremendously in the developed countries, in some countries reaching up to 50% of the countries' need for lubricating oil (Thrash, 1991) and there are different methods developed by different countries in the western countries, on how to refine used lubricating oil for reuse. The basic principle remains the same and utilizes many of the following basic steps:

Removal of water and solid particles by settling.
Sulphuric acid treatment to remove gums, greases, etc.
Alkaline treatment to neutralize acid.
Water washing to remove "soap".
Clay contacting to bleach the oil and absorb impurities.
Striping to drive off moisture and volatile oils.
Filtering to remove clay and other solids.
Blending to specification.

The objective of this research is centered mostly on obtaining a high quality production of lubricating oil from used lubricating oil by the re-refining of used lubricating oil, thereby reducing environmental pollution and also minimizing importation of lubrication oil.

Literature survey

It is essential to recognize that, all used oils should be collected for controlled disposals. Some products, such as transformer oils and hydraulic oils, can be readily collected from large industrial concerns, regenerated to a recognized standard and returned to the original source. Oil from the automotive sources will include mono and multi-grade crankcase oils from petrol and diesel engines, together with industrial lubricants that have been inadequately segregated may also be included (Gergel and La Tour, 1977). Lubrication oil is used to provide a film between the moving parts of machine and engines to prevent wear with little or no loss of power. The conventional steps in lubricating oil manufacture are pretreatment of the crude oil charge, followed by distillation of the crude in two steps (an atmospheric tower and vacuum tower), deasphalting (as required by the nature of the crude oil charge), dewaxing, solvent extraction, filtering and blending including mixing various additives with the final lubricating oil (Bromilow, 1990).

The prime objective in the production of lubricating oil is the separation of wax distillate and cylinder stock without any decomposition or cracking of the lubrication oil fractions, thus a vacuum distillation unit is used to separate the wax distillate and the bottom stock at a lower temperature. The properties which make the high boiling paraffin hydrocarbon suitable for lubricating manufacture include stability at high temperatures, fluidity at low temperature, only a moderate change in viscosity over a broad temperature range and sufficient adhesiveness to keep it in place under high shear forces. The desired fractions for the manufacture of the lubricating oil have high boiling points and its separation into various boiling points range cuts must be accomplished under reduced pressure. The vacuum tower produces some fuel oil overhead which is sold as a separate product or sent to another area of the refinery for further processing and blending. The two main products from the vacuum tower are wax distillate and cylinder stock which is the bottom product. Both streams contain desirable lubricating oil constituents as well as by-products. The wax distillate is charged directly to the dewaxing unit. The vacuum tower bottoms, or cylinder stock are charged to deasphalting unit. These two fractions from the basic stock for lubricating oil manufacture (Hamad, 2005).

General characteristics of lubricating oils

All lubricants are characterized by some properties,

A comparative study of recycling of used lubrication Oils using distillation, acid and activated charcoal...

7

which are very peculiar to them. Such properties include: viscosity index, cloud point, flash point, pour point, total base number (TBN), ash content, water content, corrosive properties, relative density, insolubles, total acid number (TAN) etc.

Viscosity

Viscosity is defined as the force acting on a unit area where the velocity gradient is equal at a given density of the fluid. Viscosity is strongly depending on the temperature. With increasing temperature, the viscosity has to be stated for a certain temperature. The most important fluid characteristic of a lubricant is its viscosity under the operation condition to which it is subjected in the unit. It is the characteristic of a liquid which relates a shearing stress to the viscosity gradient it produces in the liquid.

Lubrication oils are identified by Society of Automotive Engineers (SAE) number. The SAE viscosity numbers are used by most automotive equipment manufacturers to describe the viscosity of the oil they recommend for use in their products. The greater or higher the SAE viscosity numbers, the more viscous or heavier is the lubricating oil (Scapin, 2007). Viscosity numbers are given in terms of saybolt second universal, SSU. The addition of certain additives is for the improvement of viscosity-temperature characteristics.

Specific gravity (Density)

From ordinary theory, we know that density of a substance is equal to the mass of a substance divided by the volume of the substance, that is:

$$\text{Density (d)} = \frac{\text{Mass (m)}}{\text{Volume (v)}}$$

Specific gravity is the ratio of the density of the material to density of the equal volume of water. The temperature at which the density is been measured must be known for density changes as temperature changes.

Cloud point and pour point

Cloud and pour points, ASTM D97 – 47. Cloud point is the temperature at which paraffinic wax and other oil is cooled under a given condition. Pour point is not a measure of the temperature at which the oil ceases to flow under service conditions of a specific system. It is very important to users of lubricants in low temperature environment.

Water content

Water found in lubricating oil in service depends on

where the automobile is being used. In almost all system, traces of water in the lubricant are unavoidable, arising from such sources as leaking oil coolers, engine cooling system leaks and in all types of machinery, from atmospheric condensation. Accordingly, the water content must not exceed the "action" levels (more than 0.5) recommended for the different grades of oil and application. In places where there are bad roads and drainage system, one is bound to see water as part of contaminants of the oil. The water in the radiator may also contribute to the presence of water in lubricating oils in use. The presence of excessive water contamination will affect the viscosity of the oil and this may give rise to emulsion formation and can also lead to gear tooth and bearing problems.

Flash point

Flash point is the minimum temperature at which an oil gives off sufficient vapours to form an explosive mixture with air. The flash point test gives an indication of the presence of volatile compounds in oil and is the temperature to which the oil must be heated under specific conditions to give off sufficient vapour to form a flammable mixture with air. There are various methods of determining flash point of oils as contained in ASTM (Art, 2010). Flash point, open cup is the temperature at which a flash appears on the surface of the sample when a small flame of specified size is passed across the cup at regular temperature intervals while the oil in the cup is being heated at a specified rate.

Fire point

Fire point and flash point are significant in cases where high temperature operations are encountered. The fire point, open cup is the temperature at which the oil ignites and continues to burn for at least 5 seconds. Fire point is obtained as a continuation of the flash point test. The fire and flash points by Cleveland open cup, ASTM D92 – 56.

METHODOLOGY

The following materials were used in the recovery of the used lubricating oil vis-à-vis concentrated sulphuric acid, caustic soda, clay, activated charcoal, ethyl acetate. The apparatus and equipment used include the following:

Bunchner funnel
Filtering flask
Seperating funnels
Distillation column
Flash point tester
Thermometer (360°C)
Erlenmeyer flask (250 ml)
Beaker
Measuring cylinder

Vacuum pump
Sulphur content analyzer
Atomic absorption spectroscopy (AAS)
Coiled condenser
Ubbelohde viscometer
Erlenmeyer flask
Round bottom flask

The methods used in the re-refining of the used lubricating oils include methods such as, filtration to remove impurities, acid treatment, acid/clay treatment, and distillation/clay treatment and activated charcoal/clay treatment.

Collection of test samples

The test samples of used lube oils (premium motor oil SAE-40) and (Quartz 2000-SAE 40) were collected from an oil service station. The used lubrication oil were collected from used oil dumps of a car mechanic.

Experimental procedure

The procedure for the purification of the used lubrication oil consist of filtration of the oil before subjecting to treatment using: acid/clay, distillation/clay, acid and activated charcoal/clay treatment methods. The used lubricating oil was filtered to remove impurities such as metal chips, sand, dust, particles, micro impurities, that are contained in the lube base oil. This was done using a funnel with a filter paper placed in it, then a vacuum pump was connected to the filtering flask to which the funnel was fixed with the aid of a rubber stopper. 2 L of the used lube oil were filtered for the two samples collected respectively. For the acid method, the used lubricating oil was stirred thoroughly to promote homogeneity from the stock 450 ml was measured out and transferred to a beaker and 150 ml of gasoline was also measured and added to the oil.

The lube base oil mixture was transferred to the bucker of the centrifuge and centrifuged at 1500 rpm for minutes. It was then left to settle for another 10 min before decanting into a beaker. The decanting liquid mixture was distilled; to remove water, gasoline and any other liquid that may be present. The content in the flask was cooled and treated with 10 ml of 98% conc.H_2SO_4 in a separating funnel with the mixture strongly agitated. It was then allowed to settle for 48 h after which two layer/phases were formed. The sludge was removed from the bottom of the separating funnel. After which 100 ml solution of 10% NaOH was added to neutralize the acid. It was then allowed to settle for about 30 min without agitation. The alkaline phase, which is formed at the bottom, was removed and the lube oil washed with hot water 2 times (15 ml). The oil was heated with an elemantle burner while connected with a vacuum pump.

For the distillation/clay method, the sample was made moisture free by first carrying out atmospheric distillation. 200 ml of the used oil was poured into the vacuum distillation flask. The pressure was increased slightly to subside the foam encountered. Mild heating was gently applied to remove the dissolved gases. The distillate obtained was weighed leaving a very dark and waxy residue. This lube oil stream was treated by packing 30 g of clay in a plugged funnel. (the funnel inserted with a filter paper) and allowing the lube oil to pass through the bed of clay was disposed off.

For the acid/clay method, the used lube oil was allowed to settle for 12 h. The sample was further filtered by centrifugation for 20 min at 1000 rpm. The suspended particles settled in the used oil at the bottom of the flask and the liquid portion was decanted off. The decanted liquid was thermally pretreated to degrade some of the additives and reduce the workload of the acid. 100 ml of pretreated oil was measured in a separating funnel and treated with 10 ml of

98% conc. H_2SO_4 (see details in acid treatment method). The remaining oil was clay treated using 30 g clay packed in a funnel with filter paper, which was then neutralized. In the activated charcoal/clay method, the waste base lube oil was stirred homogenously and 200 ml of the used lube oil was measured into a separating funnel containing 100 ml of ethyl acetate. The solution was left at room temperature for 24 h. After decanting from the paste the solution was treated with clay and activated charcoal. Filtration and evaporation was carried out under reduced pressure (see details in acid treatment).

Quality test

Flash point (ASTM D92)

10 ml of the re-refined lube oil was introduced into a 100 ml beaker and then a thermometer inserted. A beaker was placed on a bunsen burner. A flame source was brought at intervals to determine the temperature at which a flash appears on the surface of the sample while the lube oil in the beaker was heated.

Pour point (ASTM D97)

20 ml of the lube oil sample was introduced into a container. The lube oil sample was chilled at specific rate; certain paraffin hydrocarbon (in the form of wax) will begin to solidify and separate out in crystalline form. The temperature at which this occurs is known as cloud point. Further chilling was continued until lube oil stop to flow. The temperature this occurred, is called the pour point temperature.

Specific gravity (ASTM D941-55)

Specific gravity is the ratio of the density of the material to the density of equal volume of water. This was measured using the hydrometer. The density was observed at 60°F and the value recorded.

Metal content

Metal content in re-refined lube oil was determined using the atomic absorption spectrometry (AAS). Sample containing the metal to be analysed was dissolved in water, if insoluble, digesting in the acid dissolved it. The cathode lamp for the element was put in position and the element characteristic wavelength selected using wavelength selector standard solutions of the element to be determined are first prepared and their absorbency measure at selected wavelength.

Viscosity (ASTM D445)

This was obtained with the viscometer. The re-refined samples obtained by various methods. The fresh oil and the used lube oil were heated one after the other to attain a temperature of 100°C. The bulb of the clean viscometer was then filled with the hot oil (at 100°C) to the mark while immersed in a thermostat.

Sulphur content

This test is known or called sulphur analysis. 1 ml of the test sample was introduced into a burette and attached to the spectrophotometer clip. When the position is achieved, the radiation bottom

Table 1. Results of tests using the various refining methods.

Parameter	Fresh lube oil	Used lube oil	Distillation/clay treatment	Acid/clay treatment	Acid treatment method	Activated charcoal/clay treatment
Water content v/v	< 0.20	13.70	0.66	0.40	0.60	0.47
Specific at gravity at 60 °F	0.90	0.91	0.86	0.88	0.86	0.86
KV at 100 °F(Cs)	82.20	61.60	84.10	82.00	84.20	80.20
Viscosity index	92.80	21.10	85.80	88.90	84.40	86.80
Flash point °C	188.00	120.00	168.00	182.00	170.00	178.00
Pour point	-9.00	-35.00	-16.00	-11.00	-15.00	-13.00
Sulphur content	-	0.80	0.046	0.04	0.043	0.042
Iron (ppm)	-	22.50	10.30	2.60	10.50	9.50

is allowed to emit its rays from X ray tubes. The excitation effect is proportional to the concentration of sulphur in the sample.

RESULTS

The results of acid treatment, distillation/clay treatment and activated charcoal/clay treatment with their respective quality test and discussion of these results are given here.

Flash point

The flash points of this oil are 188°C for fresh oil, 100°C for used oil, acid treatment 168°C, 182°C for distillation/clay treatment, 170°C for acid/clay treatment and 178°C for activated charcoal/clay treatment. The decrease in value of flash point for the used oil could be as result of the presence of light ends of oils (Rincon, 2005). In essence, after undergoing combustion and oxidation at high temperature of the combustion engine, the oil breaks down into component parts, which include some light ends.

From Table 1, the pour point of fresh lubricating oil is 188°C, while those obtained using: acid treatment methods, distillation/clay, activated charcoal/clay, are 184,176,172 and 180°C. The decrease in flash point for the used oil, was as a result of distillation with fuel; that is, for an automobile with bad piston rings, the flash point will decrease because of distillation with fuel (Firas and Dumitru, 2006). Hence, the flash point for various recovery methods used here are acceptable concerning the reference standard.

Specific gravity

The specific gravity of the used lube oil is higher than the refined one from the above results.

The results for the fresh and used lubricating oils are 0.8966 and 0.9006 respectively, while those re-refined oil obtained by the various methods were: 0.8595 for distillation, 0.8779 for acid/clay treatment, 0.8564 for acid

treatment and 0.8607 for activated clay respectively. The specific gravity for the used oil was higher than that for the refined oil. The specific gravity of a contaminated oil could be lower or higher than that of its virgin/fresh lube oil depending on the type of contamination (Chevron Lubricating oil FM ISO 100). If the used lube oil was contaminated due to fuel dilution and/or water originating from fuel combustion in the engine and accidental contamination by rain, its specific gravity will be lower than that of its fresh lube oil or the re-refined one.

Viscosity

Viscosity increase can occur due to oxidation or contamination with insoluble matter, from the table, we can see a decrease in kinetic viscosity of the used oil ; this is due to contamination in form of sludge in the used oil. In general, oil is considered unfit for service, if the original viscosity increases or decreases to the next SAE number. Viscosity increase can occur due to oxidation or contamination (Scapin, 2007). Viscosity decrease can be caused by dilution with light fuel. The result of the viscosity test shows that, the used lube oil has lost most of its viscosity due to contamination. However, treatment has restored most of its viscosity. The result shows also that refining using acid/clay method gave the highest viscosity. This can be attributed to the possible conversion of possible contaminants by the acid and removal by the clay from the lube oil. In view of the desirability of the oil to act as a coolant or heat transfer medium, it must be able to retain adequate body at elevated film temperature, yet adequate fluidity elsewhere in the system. This is ensured when the viscosity is above 80. Oil as treated using all four methods meet this specification, but the acid/clay method has an advantage over others.

Metal content

The engine block is made of aluminum, iron and lead, hence during combustion of fuel in the engine chamber, the wear of these metals in parts per million (ppm) are

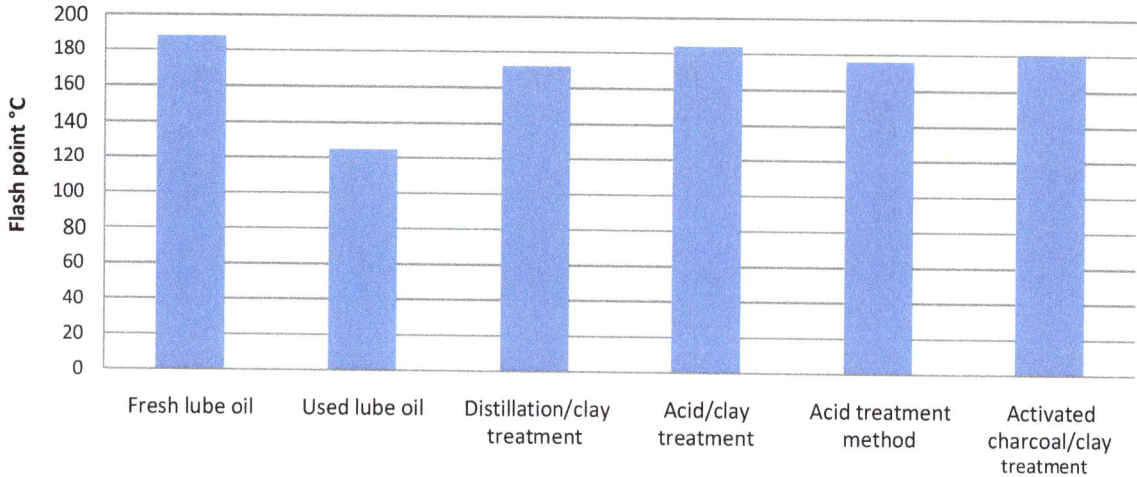

Figure 1. Effect of various refining methods on flash point of used lube oil.

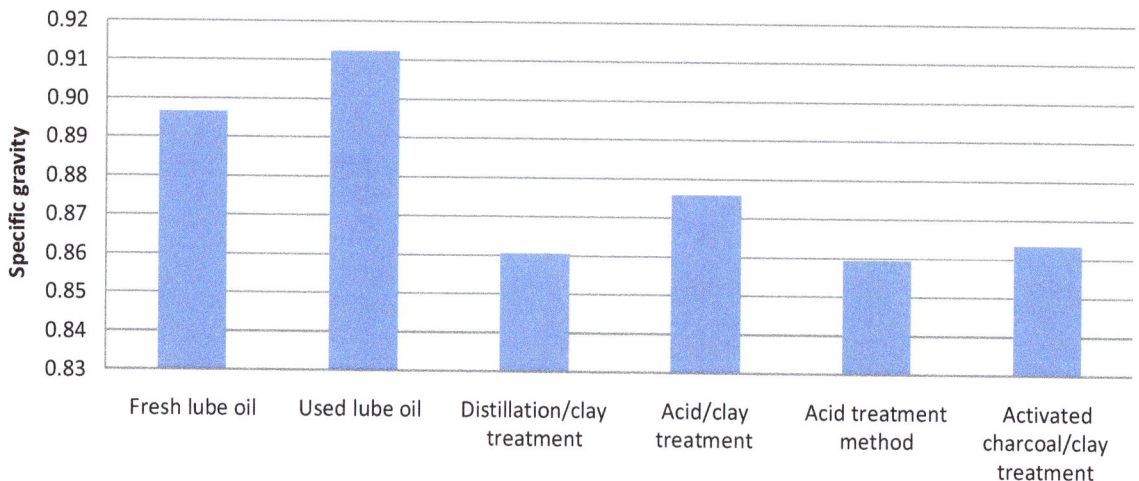

Figure 2. Effect of refining of used lube oil on its specific gravity.

found in the used oil. The wear of these metals is due to the corrosion caused by the presence of water and aided by fuel dilution due to bad piston rings (Art, 2010). One can say that, the recovery methods proved better yield when compared with that of acceptable refined base oil standard of individual metal contents. The result shows that, the acid/clay method had the least metal contents, likely due to conversion of these metals to other harmless products.

Sulphur content

The Figures 1 to 6 shows the values of sulfur content for the used lubricating oil and refined oil by the various methods. The fresh lube oil has an inherent anti-oxidant capacity. The sulphur content of used lube oil are high, this is due to the presence of wear caused between moving parts. Sulfur reacts with the metal to form

compounds of low melting point that are readily sheared without catastrophic wear. Corrosion in engines is caused by mineral acids formed by the oxidation of sulphur compounds in fuel in internal combustion engines with refined oils; those hydrocarbons that were inherently unstable will have been oxidized during use (Rincon, 2005). The function of sulphur content of refined base oil is not clear, more work is required to determine the concentration of sulphur required to meet minimum performance standards.

Pour point

From the results obtained for the used oil, pour point for the used lube oil is high. This is because of the degradation of additive in the lube oil. Pour point especially is of interest when an oil must be under relatively cold condition. Pour point will vary widely

A comparative study of recycling of used lubrication Oils using distillation, acid and activated charcoal...

11

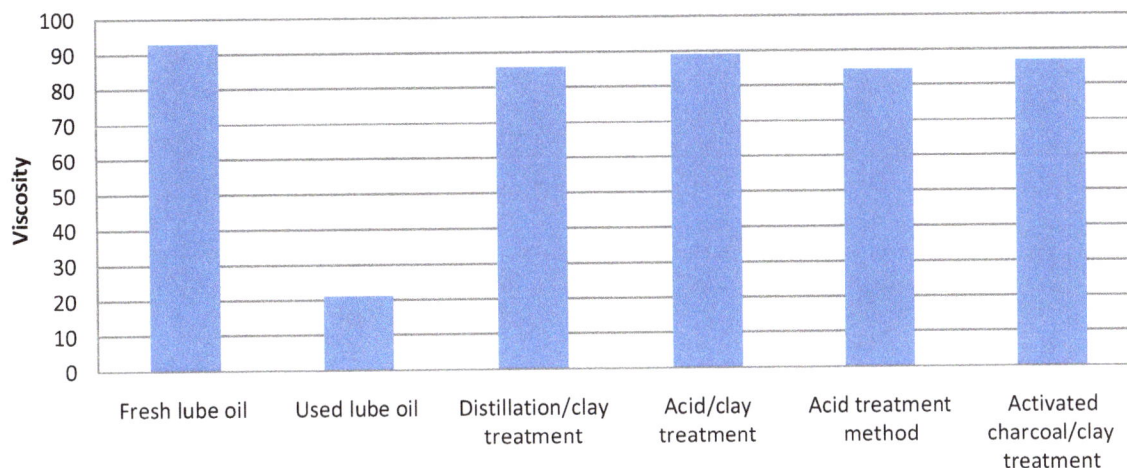

Figure 3. Effect of refining of used lube oil on its viscosity.

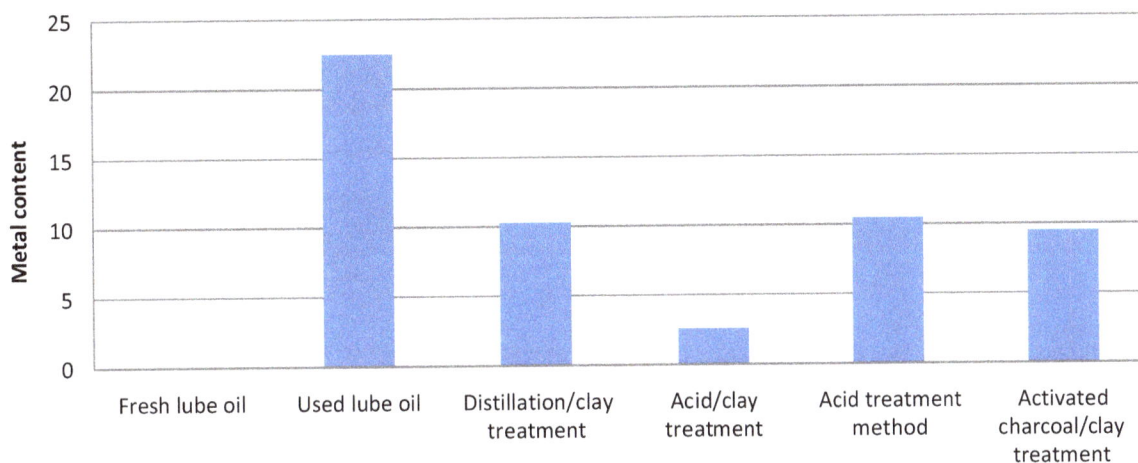

Figure 4. Effect of refining of used lube oil on metal contents.

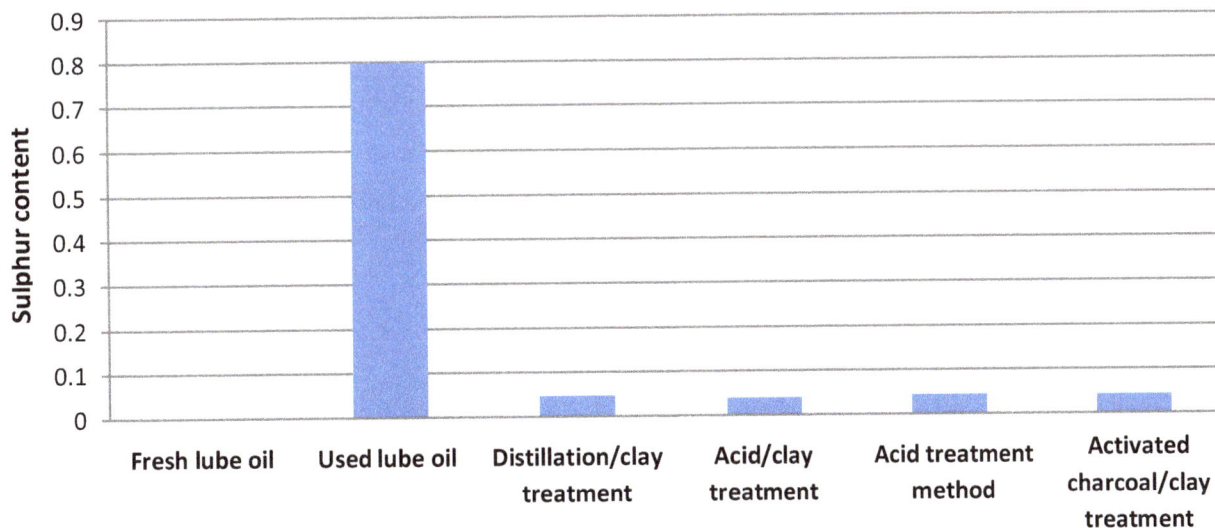

Figure 5. Effect of refining of used lube oil on sulfur contents.

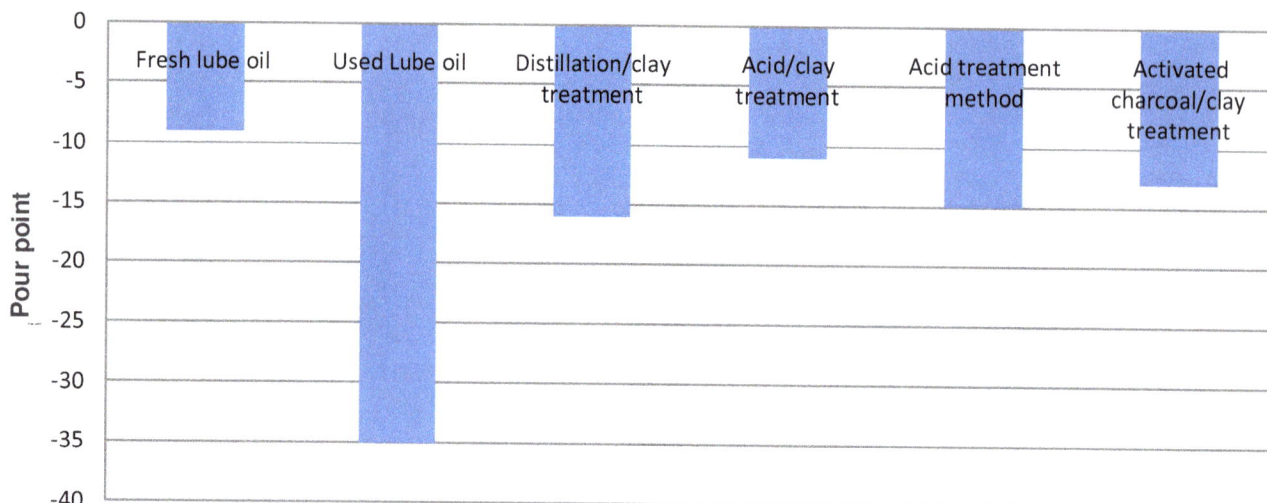

Figure 6. Effect of refining of used lube oil on its pour point.

depending on the base, the source of the lube oil and the method of refining, especially if dewaxing has been done (Firas and Dumitru, 2006; Chevron Lubricating oil FM ISO 100). Here, these recovery methods proved alright when compared with that of fresh lube oil.

Conclusion

From the results presented, it is clear that the four methods effectively removed contaminants from used lubricating base oil and returned the oil to a quality essentially equivalent to oils produced by fresh lube oil stocks. Overall, the refining method using acid/clay proves the best option, judging from the results obtained. It is envisaged that, it will cost less to procure the used lube oil and has proved to yield about 80% while the yield from crude oil is between 5 to 10%. The recycling of used lubrication oil will reduce environmental concern which it has so far posed. It will also reduce the demand for lubricant rich crude which is a finite resource.

REFERENCES

Cutler ET (2009). "Conserve Lube Oil: Re-refine" Hydrocarbon processing Vol.265, March.
Concawe (1985). "The collection, Disposal and Regeneration of waste oil and related Materials. Concawe Report". No. 85/53.
Cooke VB (1982) "The Role of Additive in the Automobile Industry, ASLE.
Coyler CC (2000). "Gasoline Engine Oils: Performance, Evaluation and Classification", 10th World petroleum Congress, Moscow, p. 112.
Whisman ML, Reynolds JW, Goetzinger JE, Cotton FO (1978). "Re-refining Makes quality Oils". Hydrocarbon Process., pp. 141-145

Bennet AY (1960) "An Engine Test for predicting the performance of engine Lubricants".
Georgel WC, La TGG (1977). "Extended Engine Oil Life Through New Technology", Seminar paper at the National Petroleum Refiners Association 1977 Annual meeting, Mar 27 – 29, San Francisco, Carlifornia.
Iarc (1984) Iarc Monographs "Risk of chemicals to humans", vol 33, IARC, Lyon, France.
Asseff PA (1961). "Lubricating Oil Additive, Description and Utilization, Lubricol Corp, Wicklif, Ohio, pp. 140-142
Mortier RM, Orszulik ST (1994). "Chemistry and technology of lubricant VCH publishers, Inc. New York, pp. 108-112
Thrash LA (1991). "Annual Refinning Survey: Oil and Gas J. (March 18), p. 84.
Gergel WC, La TGG (1977). "Extended Engine oil Life through New technology", Seminar paper at the National Petroleum Refiners Association 1977 Annual meeting, March 27 – 29, San Francisco, California.
Bromilow LG (1990). "Supply and Demand of Lube Oil: An update of the Global Perspective". Am 20 – 27 Presented at the 1990 NPRA Annual meeting.
Hamad A (2005). Used lubricating oil recycling using hydrocarbon solvents, linking hub. Elservier.com/pii/56301479704.
Scapin MA (2007). Recycling of used lubricating oils by ionizing, linking hub, elservier.com/retrieval/pii/30969806X0700182X
Art Jones (2010). Lubricating oil through the process of refining used motor oil, ww.articlealley.com/article.
Rincon J (2005). Regeneration of used lubricant oil by polar solvent extraction, pubs.acs.org/doi/abs/10.1021/ie040254.
Firas A, Dumitru P (2006). Design aspects of used lubricating oil re-refining, books.google.com.ng/books?isbn= 044452228X, p. 114.
Chevron Lubricating oil FM ISO 100.

Pyrolysis behaviour and kinetics of Moroccan oil shale with polystyrene

A. Aboulkas[1,2*], K. El harfi[1,2], M. Nadifiyine[1] and M. Benchanaa[1]

[1]Laboratoire de Recherche sur la Réactivité des Matériaux et l'Optimisation des Procédés «REMATOP», Département de chimie, Faculté des Sciences Semlalia, Université Cadi Ayyad, BP 2390, 40001 Marrakech, Maroc, Morroco.
[2]Laboratoire Interdisciplinaire de Recherche en Sciences et Techniques, Faculté polydisciplinaire de Béni-Mellal, Université Sultan Moulay Slimane, BP 592, 23000 Béni-Mellal, Maroc, Morroco.

Pyrolysis of oil shale/polystyrene mixture was performed in a thermogravimetric analyzer (TGA) from room temperature of 1273K, at heating rates of 2, 10, 20, 50 and 100 K/min. The global mass loss during oil shale/polystyrene pyrolysis was modelled by a combination of mass-loss events for oil shale and polystyrene volatiles. TGA results indicate that mixture pyrolysis can be identified in three phases. The first is attributed to the drying of absorbed water; the second was dominated by the overlapping of organic matter and plastic pyrolysis, while the third was linked to the mineral matter pyrolysis, which occurred at much higher temperatures. Discrepancies between the experimental and calculated TG/DTG profiles were considered as a measurement of the extent of interactions occurring on co-pyrolysis (10% of the difference between experimental and calculated curves in the temperature range 600 to 900K). The maximum degradation temperature of each component in the mixture was higher than those of the individual components. The calculated residue was found to be higher than experimental. These experimental results indicate a significant synergistic effect during pyrolysis of mixture of oil shale and polystyrene. The kinetic studies were performed using Flynn-Wall-Ozawa (FWO) method. The overall activation energies were 87 kJ/mol for organic matter of oil shale, 169 kJ/mol for polystyrene, and 161 kJ/mol for the mixture. Thus, it has been found that there exists an overall synergy, when two materials were pyrolysed together.

Key words: Thermogravimetric, pyrolysis, kinetics, oil shale, polystyrene.

INTRODUCTION

Our modern society is unimaginable without plastics. Nowadays, both the consumption and production of polymers are increasing, but the increasing amount of polymer wastes from them generates further environmental problems. As only a small amount of these wastes is recycled and most of the plastics are not biodegradable, these wastes need to be treated adequately to prevent environmental problems and make possible a sustainable development of modern society. Incineration with energy recovery is widely applied in some European countries. However, the generation of highly toxic chlorinated organic compounds makes this technology highly controversial

and expensive to operate (Tukker, 2002). Mechanical recycling involves the melting and re-moulding of used thermoplastics. However, the growth potential for this technology is already limited by the low quality of the plastic produced (Tukker, 2002; Association of Plastics Manufacturers in Europe (APME), 2003; Aguado and Serrano, 1999).

Utilization of waste plastics mixed with oil shale is a more attractive way for recycling the waste plastics and generating the necessary energy to supply the increasing energy demand. The pyrolysis of the mixture of waste plastics with oil shale could play an important role in converting these solid fuels into economically valuable hydrocarbons, which can be used either as fuels or as feedstock in the petrochemical industry (Ballice, 2001; Ballice et al., 2002; Tiikma et al., 2004; Gersten et al., 1999; Ballice and Reimert, 2002).

*Corresponding author. E-mail: a.aboulkas@yahoo.fr.

In Morocco, 90% of the energy consumed is dependent on imported oil. Thus, an intensive programme was commenced for the mobilization of indigenous energy sources, especially the local oil shales. Morocco is very rich in Upper Cretaceous oil shale deposits; the main sites are located at Timahdit (Middle Atlas Mountains) and Tarfaya (South Morocco) (Nuttall et al., 1983; Alpern, 1981). The oil shale deposits in Morocco represent about 15% of known oil shale resources in the world (Bekri and Ziyad, 1991). Oil shale deposits can be considered as interesting potential sources of carbon or of organic molecules which could be exploited diversely in the future (Ambles et al., 1994; Halim et al., 1997).

Thermogravimetric analysis is an analytical method to determine the decomposition rate of reactions resulting from thermal effects and the kinetic parameters of these reactions (Bagc and Kok, 2004; Barkia et al., 2003; Aboulkas et al., 2008a, b; Aboulkas and El harfi, 2008). The kinetic analysis in the thermal decomposition is the most important tool in the study of the complex pyrolysis mechanism (Aboulkas and El harfi, 2008; Aboulkas et al., 2008). In several publications, it has been reported that weight loss occurs in relation with the temperature, and kinetic analysis is used in thermal analysis processes, such as pyrolysis of oil shale, plastic or their mixtures (Aboulkas et al., 2008a, b; Aboulkas and El harfi, 2008; Yamur and Durusoy, 2006; Kok and Iscan, 2007; Kok and Pamir, 1999, 2000; Jaber and Probert, 2000; Williams and Nasir, 2000; Heikkinen et al., 2004; Encinar and Gonzalez, 2008; Sorum et al., 2001; Wu et al., 1993; Aboulkas et al., 2007; Gersten et al., 2000; Degirmenci and Durusoy, 2005). The thermal decomposition of oil shale has been studied (Yamur and Durusoy, 2006; Kok and Iscan, 2007; Kok and Pamir, 1999, 2000; Jaber and Probert, 2000; Williams and Nasir, 2000). Two or three peaks that appear in thermogravimetric curves are due to organic and mineral matter. Numerous studies on the thermal decomposition of polyolefin and, in particular, polystyrene have been carried out, especially in inert atmosphere (Heikkinen et al., 2004; Encinar and Gonzalez, 2008; Sorum et al., 2001; Wu et al., 1993).

Knowledge of the thermal behaviour of mixtures based on fossil fuel and polymers is of great importance from the processing point of view. In this sense, many reports in literature were devoted to the analysis of the effect of co-pyrolysis of fossil fuel and polymer mixtures ((Aboulkas et al., 2008; 2007; Gersten et al., 2000; Degirmenci and Durusoy, 2005; Cai et al., 2008; Vivero et al., 2005). Aboulkas et al. (2008) performed thermal degradation processes for a series of mixtures of oil shale/plastic using thermogravimetric analysis (TGA) at four heating rates of 2, 10, 20 and 50K min^{-1} from ambient tem-perature to 1273K. High density polyethylene (HDPE), low density polyethylene (LDPE) and polypropylene (PP) were selected as plastic samples. The overlapping degradation temperature of oil shale and plastic in TG/DTG curves of the mixture may provide an opportunity for free radicals from oil shale pyrolysis to participate in reactions of plastic decomposition. Gersten et al. (2000) investigated the thermal decomposition behaviour of polypropylene, oil shale and a 1:3 mixture of the two in a TG/DTG reaction system in an argon atmosphere. Experiments were conducted at three heating rates in the temperature range of 300 to 1173K. The results indicated that the characteristics of the process depend on the heating rate, and the polypropylene acts as a catalyst in the degradation of oil shale in the mixture. Degirmenci and Durusoy (2005) used the thermogravimetry analysis to obtain kinetics of the pyrolysis of oil shale, polystyrene and their mixtures. Experiments were carried out at non-isothermal decomposition conditions under argon atmosphere from 298 to 1173K at heating rate values of 10 and 60K min^{-1}. An increase was observed in the total conversion values of the blends with the increase in the blending ratio of polystyrene to oil shale. When a blend in any proportion of polystyrene to oil shale was degraded, an increase in maximum decomposition rate and a decrease in the temperature of maximum decomposition rate with the increase in polystyrene content of the sample were observed. The main conclusion is that the polystyrene accelerates the decomposition of the organic matter in the oil shale. Thermogravimetric analysis and kinetics of coal/plastic blends during co-pyrolysis in nitrogen atmosphere were investigated by Cai et al. (2008). The results indicated that plastic was decomposed in the temperature range 711 to 794K, while the thermal degradation temperature of coal was 444 to 983K. The overlapping degradation temperature interval between coal and plastic was favourable for hydrogen transfer from plastic to coal. The difference of weight loss between experimental and theoretical ones, calculated as an algebraic sum of those from each separated component, was 2.0 to 2.7% at 823 to 923K. These experimental results indicated a synergistic effect during plastic and coal co-pyrolysis at the high temperature region. The overlapping degradation temperature of coal and plastic in TG/DTG curves of the mixture may provide an opportunity for free radicals from coal pyrolysis to participate in reactions of plastic decomposition. Vivero et al. (2005) studied the thermal decomposition of blends of coal and plastic such as high density polyethylene and polypropylene using the thermogravimetric method. It was shown that plastic wastes have a strong influence on the thermoplastic properties of coal as well as the structure and thermal behaviour of the semicokes.

The goal of this paper is to examine oil shale/polystyrene mixture by thermal analysis. The pyrolysis behaviour of pure components and the mixture is measured by a thermogravimetric analyzer (TGA) and the obtained mass loss curve (TG) and its derivative (DTG) is used as a fingerprint of each component. The calculation of apparent activation energies was based on the application of the isoconversional Flynn-Wall-Ozawa method.

Table 1. Some average physicochemical characteristics of the Tarfaya oil shales.

Parameter	(wt %)
Proximate analysis	
Volatile matter	40.09
Ash	52 .83
Moisture (as received)	5.15
Fixed carbon	01.10
Elemental analysis	
C	17.60
H	1.78
N	0.70
S	0.37
Composition	
Carbonate mineral (Calcite)	70.0
Silicate mineral (Quartz, Kaolinite)	10.0
Bitumen	0.8
Pyrite	1.0
Kerogen	17.0

Table 2. Some characteristics of polystyrene.

Proximate analysis (wt %)			Elemental analysis(wt.% dry ash free)	
Volatiles	**F.C**	**Ash**	**C**	**H**
99.6	0.4	0.00	91.5	8.5

METHODOLOGY

Materials

The oil shale used in this work was from the Tarfaya deposit located in the south of Morocco. This deposit consists of several layers that are in turn subdivided in sub-layers, each having a different amount of organic matter. The samples were obtained from the R3 sub-layer, characterized by its high content of organic matter (Bekri and Ziyad, 1991). The results of the analysis of these samples are given in Table 1.

A sample from Tarfaya oil shale was obtained from the Moroccan "Office National de Recherche et d'Exploitation Pétrolière (ONAREP)". The organic matter belongs to Type II kerogen and covers a relatively wide range of maturity with $R_0 = 0.32 \pm 0.04\%$ (vitrinite reflectance). The kerogen was prepared by the following procedure: the dried oil shales were treated with chloroform to extract the bitumen. The solution was then filtered and the solvent eliminated in a rotary apparatus at reduced pressure. Pre-extracted samples were treated with diluted HCl, HF and HCl successively to eliminate carbonates and silicates. The isolated solid was washed with distilled hot water until the silver nitrate test for chlorides was negative. The pyrite was removed by the method of density difference. The H/C and O/C atomic ratios (1.62 and 0.14, respectively) correspond to a low maturity type II kerogen (Tissot and Welte, 1978; Durand and Monin, 1980).

The samples of polystyrene were provided by Plador, Marrakech, Morocco (particle size of 0.1 to 0.2 mm). The results of characterization of these materials are given in Table 2. Raw oil shale samples were ground and sieved to give particle size of 0.1 mm. Oil shale/ polystyrene mixture (1:1 in mass) were blended by tumbling for 30 min in order to achieve homogeneity. In all experiments, samples of around 20 mg with particle sizes ranging from 0.1 and 0.2 mm were placed in the platinum crucible of a thermobalance.

Experimental techniques

Raw oil shale, polystyrene and their mixture samples were subjected to thermogravimetric analysis (TGA) in an inert atmosphere of nitrogen. Rheometrix Scientific STA 1500 TGA analyzer was used to measure and record the sample mass change with temperature over the course of the pyrolysis reaction. Thermogravimetric curves were obtained at four different heating rates (2, 10, 20, 50 and 100K min^{-1}) between 300 and 1273K. Nitrogen gas was used as an inert purge gas to displace air in the pyrolysis zone, thus avoiding unwanted oxidation of the sample. A flow rate of around 60 ml min^{-1} was fed to the system from a point below the sample and at a purge time of 60 min (to be sure the air was eliminated from the system and the atmosphere is inert). The balance can hold a maximum of 45 mg; therefore, all sample amounts used in this study averaged approximately 20 mg. The experimental data presented in this paper corresponding to the different operating conditions are the mean values of runs carried out two or three times.

KINETIC MODELING

Non-isothermal kinetic study of weight loss under pyrolysis of

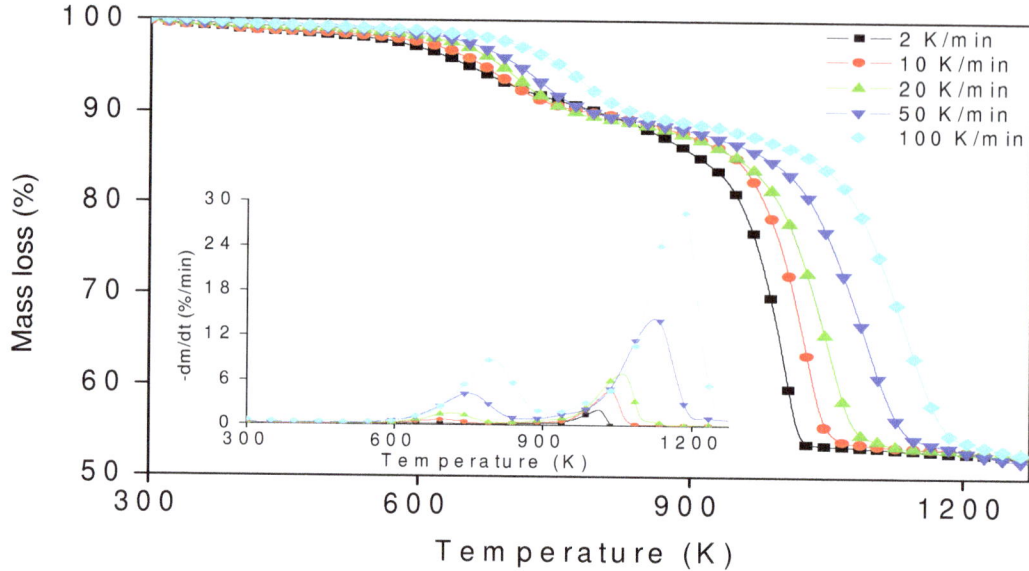

Figure 1. TG curves of oil shale at different heating rates. Inset: corresponding DTG curves.

carbonaceous materials is an extremely complex task because of the presence of numerous complex components and their parallel and consecutive reactions.

The extent of conversion or the fraction of pyrolysed material, x, is defined by the expression:

$$x = \frac{m_0 - m}{m_0 - m_f} \qquad (1)$$

where m is the mass of the sample at a given time t; m_0 and m_f refer to values at the beginning and the end of the mass event of interest. The rate of the kinetic process can be described by Equation 2:

$$\frac{dx}{dt} = K(T) f(x) \qquad (2)$$

where K(T) is a temperature-dependent reaction rate constant and f(x) is a dependent kinetic model function. There is an Arrhenius type dependency between K(T) and temperature according to Equation 3:

$$K(T) = A \exp\left(-\frac{E}{RT}\right), \qquad (3)$$

where A is the pre-exponential factor (usually assumed to be independent of temperature), E the apparent activation energy, T the absolute temperature and R is the gas constant.

For non-isothermal conditions, when the temperature varies with time with a constant heating rate, $\beta = dT/dt$, Equation 2 is modified as follows:

$$\beta \frac{dx}{dT} = A \exp\left(-\frac{E}{RT}\right) f(x) \qquad (4)$$

The use of Equation 4 supposes that a kinetic triplet (E, A, f(x)) describes the time evolution of a physical or chemical change. Upon integration, Equation 4 gives:

$$g(x) = \int_0^x \frac{dx}{f(x)} = \frac{A}{\beta} \int_0^T \exp\left(-\frac{E}{RT}\right) dT \equiv \frac{AE}{\beta R} p\left(\frac{E}{RT}\right) \qquad (5)$$

where T_0 is the initial temperature, g(x) the integral form of the reaction model and p(E/RT) is the temperature integral, which does not have an analytical solution. If T_0 is low, it may be reasonably assumed that $T_0 \to 0$, so that the lower limit of the integral on the right-hand side of Equation 5, T_0, can be approximated to be zero.

The isoconversional integral method suggested independently by Flynn and Wall (1966) and Ozawa (1965) uses Doyle's approximation (1961) for the temperature integral:

$$\ln p\left(\frac{E}{RT}\right) = -5.331 - 1.052 \frac{E}{RT} \qquad (6)$$

Relations (5) and (6) lead to:

$$\ln \beta = \ln \frac{AE}{Rg(x)} - 5.331 - 1.052 \frac{E}{RT} \qquad (7)$$

Thus, for x = const., the plot ln β vs. (1/T), obtained from thermograms recorded at several heating rates, should be a straight line whose slope can be used to evaluate the apparent activation energy.

RESULTS AND DISCUSSION

Thermogravimetric analysis of oil shale

Thermal degradation of oil shale

TG and DTG curves at different heating rates for thermal degradation of oil shale are shown in Figure 1. The oil shale degradation occurs in three steps. The first step which is from 300 to 430K corresponds to drying of the

Table 3. Characteristic temperatures of the pyrolysis of oil shale, polystyrene and their mixture.

Variable	2 K min^{-1}		10 K min^{-1}		20 K min^{-1}		50 K min^{-1}		100 K min^{-1}	
	Pyrolysis range (K)	Peak temperatures* (K)	Pyrolysis range (K)	Peak temperatures (K)	Pyrolysis range (K)	Peak temperatures (K)	Pyrolysis range (K)	Peak temperatures (K)	Pyrolysis range (K)	Peak temperatures (K)
Oil shale	537-740	664	580-780	687	602-800	716	626-848	733	660-900	770
Polystyrene	587-701	682	621-745	708	642-759	718	682-806	736	710-826	769
Oil shale/polystyrene	617-719	688	646-748	716	658-766	729	675-792	748	704-828	775

*Temperature at which the peak rate of mass loss (-dm/dt) occurs.

oil shale. The second step, at about 566 to 915K, has mass loss of 7 to 9% (depending on the heating rate) due to the degradation of the organic matter content of the oil shale. The last step, which begins at 800 to 930K (depending on the heating rate), presents a mass loss of 36% due to the transformation of the mineral matter in the oil shale (calcite, quartz, kaolinite and pyrite). The characteristic temperatures are listed in Table 3.

As the heating rate is increased, Table 3 shows that there was a lateral shift to higher temperatures at wich the peak rate of mass loss (T_{max}). The lateral shift is also illustrated in Figure 1. The rate of weight loss also reflects the lateral shift with an increase in the rate as the heating rate was increased from 2 to 100K min^{-1}. The residual weight seemed to reach some constant values after 1200K. The values of residual mass were calculated to be about 52.9%. The lateral shift to higher temperatures for the maximum region of mass loss rate has also been observed by other workers using TGA to investigate the pyrolysis of oil shales. For example, Gersten et al. (2000) showed a shift in the temperatures of maximum mass loss rate of 38°C towards higher temperatures as the heating rate was increased from 5 to 15K min^{-1} for Israel oil shale. Williams and Nasir (1999) and Jaber and Probert (2000) also showed a lateral shift in the maximum rate of weight loss for the TGA of oil-shale samples. Williams and Nasir (2000) suggested that the shift to higher temperatures of degradation represented differences in the rate of heat-transfer to the sample as the heating rate was varied.

Kinetic study of oil shale

Applying Equation 7 on the TGA data, a plot of $\ln \beta$ vs $1/T$ for oil shale was obtained. The values of activation energy were determined from the best-fit lines as explained earlier (0.99 correlation coefficient). The activation energies determined from the slope of $\ln \beta$ vs $1/T$ plots from 4 to 18% conversion for degradation of organic matter of oil shale are listed in Table 4. It was found that the activation energies remain relatively constant between 6 and 16% conversion, which reveal that there is one dominant kinetic process. The mean value of the activation energies was 87 kJ/mol. Comparison with literature data shows that the kinetic parameters are unique to each individual case of oil shale. Torrente and Galan (2001) obtained activation energy of 167 kJ/mol for non-isothermal TGA of Puertollano (Spain) oil shale. Sonibare et al. (2005) performed non-isothermal TGA on Lokpanta oil shales (Nigeria) and found the activation energies vary from 73.2 to 75 kJ/mol. Dogan and Uysal (1996), however, reported results for Turkish oil-shale, of approximately 25 kJ/mol for the lower temperature decomposition and up to 43 kJ/mol for the main stage of decomposition. The difference between the results determined by our study and literary data is probably due to the influence of process parameters, such as heating rate and particle size. In addition, oil shale, especially its kerogen, is characterized by a complex heterogeneous nature; hence, it would be difficult to obtain the same experimental results even for nominally the same sample. Therefore, the same experimental technique, including sample preparation procedure, analysis method adopted, and the kinetic model for the analysis, should be employed in order to enable a reasonable comparison to be achieved.

Thermogravimetric analysis of polystyrene

Thermal degradation of polystyrene

TG and DTG curves of thermal decomposition of polystyrene at four heating rates are represented

Table 4. Kinetic parameters for pyrolysis of oil shale, polystyrene and their mixture.

Oil shale		Polystyrene		Oil shale/Polystyrene	
x	Ea (kJ/mol)	x	Ea (kJ/mol)	x	Ea (kJ/mol)
0.04	104	0.1	158	0.15	147
0.06	72	0.2	159	0.20	153
0.08	69	0.3	163	0.25	155
0.10	77	0.4	169	0.30	158
0.12	87	0.5	173	0.35	159
0.14	92	0.6	176	0.40	161
0.16	89	0.7	176	0.45	165
0.18	109	0.8	175	0.50	168
		0.9	172	0.55	169
				0.60	167
				0.65	166
				0.70	164
Mean	87	Mean	169	Mean	161

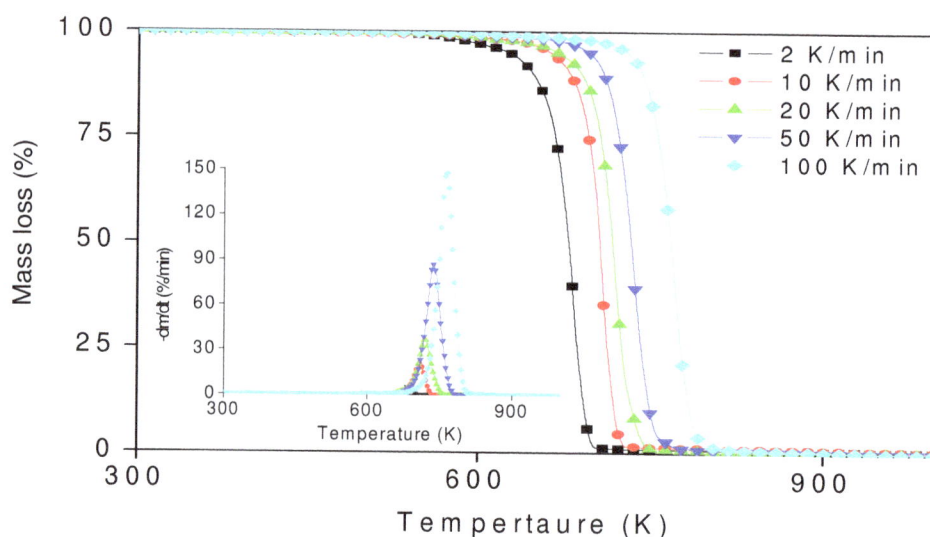

Figure 2. TG curves of polystyrene at different heating rates. Inset: corresponding DTG curves.

in Figure 2. It can be seen that the shape of the mass curves does not change with variations in heating rate, but the peak mass loss temperatures show an increase at higher heating rates. The mass loss shows that degradation occurs almost totally in one step as can be concluded by the presence of only one peak in DTG.

The TG curves show that the polystyrene thermal degradation starts at 550K and is almost complete at approximately 820K. At higher heating rates, the maximum degradation rate shifted from 682K at 2K min⁻¹ to 769K at 100K min⁻¹. The maximum degradation rate also increased from 4.4% min⁻¹ at 2K min⁻¹ to 150% min⁻¹ at 100K min⁻¹. The TG/DTG curves were displaced to higher temperature due to the rate of heat transfer increasing with increasing heating rate. The characteristic temperatures are summarized in Table 3.

Kinetic study of polystyrene

A similar kinetic study was carried out for polystyrene. The value of activation energy of degradation of these samples from 10 to 90% conversion is cited in Table 4. These values also remain relatively constant after 10% conversion, which reveal that there is one dominant kinetic process of polystyrene. It was found that the mean value of activation energy was 169 kJ/mol. The calculated apparent activation energies reported in the literature for polystyrene varied over a wide range. Similar results were obtained by Wu et al. (1993) on the pyrolysis of

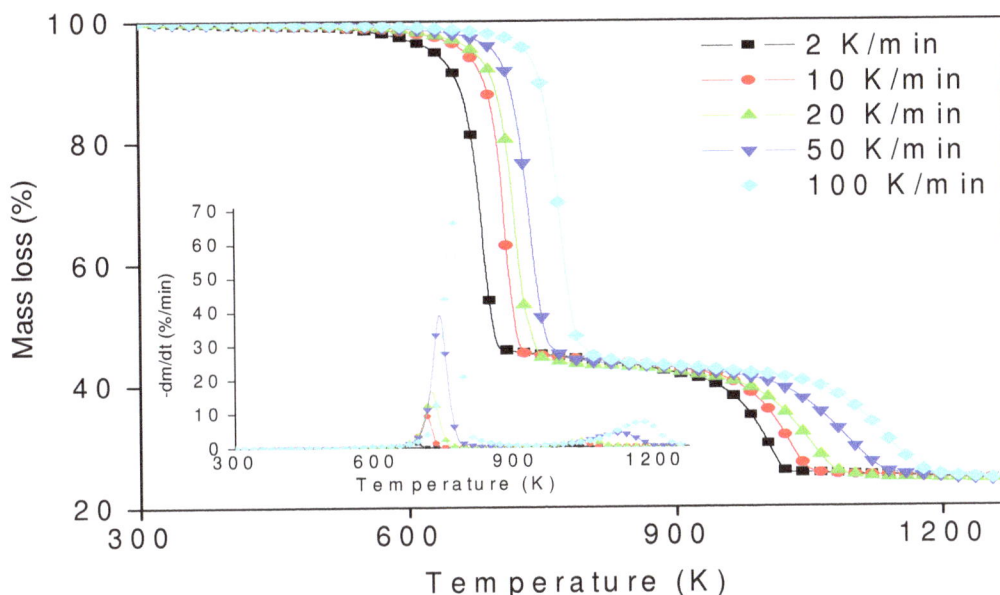

Figure 3. TG curves of Oil shale/polystyrene mixture at different heating rates. Inset: corresponding DTG curves.

plastic mixtures of municipal solid waste (MSW), the apparent activation energy of polystyrene degradation was found of 172 kJ/mol. Whereas, Encinar and Gonzalez (2008) found for polystyrene degradation an activation energy of 137 kJ/mol. Aguado et al. (2003) reported the activation energy of polystyrene degradation of 123 kJ/mol, and Sorum et al. (2001) found for polystyrene degradation an activation energy of 312 kJ/mol.

Thermogravimetric analysis of oil shale/ polystyrene mixture

Thermal degradation of oil shale/ polystyrene mixture

TG/DTG curves of thermal degradation of oil shale/ polystyrene mixture are represented in Figure 3. In general, we can note that the domains of degradation of both components are well overlapped. It can be seen that no separation of the DTG peaks of organic matter of oil shale and polystyrene exists. It should be added that the characteristic temperatures for the mixtures change in comparison with those for each component (Table 3). Both co-components increase their thermal stability as can be concluded from the shift of peak temperature toward higher values (Table 3). It should be added that the residue yield of the mixtures is lower than that of oil shale alone.

TG/DTG curves show that the degradation proceeds through three steps. The first step, obtained in the temperature range 300 to 430K, is attributed to the departure of the adsorbed water from the sample. The second step occurs between 540 and 900K and exhibit a total mass loss of 57.4 to 58.2% (depending on the heating rate), which corresponds to the overlapping of the organic matter of the oil shale and polystyrene degradations. The third step between 870 and 1250K with 18.3 to 18.9% wt (depending on the heating rate) mass loss was attributed to the degradation of mineral matter of oil shale content in the mixture. Some differences in such temperatures are observed when compared to those of pure materials. Polystyrene significantly increases its temperature at the maximum rate of mass loss (6 to 12 °C), while the variations for organic matter of oil shale are more pronounced (5 to 29 °C).

As in the case of oil shale and polystyrene, the characteristic temperatures of the process depend on the heating rate (Table 3). A higher heating rate shifted the TG/DTG curves rightward to the higher temperature range. To investigate whether interactions existed between the oil shale and polystyrene, theoretical TG/DTG curves where calculated. This curve represented the sum of individual components behaviour in the mixture:

$$m_{sum} = x_1 m_{oilshale} + x_2 m_{polystyrene}$$

where $m_{oil\ shale}$ and $m_{polystyrene}$ are mass loss as of the materials in the same operational condition, and x_1 and x_2 are the mass fractions of oil shale and polystyrene in the mixture. The calculated and experimental curves for oil shale/ polystyrene mixture at 10K min^{-1} are illustrated in Figure 4. We can note that an interaction is observed between organic matter of oil shale and polystyrene at 570 to 840K, while the degradation of mineral matter of

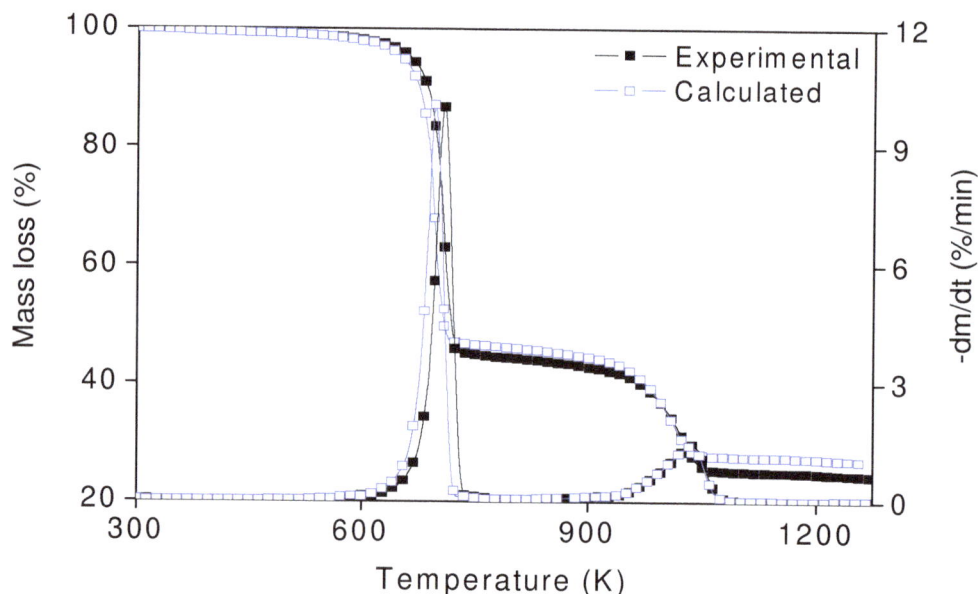

Figure 4. Comparison between experimental and calculated TG/DTG values for the oil shale/polystyrene mixture at 10 K min⁻¹.

oil shale is not affected. However, the temperature of the DTG maximum of oil shale and polystyrene shifted to higher temperatures in the mixture. The maximum temperature increased by 5 to 29°C for oil shale and 6 to 12°C for polystyrene. Comparing the TG/DTG curves, discrepancies between experimental and calculated curves are observed (10% of the difference between experimental and calculated curves in the temperature range 600 to 900K). The presence of polystyrene resulted in the increase of weight loss of mixture organic matter. Again, the residue obtained was found to be lower than the calculated value. Hence, there may be synergistic interaction between oil shale and polystyrene, when they are co-pyrolyzed. This synergistic interaction is due probably to that the products formed during polystyrene degradation participate in reactions of the oil shale degradation. These results are in good agreement with those of Gersten et al. (2000) and Degirmenci et al. (2005), who concluded that plastics accelerate the decomposition of the organic matter of oil shale. Cai et al. (2008) found that the reaction of hydrogen transferring from a polyolefinic chain to coal radicals will stabilize the primary products from coal thermal degradation. This would result in higher weight loss and lower yield of char.

Kinetic study of oil shale/polystyrene mixture

Flynn-Wall-Ozawa method was also applied to the study of the mixture of oil shale with polystyrene. The relationship between the activation energy and the conversion are shown in Table 4. It is interesting to note that the activation energy remains relatively constant between 15

and 75% conversion (between 147 and 169 kJ/mol). The average value of activation energy of oil shale/polystyrene process is 161 kJ/mol.

The obtained activation energy for the overlap of organic matter of oil shale and polystyrene decomposition in the mixture (161 kJ/mol) is higher than that of the oil shale (87 kJ/mol) and close to that of polystyrene (169 kJ/mol). When comparing these results with those of the thermal decomposition of the pure materials, it is noteworthy that the activation energy of polystyrene in the mixture was lower than the activation energy of polystyrene and the activation energy of the organic matter of oil shale in the mixture is higher than for the oil shale. It is possible to think that polystyrene acted as a catalyst in the pyrolysis of oil shale in the mixture while the mineral matter of oil shale acted as inhibitor in the pyrolysis of polystyrene in the mixture (Aboulkas et al., 2008; Gersten et al., 2000; Degirmenci and Durusoy, 2005). The mechanism of the synergistic effect between oil shale and polystyrene during co-pyrolysis is not very clear. According to the common views, the thermal degradation of polyolefins proceeds as a radical chain process, including the steps of radical initiation, chain propagation and radical termination. The mechanisms of this process and composition of formed products were studied in a number of articles (Faravelli et al., 2001; Sharypov et al., 2003; Ishaq et al., 2006; Karaduman, 2002). It is reasonable to explain the influence of oil shale on the thermal degradation of polystyrene in the framework of the well-known mechanism of polystyrene decomposition (Faravelli et al., 2001). In the first stage of the co-pyrolysis process, oil shale components can start the radical formation initiating the scission of the synthetic polymer chain. Then the reaction of hydrogen transferring from a polyolefinic chain to oil shale-derived radicals will probably stabilize the primary products from oil shale thermal degradation. This would result in higher weight loss and lower yield of char, as observed in our experiments.

Conclusions

A thermogravimetric study revealed that three stages of weight loss are involved during oil shale/polystyrene decomposition. The first step, obtained in the temperature range (300 to 430K), is attributed to the departure of the adsorbed water from the sample. The second step occurs between 570 and 840K, which corresponds to the overlapping of the organic matter of the oil shale and plastic degradations. The third step between 870 and 1250K was attributed to the degradation of mineral matter of oil shale content in the mixture (carbonate and silicate).

Comparing the TG/DTG curves, discrepancies between experimental and calculated curves are observed (10% of the difference between experimental and calculated curves in the temperature range 600 to 900K). Both co-components increase its T_{max} toward higher values. The presence of polystyrene resulted in the increase of weight loss of mixture organic matter. Furthermore, the residue obtained (24%) was found to be lower than the calculated value (27.7%). Hence there may be synergistic interaction between oil shale and polystyrene, when they are co-pyrolyzed.

The obtained activation energy of overlap organic matter of oil shale and polystyrene decomposition in the mixture (161 kJ/mol) is higher than that of the oil shale (87 kJ/mol) and close to that of polystyrene (169 kJ/mol). The overlapping degradation temperature of oil shale and polystyrene in TG/DTG curves of the mixture may provide an opportunity for free radicals from oil shale pyrolysis to participate in reactions of polystyrene decomposition.

REFERENCES

Aboulkas A, El harfi K (2008). "Study of the kinetics and mechanisms of thermal decomposition of Moroccan Tarfaya oil shale and their kerogen." Oil shale, 25: 426-443.

Aboulkas A, El harfi K, El bouadili A (2008). "Kinetic and mechanism of Tarfaya (Morocco) oil shale and LDPE mixture pyrolysis," J. Mater. Process. Technol., 206: 16-24.

Aboulkas A, El Harfi K, El Bouadili A, BenChanâa M, Mokhlisse A (2007). "Pyrolysis kinetics of polypropylene, Morocco oil shale and their mixtures". J. Therm. Anal. Calorim., 89: 203-209.

Aboulkas A, El harfi K, Nadifiyine N, El bouadili A (2008). "Investigation on pyrolysis of Morocco oil shale/plastic mixtures by thermogravimetric analysis," Fuel Process. Technol., 89:1000-1006.

Aguado J, Serrano D (1999) "Feedstock Recycling of Plastic Wastes Series (The Royal Society of Chemistry," Cambridge, UK.

Aguado R, Olazar M, Gaisán B, Prieto R, Bilbao J (2003). "Kinetics of polystyrene pyrolysis in a conical spouted bed reactor". Chem. Eng. J., 92: 91-99.

Alpern B (1981). "Les schistes bitumineux: constitution, réserves, valorisation," Bulletin des Centres de Recherches Exploration

Production, Elf Aquitaine. Bull. Centre Rech. Explor. Prod. Elf-Aquitaine, 5: 319.

Ambles A, Halim M, Jacquesy JC, Vitorovic D, Ziyad M(1994). "Characterization of kerogen from Timahdit shale (Y-layer) based on multistage alkaline permanganate degradation." Fuel, 73: 17-24.

Association of Plastics Manufacturers in Europe (APME) (2003), "An analysis of plastics consumption and recovery in Europe. Association of Plastics Manufacturers in Europe 9APME," Brussels, Belgium,

Bagc S, Kok MV (2004). "Combustion Reaction Kinetics Studies of Turkish Crude Oils." Energy Fuels, 18: 1472-1481.

Ballice L (2001). "Classification of volatile products evolved from the temperature-programmed co-pyrolysis of Turkish oil shales with atactic polypropylene (APP)," Energy Fuels, 15: 659-665.

Ballice L, Reimert R (2002). "Temperature-programmed co-pyrolysis of Turkish lignite with polypropylene." J. Analytical Appl. Pyrol., 65: 207-219.

Ballice L, Yuksel M, Saglam M, Reimert R, Schulz H (2002). "Classification of volatile products evolved during temperature-programmed co-pyrolysis of Turkish oil shales with low density polyethylene." Fuel, 81: 1233-1240.

Barkia H, Belkbir L, Jayaweera SAA (2003). "Oxidation kinetics of Timahdit and Tarfaya Moroccan

Bekri O, Ziyad M (1991). "Synthesis of oil Shale R & D Activities in Morocco," Proceedings of the 1991 Eastern Oil Shale Symposium, Lexington, Kentucky.

Cai J, Wang Y, Zhou L, Huang Q (2008). "Thermogravimetric analysis and kinetics of coal/plastic blends during co-pyrolysis in nitrogen atmosphere." Fuel Process. Technol., 89: 21-27.

Degirmenci L, Durusoy T (2005) "Thermal degradation kinetics of Göynük oil shale with polystyrene." J. Therm. Anal. Calorim., 79: 663-668.

Dogan OM, Uysal BZ (1996). "Non-isothermal pyrolysis kinetics of three Turkish oil shales." Fuel, 75: 1424-1428.

Doyle C (1961). "Kinetic analysis of thermogravimetric data." J. Appl. Polym. Sci., 5: 285-292.

Durand B, Monin JC (1980). "Elemental analysis of kerogen. In: Durand, B. (Ed.), Kerogen," Technip, Paris.

Encinar JM, Gonzalez JF (2008). "Pyrolysis of synthetic polymers and plastic wastes. Kinetic study," Fuel Process. Technol., 89: 678-686.

Faravelli T, Pinciroli M, Pisano F, Bozzano G, Dente M, Ranzi E (2001). "Thermal degradation of polystyrene." J. Anal. Appl. Pyrol., 60: 103-121.

Flynn J, Wall LA (1966). "Quick direct method for the determination of activation energy from thermogravimetric data." Polym. Lett., 4: 323-328.

Gersten J, Fainberg V, Garbar A, Hetsroni G, Shindler Y (1999). "Utilization of waste polymers through one-stage low-temperature pyrolysis with oil shale." Fuel, 78: 987-990.

Gersten J, Fainberg V, Hetsroni A, Shindler Y (2000). "Kinetic study of the thermal decomposition of polypropylene, oil shale, and their mixture." Fuel, 79: 1679-1686.

Halim M, Joffre J, Ambles A (1997). "Characterization and classification of Tarfaya kerogen (South Morocco) based on its oxidation products." Chem. Geol., 141: 225-234.

Heikkinen JM, Hordijk JC, Jong W, Spliethoff H (2004). "Thermogravimetry as a tool to classify waste components to be used for energy generation." J. Anal. Appl. Pyrol., 71: 883-900.

Ishaq M, Ahmad I, Shakirullah M, Khan MA, Rehman H, Bahader A (2006). "Pyrolysis of some whole plastics and plastics-coal mixtures." Energy Convers. Manag., 47: 3216-3223.

Jaber JO, Probert SD (2000). "Non-isothermal thermogravimetry and decomposition kinetics of two Jordanian oil shales under different processing conditions," Fuel Process. Technol., 63:57-70.

Karaduman A (2002). "Pyrolysis of Polystyrene Plastic Wastes with Some Organic Compounds for Enhancing Styrene Yield." Energy Sources, 24: 667-674.

Kok MV, Iscan AG (2007). "Oil shale kinetics by differential methods." J. Therm. Anal. Calorim., 88: 657-661.

Kok MV, Pamir R (1999). "Non-Isothermal Pyrolysis and Kinetics of Oil Shales." J. Therm. Anal. Calorim., 56: 953-958.

Kok MV, Pamir R (2000)."Comparative pyrolysis and combustion kinetics of oil shales." J. Anal. Appl. Pyrol., 55:185-194.

Nuttall HE, Guo TM, Schrader S, Thakur DS (1983). ACS Symposium Series 230, American Chemical Society, Washington, DC.

Ozawa T (1965). "A New Method of Analyzing Thermogravimetric Data." B. Chem. Soc. Jpn., 38: 1881-1886.

Sharypov VI, Beregovtsova NG, Kuznetsov BN, Membrado L, Cebolla VL, Marin N (2003). "Co-pyrolysis of wood biomass and synthetic polymers mixtures. Part III: characterisation of heavy products." J. Anal. Appl. Pyrol., 67: 325-340.

Sonibare OO, Ehinola OA, Egashira R (2005). "Thermal and geochemical characterization of Lokpanta oil shales, Nigeria." Energy Convers. Manage., 46: 2335-2344.

Sorum L, Grønli MG, Hustad JE (2001) "Pyrolysis characteristics and kinetics of municipal solid wastes," Fuel, 80: 1217-1227.

Tiikma L, Luik L, Pryadka N (2004). "Co-pyrolysis of Estonian shales with low-density polyethylene." Oil Shale, 21: 75-85.

Tissot BP, Welte DH (1978). "Petroleum Formation and Occurrence," Springer-Verlag, Berlin.

Torrente MC, Galan MA (2001). "Kinetics of the thermal decomposition of oil shale from Puertollano (Spain)." Fuel, 80: 327-334.

Tukker A (2002). "Plastic waste feedstock recycling, chemical recycling and incineration," Rapra Review Reports, Report 148, Rapra Technology Ltd., Shropshire, UK, 3(4).

Vivero L, Barriocanal C, Alvarez R, Diez MA (20050). "Effects of plastic wastes on coal pyrolysis behaviour and the structure of semicokes." J. Anal. Appl. Pyrol., 74: 327-336.

Williams PT, Nasir A (1999). "Influence of process conditions on the pyrolysis of Pakistani oil shales." Fuel, 78: 653-662.

Williams PT, Nasir A (2000). "Investigation of oil-shale pyrolysis processing conditions using thermogravimetric analysis." Appl. Energy, 66: 113-133.

Wu CH, Chang CY, Hor JL, Shih SM, Chen LW, Chang FW (1993). "On the thermal treatment of plastic mixtures of MSW: Pyrolysis kinetics," Waste Manage., 13: 221-235.

Yamur S, Durusoy T (2006). "Kinetics of the pyrolysis and combustion of göynük oil shale." J. Therm. Anal. Calorim., 86: 479-482.

4

Nitrogen fixing capacity of legumes and their Rhizospheral microflora in diesel oil polluted soil in the tropics

M. A. Ekpo* and A. J. Nkanang

Department of Microbiology, University of Uyo, Uyo, Akwa Ibom State, Nigeria.

The nitrogen fixing capacity of legumes cowpea *(Vigna unguiculata)* and groundnut *(Arachis hypogea)* and their micrflora grown in diesel oil simulated utisol was investigated. Result revealed that concentration as low as 1% v/w of diesel oil significantly affected the densities of nitrogen fixing bacteria, bacteriods, actinomycetes and fungi associated with the legumes. The heterotrophic bacteria count in the rhizosphere of cowpea reduced from $2.46 \pm 0.72 \times 10^7$ to $1.5 \pm 0.37 \times 10^7$ cfu/g after a growth duration of 12 weeks while it reduced from $3.4 \pm 1.25 \times 10^7$ to $1.52 \pm 0.36 \times 10^7$ cfu/g for groundnut in the same growth period. Nitrifying bacteria count reduced from $3.25 \pm 1.19 \times 10^4$ to $4.5 \pm 0.18 \times 10^3$ cfu/g for cowpea and $3.43 \pm 1.23 \times 10^4$ to $2.7 \pm 0.21 \times 10^3$ cfu/g for groundnut. Bacteriods count also significantly ($P > 0.05$) reduced from $3.85 \pm 2.30 \times 10^5$ cfu/g for the control treatment to $1.25 \pm 2.23 \times 10^5$ cfu/g in 1% level of pollution with no bacteriod formed in both 4 and 8% pollution due to inhibition of nodule formation by the diesel oil. Significant reduction ($P > 0.05$) was also observed in fungal and actinomycetes counts. Generally, organisms in the rhizosphere of groundnut exhibited more tolerance to diesel oil pollution than those found in the rhizosphere of cowpea. This study revealed that diesel oil adversely affected nitrogen fixing bacteria and bacteriods and consequently the nitrogen fixation in the soil.

Key words: *Vigna unguiculata*, *Arachis hypogea*, bacteriods, rhizosphere, actinomycetes.

INTRODUCTION

Legumes are plants that produces nitrogen fixing root or stem nodules which forms symbiotic association with Rhizobium. They include beans, peas, clovers and soybean (Harrison, 2003). The emergence of crude oil industries has contributed immensely to changing the state of Nigeria economy and the environment. The oil industry is a major source of environmental pollution and its adverse ecological impacts have been reported (Ibia et al., 2002; Ekpo and Thomas, 2007).

This is widely spread with specifically more serious damage on the oil producing areas. The most obvious area which has generated a lot of concern is spillage resulting from oil well blowout or pipeline leakages with

each major spill incident increasing the vulnerability of our fragile environment (Ibe, 2000; Ekpo and Nwankpa, 2005). The impact of petroleum exploration could alter essential microbial biogeochemical cycling processes, resulting in altered productivity of affected ecosystem (Caravaca and Rodan, 2003).

Pollution generally, can lead to a succession or total extinction of species in the affected habitat (Budny et al., 2002; Delille and Pelletier, 2002). This is because most spills are often toxic and generally cause deficiency in essential plant nutrients. Apart from phytotoxicity, nitrogen is the major element limiting plant growth in most spills and hydrocarbon contaminated sites (Wyszkowski et al., 2004).

This investigation was to assess the effects of diesel oil pollution on the rhizospheral microflora and the nitrogen fixing capacity of legumes in the tropics.

*Corresponding author. E-mail: emacopron@yahoo.com.

MATERIALS AND METHODS

Soil analysis

The soil used was an acidic sandy loamy soil classified as ferralitic sandy loam utisol (D'Hoore, 1964) collected from the botanical garden of University of Uyo, Nigeria. The soil had no previous exposure to petroleum hydrocarbons. The physico-chemical characterization of the soil samples were carried out using standard methods. The initial soil sampling was carried out for routine analysis. The soil samples were collected at a depth of 0 – 15 cm, air dried and sieved through a 2 mm sieve. The particles size distribution was determine by the pipette method of (Gee and Bauder, 1986). Soil pH was done using 1:2 soil/water volume ratio on a digital pH meter (model EQ-610.). Organic carbon in the soil was determine by the dichromate oxidation method of Walkey-Black wet oxidation as describe by Nelson and Sommers (1982). Total nitrogen was estimated by macro-Kjeldhal procedure (Jacson, 1962). Available phosphorus was estimated using the method of Bray and Kurtz (1945). Exchangeable bases Ca, Mg, Na and K were extracted using Batch method with 1 M NH_4OH. Ca and Mg were determined using Ethylene Diamine Tetra Acetic Acid (EDTA) and measured using a flame photometer. Exchangeable acidity in the soil was extracted with 1 M KCl and determined for Aluminum by titration method with 0.05 M NaOH. Fe, Cu, Zn and Mn were extracted with sodium bicarbonate and their concentrations read from Atomic Absorption Spectrophotometer (AAS). The total hydrocarbon content (THC) of the soil was estimated calorimetrically at 400 nm using Fisher's Electro-photometer after extraction with Carbon tetrachloride.

Soil treatment

The soil collected from botanical garden of University of Uyo, Nigeria, was polluted with diesel oil. Each experimental unit had 3 kg of soil weighed into pots. The diesel oil was applied evenly using a fine hose and properly worked into the soil in each of the treatments (0, 30.0, 120 and 240 ml) and replicated six times. These gave a percentage pollution of 0, 1, 4 and 8%v/w. The pots were then labelled and exposed to ambient conditions for one week before planting the test crops. Pots containing garden soil with no oil supplements served as control.

Cultivation of test crops

The legumes *Vigna unguiculata* (cowpea) of the variety IT 97 400-3 and *Arachis hypogea* (groundnut) of the variety Rm P12 belonging to the families papilioceae were obtained form the University of Agriculture, Umudike, Nigeria. Apparently, healthy seeds were sorted out and 4 seeds of each test crops were planted in the oil simulated soils. The seeds on germination were thinned to 2 seedlings per pot to create space for vigorous growth.

Enumeration of microorganisms from rhizosphere and nodules

The rhizosphere microorganisms of legumes were enumerated by the viable plate count method of Collins and Lyne (2004). Two samples of the rhizosphere soil (plant root plus adhering soil) were carefully obtained per plant for each treatment (oil contamination and uncontaminated (control) utisols. 10 g of the soil sample was uniformly mixed in 100 ml of sterile water. Ten-fold serial dilutions ranging from 10^{-1} to 10^{-7} of the oil contaminated and uncontaminated soil samples were prepared. From each dilution, 0.1 ml was plated out on appropriate microbial media. Bacto- plate count agar (Difco) plates, into which three drops of fungizone (50 μgml^{-1}) had been incorporated to inhibit fungal growth, were used for the estimation of heterotrophic bacteria. Duplicates plates from dilution 10^{-6} were prepared and incubated at 28 ± 2°C for 48 h before enumeration.

Bacto–Actinomycete agar (Difco) was used for the enumeration of actinomycetes in the rhizosphere of the crops. The pH of the medium was adjusted to 5.5 by incorporation of lactic acid to suppress the growth of non-filamentous bacteria. The plates were incubated at 28 ± 2°C for 48 h before enumeration. The number of diazotrophic bacteria was estimated on Bacto- nitrate agar (Difco). Inoculated nitrate agar plates were incubated at 37°C for 4 days before enumeration. Uninoculated nitrate agar plates were also incubated to serve as control. The rhizosphere fungal flora of the legumes were enumerated on Sabouraud dextrose agar (Difco) plates into which three drops of streptomycin (50 μgml^{-1}) was incorporated to suppress bacterial growth. The fungal plates were incubated at ambient temperature for 5 - 7 days before enumeration.

The bacteriods were enumerated using the method of Agboola (1986). Nodules were carefully obtained per plant for each treatment (oil contaminated and uncontaminated soil). The nodules were held in a Gooch crucible and washed thoroughly. The crucible containing the nodules was then dipped into a Petri dish containing diluted mercuric chloride solution (ratio 1:1000) for 5 min. These were then immersed into successive dishes of sterile water, using sterile forceps and were crushed in a sterile moter. 1 ml of sterile water was then put into the crushed nodule to obtain 10^{-1} dilution. From this a serial dilution to a dilution of 10^{-4} for the oil contaminated and uncontaminated nodule sample were prepared. 0.1 ml of the forth dilution was plated on yeast mannitol agar. Bacteriods plates were incubated at ambient temperature for 5 - 7 days before enumeration.

Isolation of microorganisms from rhizosphere and nodules

Discrete bacterial colonies which developed on the plates were randomly picked and purified by sub-culturing into fresh nutrient agar plates using streak plate techniques. Discrete colonies which appeared on the plates were then transferred into nutrient agar slants and stored as stock cultures for further test.

The fungi present in the rhizosphere of the legumes were isolated using the same procedure for bacterial isolation above but employing Sabauraud dextrose agar plates fortified with streptomycin. A portion of each fungal colony which developed was picked using sterile inoculating needle and aseptically sub-cultured into fresh Sabauraud dextrose agar plate. The plates were kept as stock cultures for identification tests.

Characterization and identification of isolates

The bacterial isolates were gram-stained and examined morphologically. Biochemical characterization was carried out including the isolates' ability to assimilate sugars. The probable identities of the isolates were determined using methods described by Holt et al. (1994), Buchanan and Gibbons (1974). The fungal isolates were examined macroscopically and then identified according to the method described by Domsch et al. (1980), Huntar and Benneth (1973) and Larone (1976). Bacteriods isolates were identified based on their cultural characteristics and cell morphology (Agboola, 1986).

RESULTS

The physico-chemical properties of the soil and the

Table 1. Some physicochemical properties of utisoil investigate.

Soil properties	Utisol before pollution				Utisol treated with diesel oil (12 weeks after pollution)			
	Cowpea (% v/w)				Groundnut (% v/w)			
Particle size (Depth 0 - 15 cm)	0	1	4	8	0	1	4	8
Sand	85.6	81.92	79.92	79.92	81.92	83.92	81.92	79.92	79.92
Silt	10.2	8.00	8.00	10.00	10.00	8.00	10.00	8.00	10.00
Clay	12.8	10.08	12.08	10.08	8.08	8.08	10.08	12.08	12.08
Chemical properties								
pH	6.16	6.1	6.40	6.30	6.00	6.10	6.10	6.20	6.20
Electrical Conduct. (ds/m)	0.047	0.327	0.168	0.132	0.092	0.024	0.100	0.136	0.107
Organic carbon (%)	1.84	1.79	2.31	2.69	4.11	1.10	2.89	2.02	4.86
Total Nitrogen (%)	0.134	0.05	0.10	0.12	0.18	0.06	0.13	0.11	0.21
Available P.(mg/kg)	69.93	65.33	56.66	52.99	47.33	68.66	52.66	50.66	47.99
Exchangeable								
- Ca (mol/kg)	2.27	2.56	2.40	2.60	2.00	2.34	2.40	2.70	2.48
- Mg (mol/kg)	1.20	1.10	1.10	1.20	1.00	1.13	1.07	1.20	1.30
- Na (mol/kg)	0.06	0.06	0.04	0.05	0.05	0.05	0.04	0.05	0.04
- K (mol/kg)	0.08	0.07	0.07	0.06	0.07	0.08	0.07	0.07	0.06
Exchangeable acidity								
(mol/kg)	1.56	1.44	1.60	1.76	1.40	1.50	2.64	1.60	1.90
Cation Exchange								
Capacity (cmol/kg)	5.17	5.23	5.11	5.68	4.52	5.37	6.22	5.62	5.78
Base saluation (%)	69.83	68.07	68.69	69.04	69.03	67.35	57.56	71.53	67.13
Total Hydrocarbon (mg/kg)	0.00	0.00	30.0	50.0	300.0	0.00	0.00	50.0	100.0
Heavy metals								
Fe	923.5	1313	1251	1251	1042	1396	1021	1125	1084
Cu	18.20	20.3	18.4	19.35	20.85	22.15	24.35	24.65	22.1
Zn	67.55	82.6	84.7	80.75	87.5	83.6	100.3	102.6	85.35
Mn	253.6	363.2	370.8	373.6	371.6	383.7	408.1	413.8	332.1
Cd		1.05	0.8	1.20	0.55	1.80	1.45	0.95	1.10
Cr		43.75	85.45	55.85	48.8	58.9	77.05	655.5	82.1
V		3.55	4.95	2.40	3.75	2.90	5.45	4.30	2.45

experimental plot before and after pollution are presented in Table 1. The soil, an acidic sandy loamy soil (utisol) with no hydrocarbon content, had originally considerable amount of nitrogen (0.134%), available phosphorus (69.93 mg/kg) and organic carbon (1.84%). On simulation with diesel oil, organic carbon content of the utisol increased with increased level of pollution while available phosphorus reduced to 47.33 and 47.99 mg/kg in the 8% soil treatment in cowpea and groundnut respectively. Also the total nitrogen content were slightly increased in 8% polluted soil to 0.18 and 0.21% for cowpea and groundnut respectively within 12 weeks after planting. The simulated soil had THC of 30 and 50 mg/kg in cowpea and groundnut respectively at 1% pollution level.

Microbial analysis of the rhizosphere and nodules of the crop cultivated in the oil contaminated and uncontaminated utisol are presented in Tables 2, 3, 4, 5 and 6. Table 2 shows the total heterotrophic bacteria count (THBC) in the rhizosphere of the legumes. There was a gradual increase in the THBC up to the 8th weeks after planting (WAP) and thereafter reduction in some of the treatments. It was however, observed that there was a significant decrease in count in the rhizosphere of both the cowpea and groundnut in the soil with higher levels of pollution.

The actinomycetes count in the rhizosphere of legumes is presented in Table 3. It was observed that the highest pollution level of diesel oil significantly reduced the mean count to $6.5 \pm 0.37 \times 10^2$ cfu/g for cowpea and $1.9 \pm 0.06 \times 10^2$ cfu/g for groundnut compared to the control with 1.6

Table 2. Heterotrophic bacteria count ($\times 10^7$ cfu/g) in the rhizosphere of legumes grown in oil contaminated and uncontaminated ultisol

Crop	Treatment	Age (weeks after planting)						
		2	4	6	8	10	12	mean
Cowpea	0	1.51	1.95	2.31	2.5	3.01	3.51	2.46±0.72
	1	1.32	1.81	1.91	2.0	2.54	2.71	2.04±0.50
	4	1.21	1.5	1.7	1.82	2.02	2.22	1.75±0.36
	8	1.03	1.21	1.41	1.56	1.81	2.03	1.50±0.37
Groundnut	0	2.0	3.0	5.02	4.7	3.53	2.2	3.40±1.25
	1	1.51	2.82	4.0	3.51	3.11	2.82	2.16±0.84
	4	1.31	1.51	3.51	3.0	2.72	2.52	2.42±0.85
	8	1.0	1.31	2.01	1.81	1.61	1.41	1.52±0.36

Cowpea, $S_x = 0.69$. $LSD_{0.05} = 0.51$. *Significant at $P > 0.05$, Groundnut, $S_x = 1.09$. $LSD_{0.05} = 0.88$. *Significant at $P > 0.05$.

Table 3. Actinomycetes count ($\times 10^3$ cfu/g) in the rhizosphere of legumes grown in oil contaminated and uncontaminated utisol

Crop	Treatment	Age (weeks after planting)						
		2	4	6	8	10	12	mean
Cowpea	0	0.51	1.5	1.8	1.92	2.01	2.21	1.65±0.61
	1	0.43	1.3	1.51	1.72	1.8	2.0	1.46±0.55
	4	0.21	1.01	1.31	1.51	1.71	1.82	1.26±0.59
	8	0.12	0.33	0.61	0.8	0.94	1.12	0.65±0.37
Groundnut	0	0.7	1.3	2.8	3.0	3.2	3.9	2.48±1.22
	1	0.5	1.2	2.5	2.0	2.2	2.5	1.81±0.82
	4	0.4	1.0	1.5	1.0	1.2	1.5	1.10±0.41
	8	0.3	0.23	0.2	0.13	0.06	0.14	0.19±0.06

Cowpea, $S_x = 0.65$. $LSD_{0.05} = 0.58$. *Significant at $P > 0.05$. Groundnut, $S_x = 1.12$. $LSD_{0.05} = 0.75$. *Significant at $P > 0.05$.

Table 4. Nitrifying bacteria count ($\times 10^4$ cfu/g) in the rhizosphere of legumes grown in oil contaminated and uncontaminated ultisol.

Crop	Treatment	Age (weeks after planting)						
		2	4	6	8	10	12	mean
Cowpea	0	1.52	2.5	3.01	3.73	4.02	4.91	3.28±1.19
	1	1.01	1.51	2.01	1.5	1.32	1.01	1.39±0.37
	4	0.6	1.4	1.0	0.8	0.61	0.4	0.80±0.35
	8	0.41	0.61	0.71	0.51	0.3	0.21	0.45±0.18
Groundnut	0	1.81	2.4	3.11	3.8	4.5	5.01	3.43±1.23
	1	1.22	1.0	0.71	0.51	0.3	0.2	0.65±0.39
	4	0.81	0.6	0.5	0.3	0.21	0.11	0.42±0.26
	8	0.6	0.42	0.31	0.21	0.09	0.04	0.27±0.21

Cowpea, $S_x = 1.27$. $LSD_{0.05} = 0.65$. *Significant at $P > 0.05$. Groundnut, $S_x = 1.46$. $LSD_{0.05} = 0.65$. *Significant at $P > 0.05$.

± 0.61 × 10^3 cfu/g and 2.48 ± 1.22 × 10^3 cfu/g respectively. Table 4 shows the nitrifying bacteria count in the rhizosphere of legumes grown in diesel oil contaminated utisol. The nitrifying bacteria were found to be highly sensitive to the diesel oil that even 1% pollution was significantly different compared to the control.

Fungi count in the rhizosphere of legumes in soil contaminated with diesel oil is presented in Table 5. The rhizosphere of cowpea was observed to enhanced greater fungal growth compared to that of groundnut. There was however, a significant decrease caused by the diesel oil contamination of the soil. Table 6 shows the

Table 5. Fungi count ($\times 10^4$ cfu/g) in the rhizosphere of legumes grown in oil contaminated and uncontaminated ultisol.

Crop	Treatment	Age (weeks after planting)						
		2	4	6	8	10	12	mean
Cowpea	0	7.1	9.1	10	12	13	15.1	11.5±2.88
	1	1.5	2.1	4.1	5.1	7.1	8.0	4.65±2.61
	4	1.3	3.0	5.0	6.1	6.4	7.0	4.80±2.21
	8	0.4	1.1	3.0	4.0	4.4	5.2	3.02±1.90
Groundnut	0	0.4	0.61	0.7	0.91	1.61	1.7	0.98±0.54
	1	0.12	0.2	0.2	0.32	0.41	0.81	0.34±0.25
	4	0.14	0.2	0.24	0.3	0.51	0.72	0.35±0.22
	8	0.2	0.32	0.44	0.51	0.7	0.92	0.51±0.26

Cowpea, S_x = 3.86. $LSD_{0.05}$ = 2.42. *Significant at $P > 0.05$. Groundnut, S_x =.0.41. $LSD_{0.05}$ = 0.34. *Significant at $P > 0.05$.

Table 6. Bacteroids count ($\times 10^5$ cfulg) in the rhizosphere of legumes grown in oil contaminated and uncontaminated utisol.

Crop	Treatment	Age(weeks after planting)						
		2	4	6	8	10	12	mean
Cowpea	0	0	5.3	5.1	5.0	5.7	2.0	3.85±2.30
	1	0	5.5	2.0	0	0	0	1.25±2.23
	4	0	0	0	0	0	0	0
	8	0	0	0	0	0	0	0
Groundnut	0	0	5.1	4.8	7.7	5.3	6	4.8±2.57
	1	0	0	1.1	0.5	5.0	0.5	1.10±1.91
	4	0	0	0	0	0	0	0
	8	0	0	0	0	0	0	0

bacteriods count in legumes grown in contaminated soil.

The diesel oil greatly affected soil bacteriod counts. It was observed that there was no bacteriod formation in the 4 and 8% pollution of the diesel oil. Table 7 shows the microorganisms isolated from the rhizosphere of legumes planted in diesel oil contaminated and uncontaminated soil. Though the diesel oil affected some organisms which disappear during the course of the experiment, most of the organisms were able to survive till the end of the experiment.

DISCUSSION

The investigation on the impact of diesel oil pollution on the rhizosphere of legumes revealed several adverse effects both on the legumes' capacity to mutually enhance the nitrogen fixation, the microbial community of the rhizosphere and the physico-chemical properties of the utisol. Total nitrogen increased slightly to the highest pollution level (8% treatment). This may be due to the diesel oil which is known to affect the macro elements content of the soil and modifies the soil biological activities as a result of its effect on the urease, acid and alkaline phosphatase activities in the soil (Wyszkowski and Wyszkowska, 2005; Akubugwo et al., 2009). Similar findings had earlier been reported by Odu (1981) that soil organic matter and soil nitrogen increases after degradation of oil polluted soils by microbial biodegraders.

Available phosphorus reduced with increased concentration of diesel oil applied to cultivated soil. This may be due to the effect of the diesel oil on the microorganisms which affect the activities of the enzyme phosphatase, therefore inhibiting the release of phosphorus from organic materials in the soil. It could also be attributed to the phosphorus utilized by the plants since it has been reported that nitrogen-fixing leguminous plants utilize more phosphorus due to the high energy cost of nitrogen fixation and the maintenance of functional nodules (Graham, 1998). Margesin and Schinner (2001) working on diesel-oil-contaminated soil in the Alpine glacier Skiing Area, also observed a decrease in the available phosphorus content of the soil and attributed it to microbial metabolism and immo-bilization of apatite (Calcium phosphate). The pH was also observed to slightly increase in both the rhizosphere of

Table 7. Microorganism isolated from diesel oil contaminated and uncontaminated Utisol.

Bacteria	Fungi	Yeast	Bacteriods
Clotridium botilinum	Aspergillus flavus*	Sacchoromyces Cerevisa	Rhizobium phaseoli
Micrococcu s luteus	Aspergills niger*	Candida tropocalis	Rhizobium
Lystevia monocytogens	Aspergillus Fumigatus*	Candiad psevdotropocalis	leguminosarium
Actinomyces isrealii	Aspergillus terreus*		
Bacillus pumilus	Botrytis aclad		
Clostridium hastolyticum	Penicillum citrinum*		
Bacillus spharicus	Penicillum trequentans*		
Bacillus circus	Fusarium roseum*		
Pseudomonas mallci	Trichoderma horizonum*		
Actinomyces neeslundii	Absidia sp.*		
Enterococcus feacalis	Scopolariopsis candida		
Shigella dysenteriac	Cephalosporium sp.*		
Pseudomonas avriginosa			
Azotobater nigricns			
Nitrosomonas euroaea			
Nitrobacter vulgaris			
Bacillus megatorium*			
Bacillus sp.*			
Corynebacterium ovis*			
Norcardia madurae*			
Bacillus polymyxa*			
Bacillus brevis*			

cowpea and groundnut. The exchangeable cations (Ca, Mg, Na, K) showed no remarkable difference from the results before pollution. The slight decrease could be due to the effect of immobilization of nutrients by the microorganisms as previously reported by Essien and Udotong (1999).

A diversed microbial community in the rhizosphere of the legumes in the simulated diesel oil polluted soil was observed. Specifically, there was a gradual increase in the THBC upto the 10th week in most of the treatments. This increase is as a result of the utilization of the diesel oil by indigenous hydrocarbon biodegraders as their energy and carbon source. Similar findings have been reported by Roscoe et al. (1989) who noted increase in the microbial population in a crude oil contaminated soils. It has also been reported that plant rhizosphere are highly favourable for the proliferation and metabolism of microbial types, hence the increase in rhizosphere heterotrophic bacteria population observed corresponds with increase amount of nutrient accumulation in the site (Brown, 1995). Increase in soil nutrient as a result of bio-degradation has also been known to enhance microbial growth due to availability of carbon, energy, nitrogen and sulfur, which are essential in the synthesis of amino acid in the microbial system (Okpokwasili and James, 1995; Chikere and Okpokwasili, 2003). Li et al. (2007) also noted increase growth of aerobic heterotrophic bacteria in a diesel-oil-stimulated soil with increase activities of soil

dehydrogenase, hydrogen-peroxidase and polyphenol oxidase.

The relatively low counts of actinomycetes encountered in the plant rhizosphere were expected. Specifically, significant decrease (P = 0.05) was observed in 8% treatment in the rhizosphere of cowpea and in both 4 and 8% in the rhizosphere of groundnut. The differences in the actinomycetes count in the microflora of cowpea and groundnut could be attributed to differences in the type of exudates produced by the legumes. Li et al. (2007) working with diesel-oil-simulated soil observed a significant decrease of the colonies of soil actinomyces and filamentous fungi and noted that this can be taken as a sensitive biological indicator of petroleum contamination. The differences in microbial community of different plant species has also been reported by Muratova et al. (2003) and noted that the actinomycetes in the rhizosphere of alfalfa was less than those found in the rhizosphere of reed in a butiment polluted soil. This difference is said to be the result of interaction between plant roots, root exudates and microorganisms within the rhizosphere and they play important role in regulating rhizosphere microbial processes and thereby significantly affecting plant growth (Bonkowski et al., 2000).

Nitrifying bacteria was observed to reduce with increased concentration of diesel oil. This could be due to the fact that under the diesel oil environment, the nitrifying bacteria could not effectively compete with other organisms that multiplied rapidly, resulting in the exhaustion of the available inorganic nitrogen. This finding agrees with the report of Odu (1981) who noticed that aerobic nitrogen fixers became relatively more abundant than other organisms while nitrifying organisms became considerably reduced in number. It has also been reported that soil polluted with bitumen recorded reduced count of nitrifying, nitrogen fixing, denitrifying and amonifying bacteria in the rhizosphere of both alfalfa and reed (Muratova et al., 2003).

Fungal population was observed to reduce with increase concentration of diesel oil. However, it was observed that in both 4 and 8% pollution, the count of fungi in groundnut rhizosphere was lower than those in cowpea. Though both rhizosphere received the same treatment, the exudates from the cowpea root must have ameliorated the effect of the diesel oil, thus the higher count of fungi in its rhizosphere compared to those of the groundnut. Yong and Crowley (2006) working on the rhizosphere microbial community structure with the use of 165 ribosomal DNA (rDNA) fingerprints, observed that the bacteria and fungi communities in the rhizosphere were substantially different in different root zones and that a rhizosphere community may be altered by changes in root exudates composition caused by changes in plant nutritional status.

The effect of the diesel oil was more pronounced on the count of the bacteriods. The 4 and 8% pollution totally inhibited the formation of nodules with only few nodules in the 1% concentration compared to the 100% nodule formation in all the control. This is expected because hydrocarbon has been known to adversely affect the nitrogen fixing bacteria within the rhizosphere (Muratova et al., 2003).

It was also observed in this study that some bacterial species. *Bacillus pumilus*, *Pseudomonas mallci*, *Enterococcus feacalis* and *Micrococcus luteus* and all fungal species except *Scopolaropsis candida* persisted after the pollution. This could be due to the fact that these microorganisms were able to degrade the diesel oil thereby increasing nutrient status of the soil for their survival. This observation was also reported by Ekpo and Thomas (2007) noting that in a crude oil impacted soil, some bacteria and most fungi possess enhanced physiological tolerance and were able to utilize the crude oil. Some microorganisms example *Clostridium botulinum*, *Listeria monocytogen* and *Rhizobium leguminosarium* were isolated only before pollution. They could have been eliminated because of their inability to utilize hydrocarbons as their sole carbon and energy source (Yong and Crowley, 2006).

Conclusion

This study has revealed that although there was a general decrease in the microbial biomas in the rhizosphere of the leguminous plant with increase in

concentration of the diesel oil, there were variations in the plant rhizosphere. Heterothrophic bacteria and actinomycetes in the rhizosphere of groundnut (*A. hypogea*) exhibited higher resistance to the diesel oil compared to the rhizosphere of cowpea *(V. unguiculata).* On the other hand, fungi in the rhizosphere of cowpea was observed to exhibit higher resistance compared to those in the rhizosphere of groundnut. The diesel oil adversely affected the nodulation of the legumes, the nitrogen fixing bacteria and consequently the nitrogen fixing capacity of the legumes. This study has also revealed that the diesel oil affects the development of nodules and the multiplication of bacteriods inside the nodules. It is therefore important to guide against the pollution of our agricultural soils with diesel oil to enhance the fertility status of our soil.

REFERENCES

Agboola AA (1986). Laboratory Manval for Agronomic studies in soil Plant and Microbiology. University of Ibadan, Ibadon, p. 80.

Akubugwo EI, Ogbuji GC, Chinyere CG, Ugbogu EA (2009). Physicochemical properties and enzyme activity study in a refined oil contaminated soil in Isiukwuato, Abia State, Nigeria. Biokemistri., 21: 79-84.

Bonkowski M, Cheng W, Griffiths BS, Alphei J, Scheu S (2000). Microbial-fuanal interaction and effect on plant growth. Eur. J. Soil Biol., 36:135-147.

Bray RH, Kurtz LT (1945). Determination of total Organic and available form of phosphorus in soils. Soil Sci. Soc. Niger., 59: 39-45.

Brown ME (1995). Rhizosphere microorganisms opportunist bandits or benefactors. In soil Microbiology: A critical Review. Walker N(Ed) Halsted press, wiley, New york. pp. 21-38.

Budny JG, Paton GI, Campbell CD (2002). Microbial communities in different soil types do not converge after diesel contamination. J. Appl. Microbiol., 92: 276-288.

Caravaca F, Rodán A (2003). Assessing changes in physical and biological properties in a soil contami nated by oil sludges under semiarid Mediterranean conditions. Geoderma, 117: 53–61.

Chikere BO, Okpokwasili GC (2003). Enhancement of biodegradation of petrochemicals by nutrient supplementation. Niger. J. Microbiol., 17(2): 130-135.

Collins OH, Lyne PM (2004). Microbiological Methods. A Hodder Arnold publication (8th Ed.). ISBN – 13 – 978 – 0340808962).London Great Britain, p. 484.

Delille D, Pelletier E (2002). Natural attenuation of diesel-oil contamination in a subantarctic soil (Crozed Island). Polar Biol., 25: 682-687.

Hoore JD (1964). Dongne, Hertley and Watson soil classification scheme In: Soil map of Africa explanatory monograhp. p. 202

Domsch KH, Gam W, Anckson T (1980). Compendium of soil fungi. Academic press. pp. 377.

Ekpo MA, Nwankpa IL (2005). The effect of crude oil on microorganism andgrowth of ginger (*Zingiber officinale*) in the tropics. J. Sustainable Trop. Agric. Res., 16: 67-71.

Ekpo MA, Thomas NN (2007). An investigation on the state of microorganisms and fluted pumpkin (*Telfairia occidentalis*) in a crude oil impacted garden soil. Niger. J. Microbiol., 21: 1572-1577.

Essien JP, Udotong IR (1999). Variation in rhizasphere microbiology properties of vegetables grown on oil contaminated utisol. Trop. J. Environ. Sci. Technol., 3: 4-12.

Gee GW, Bauder (1986). Particle size analysis. In: Method of soil Analysis part 1 (ed. Aklote). American society of Agronomy Madison WL, VSA. pp. 387-407.

Graham PH (1998). Symbiotic Nitrogen Fixation. In: Principle and application of soil microbiology. Prentice Hall. pp. 323-347.

Harrison JA (2003). The Nitrogen cycle of Microbes and Man. " Vsion

Holt JA, King NR, Smith PA, Statey JT, Wilians SJ (1994). Bergey's manual of determinative bacteriology 9th Ed. Williams and wilkins. Baltimove p. 789.

Huntar BB, Benneth HL (1973). Deuteromycetes (fungi impertecti). In Laskin AL, lechevalier HA (eds). C. R. C. Hand book of microbiology organic microbiology C. R. press. Cleveland, Ohio. 1: 405-433.

Ibe KM (2000). The Impact Assessment of oil spill on soils and shallow Ground water in the Niger Delta. J. Biotechnol., 1: 6-8.

Ibia TO, Ekpo MA, Inyang LD (2002). Soil Characterisation, Plant disease and Microbial Survey in Gas Flaring Community in Nigeria. World J. Biotechnol., 3: 443 - 453.

Jacson WL (1962). Soil Chemistry Analysis. Prentice – Hall Inc. Englewood. New Jersey. pp. 205-226.

Larone DH (1976). Medically Important fungi. A Gvide to Identification. Hamper and Row Hagerstown. Medison, p. 325.

Li HY, Zhang I, Kravchenko HXu, Zhano C (2007). Dynamic changes in microbial activities and community structure during biodegradation of petroleum compound: A laboratory experiment. J. Environ. Sci., 19: 1003-1013.

Margesin R, Schinner F (2001). Bioremediation (Natural attenuation and biostimulation) of diesel-oil-contaminated soil in an Alpine glacier Skiing Area. Appl. Environmt. Micrbiol., 67(7): 3127-3133.

Muratova A, Naruta N, Wand H, Turkovskaya O, Kuschlc P, John R, Merbach W (2003). Rhizosphere microflora of plants used for the phytoremediation of bitument-contamination. Micrbiological Research 158: 151-161.

Nelson DW, Sommers LE (1982). Total carbon, organic carbon, organic matter. In Pages AL, Miller RH, Keeney DR (eds), Methods of soil Analysis part 2. American society of Agronomy. Madison, 2: 539-579.

Odu CTI (1981). Microbiology of soil contaminated with petroleum hydrocarbon. In Extent of contamination and some soil and microbial properties after contamination. J. Inst. Pollut., (7): 279-286.

Okpokwasili GC, James WA (1995). Microbial contamination of kerosene, gasoline and crude oil and their spoilage potentials. Mat. Org., 29: 147- 156.

Roscoe YL, Mcgill WE, Nbory MP, Toogood JA (1989). Mehtod of accelerating oil degradation in soil. In: Proceeding in workshop on reclamation of disturbed Northern Fores, research Center, Alberta, Edmonton, pp. 462-470.

Wyszkowski M, Wyszkowska J, Ziółkowska A (2004). Effect of soil contamination with diesel oil on yellow lupine yield and macroelements content. Plant, Soil Environ., 50: 218–226.

Wyszkowski MJ, Wyszkowska J (2005). Effect of enzymatic activity of diesel oil contaminated soil on the chemical composition of oat (Avena sativa L.) and maize (Zea mays L.). Plant, Soil Environ., 51: 360-367.

Yong C, Crowley DE (2006). Rhizosphere microbial community structure in relation to root location and plant iron nutrition status. Appl. Environ. Microbiol., 66 : 345-351.

5

Fast combustion assisted gravity drainage (FCAGD) process for heavy oil recovery enhancement

Mehrdad Alemi[1], Mansour Kalbasi[1,2]* and Fariborz Rashidi[1,2]

[1]Department of Petroleum Engineering, Amirkabir University of Technology, Tehran, Iran.
[2]Department of Chemical Engineering, Amirkabir University of Technology, Tehran, Iran.

"Fast combustion assisted gravity drainage" (FCAGD) process is a novel heavy oil recovery enhancement method which is a specific combination of the two *in-situ* combustion process and steam assisted gravity drainage process. In this paper, the novel FCAGD thermal heavy oil recovery enhancement process has been presented by an improved simulation with an Eclipse500 simulator and has been analytically validated with Visual Basic.net code. As a result, the oil recovery factor of the simulation of the FCAGD process has been increased to about 48%. But, in comparison to its improved one, it is less efficient because as a result of the improved FCAGD process simulation, the oil recovery factor has been increased to about 52% in the reservoir. This novel process can be substituted with other thermal heavy oil recovery enhancement methods under proper technical circumstances with a greater recovery factor of oil produced.

Key words: Fast combustion assisted gravity drainage, improved simulation, thermal heavy oil recovery enhancement methods.

INTRODUCTION

Generally, in thermal heavy oil recovery enhancement methods, the conductive heat transfer to a reservoir of heavy oil can cause the reduction of oil viscosity and density and the increase of its mobility.

In the sample heavy crude oil reservoir, the heavy crude oil properties are such as: API=7.24, Flash point=180 F, pour point=4 centigrade, viscosity=1758 cp, interfacial tension=10 mN/m.

Among the most important thermal heavy oil recovery enhancement methods, it is possible to point out to *In-situ* combustion (ISC) and steam assisted gravity drainage (SAGD) processes. The *In-Situ* Combustion process is a displacement process that oxygen (air) is injected into the reservoir which reacts with the crude oil and forms a high-temperature combustion front which moves ahead through the reservoir. This method can be carried out in different forms based on the available circumstances.

Two main types of oxidation reactions with high and low temperatures are done and each of them will be effective in the whole of the process.

The steam assisted gravity drainage process is done by means of a horizontal injection well above a horizontal production well that a high-temperature steam saturated zone is formed in the heavy oil reservoir. The heat transfer from the injected steam to the reservoir rock and fluid (the original heavy and cold oil) is in the form of conduction.

The reservoir gravity force can cause the drainage of the heated heavy oil with more mobility (the activated oil) accompanied with the produced condensed water (hot water) towards the reservoir bottom and the horizontal production well.

"Fast combustion assisted gravity drainage process" (FCAGD) as a heavy oil recovery enhancement novel method in the world is a specific combination of the two *In-Situ* Combustion process (mostly in terms of the process mechanism and the combustion reactions and also the injected gas type) and the Steam Assisted Gravity Drainage process (mostly in terms of wells configuration) with three parallel and the same depth and lateral distance horizontal wells for more recovery factor of oil.

This novel combined process production performance depends on many parameters related to the reservoir and the injection /production wells that must be improved by means of the application of some specific solution ways (improvements).

However, there have been previous researches on heavy oil recovery enhancements of the ISC and SAGD methods.

MATERIALS AND METHODS

Methodology of this research

The FCAGD process is a novel thermal heavy oil recovery enhancement method which is a specific combination of the two *in-situ* combustion process (mostly in terms of the process mechanism and the combustion reactions and the injected gas type) and the steam assisted gravity drainage process (mostly in terms of well configuration). This novel process can be substituted with other thermal heavy oil recovery enhancement methods under proper technical circumstances with a greater recovery factor of oil produced.

This novel FCAGD thermal heavy oil recovery enhancement process has been presented by an improved simulation with an Eclipse500 simulator and has been analytically validated with Visual Basic.net code.

According to the simulation with an Eclipse500 simulator (Figures 1 to 8), the important technical and economical production parameter known as produced oil "recovery factor" has been gained and compared in both the ordinary simulation of the FCAGD process and the improved simulation of the FCAGD process.

As a result, the oil recovery factor of the simulation of the FCAGD process has been gained to about 48%. But, in comparison to its improved one, it is less efficient because as a result of the improved FCAGD process simulation, the oil recovery factor has been increased to about 52% in the reservoir.

Review of previous methods

(i) Awoleke et al. (2010) have considered the reservoir heterogeneity impact on the ISC process oil recovery factor. In this research, by means of experimental data and different technical analyses, the reservoir heterogeneity impact both in micro and macro scales for the combustion front movement has been identified. As a result, the injected oxygen movement into the reservoir in high permeable zones has been much more better than in low permeable zones.

(ii) Parikshit (2009) has considered the use of a down-hole igniter impact on the ISC method oil recovery factor. As a result, this strategy could be used for wet ISI for more reservoir volumetric sweep efficiency.

(iii) Kristensen et al. (2008) have considered the impact of the ISC process phase behavior modeling on its performance. In this method, a one-dimensional model of sensitivity analysis and numerical methods have been used. As a result, one important and critical condition for the ISC process has been the maintenance of the combustion front with a high temperature. Also any changes in the combustion front phase behavior could bring about either the more growth or the extinction of the process.

(iv) Tavallali et al. (Austria, 2011) have considered the SAGD process with twin horizontal injector and producer wells configuration in the Athabasca McMurray field, Alberta, Canada. As a result, an appropriate well configuration could be a 1:1 horizontal injector and producer well with 5 m vertical distance from each other.

(v) Oskouei (2011) have considered the impact of some formation zones with initial gas saturation on the SAGD process performance in experimental evaluations with physical models. As a result, the existence of some formation zones with initial gas saturation roughly equal to 9% or more could hasten the heat distribution and retard the heat chamber growth of hot steam. So, the case could have a negative impact on the SAGD process performance.

(vi) Farajzadeh et al. (2012) have considered the impact of the use of foam with the injected steam on the SAGD method oil recovery factor. As a result, the case could cause the oil recovery factor to increase up to 30%.

(vii) Dang et al. (2010) have considered the reservoir heterogeneity impact on the SAGD method oil recovery factor. As a result, reservoir heterogeneity and the existence of some thief zones with upper or lower water zones, abundant horizontal fractures and etc... could have negative impacts on the SAGD process performance.

(viii) Thorne and Zhao (2009) have considered the wells flow pressure drops impact on the SAGD method oil recovery factor. As a result, the existence of pressure drop between the twin horizontal injector and producer wells required more steam injection otherwise the oil recovery factor would decrease.

(ix) Yucel and Yannis (2005) have considered the impact of reservoir heterogeneity on the ISC process oil recovery factor. The studies have shown that the existence of reservoir heterogeneity could cause more heat loss, temperature reduction of progressing combustion front in high permeable zones and less sweep efficiency of oil in the reservoir.

DISCUSSION

The novel FCAGD process can be substituted with other thermal heavy oil recovery enhancement methods under proper technical circumstances with a greater recovery factor of oil produced (Rahnema and Mamora, 2010).

Here, this novel FCAGD thermal heavy oil recovery enhancement process has been presented by an improved simulation with an Eclipse500 simulator and has been analytically validated with Visual Basic.net code.

Here, Figure 1 shows the FCAGD process simulation-FOPR, FOIP versus Time graph. Figure 2 shows the Improved FCAGD process simulation-FOPR, FOIP versus Time graph. Figure 3 shows the FCAGD process simulation- wells configuration. Figure 4 shows the FCAGD process simulation-Matrix pressure (psi). Figure 5 shows the FCAGD process simulation-Matrix oil saturation. Figure 6 shows the improved FCAGD process simulation- huff and puff well configuration. Figure 7 shows the improved FCAGD process simulation-Matrix pressure (psi) and at last, Figure 8 shows the improved FCAGD process simulation-Matrix oil saturation.

Here, according to the sketched 2D graphs (FOPR, FOIP-Time graphs) and also the Flovis software 3D graphs, the important technical and economical production parameter known as produced oil "recovery factor" has been gained and compared in both the ordinary simulation of the FCAGD process (without the

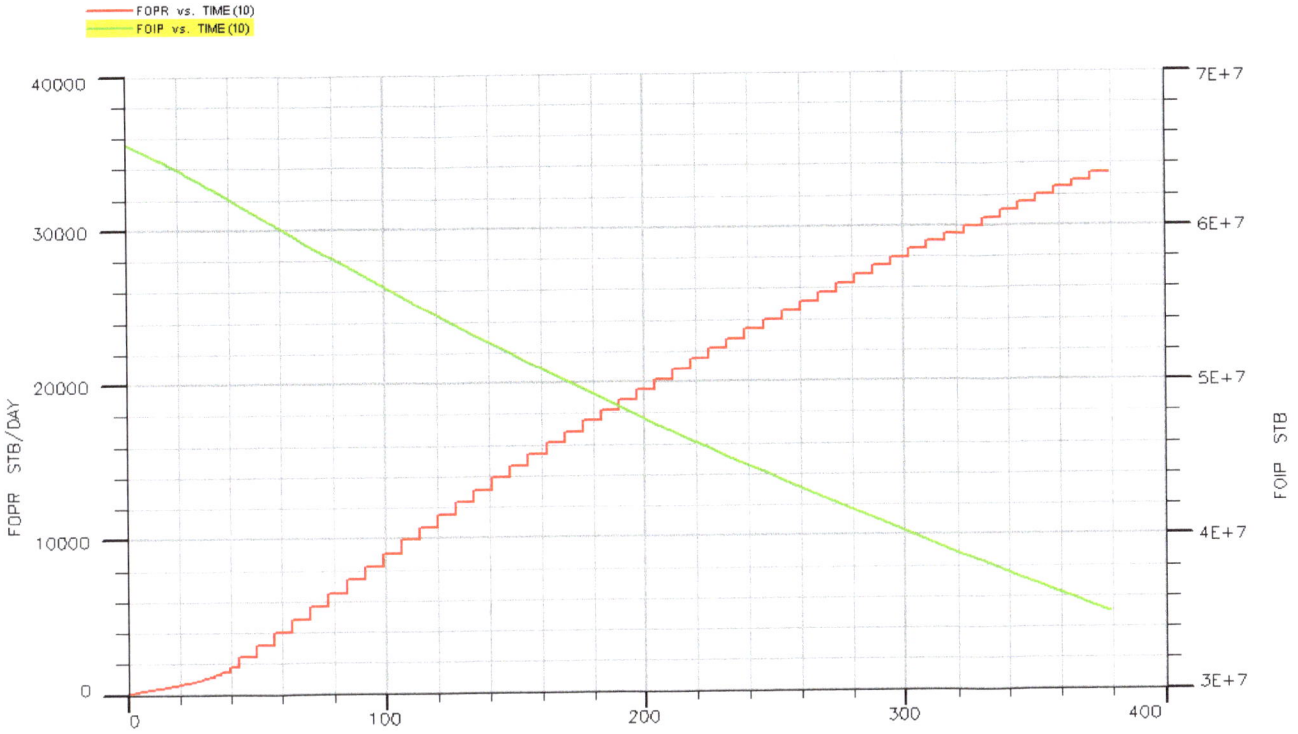

Figure 1. FCAGD process simulation - FOPR, FOIP versus Time Graph.

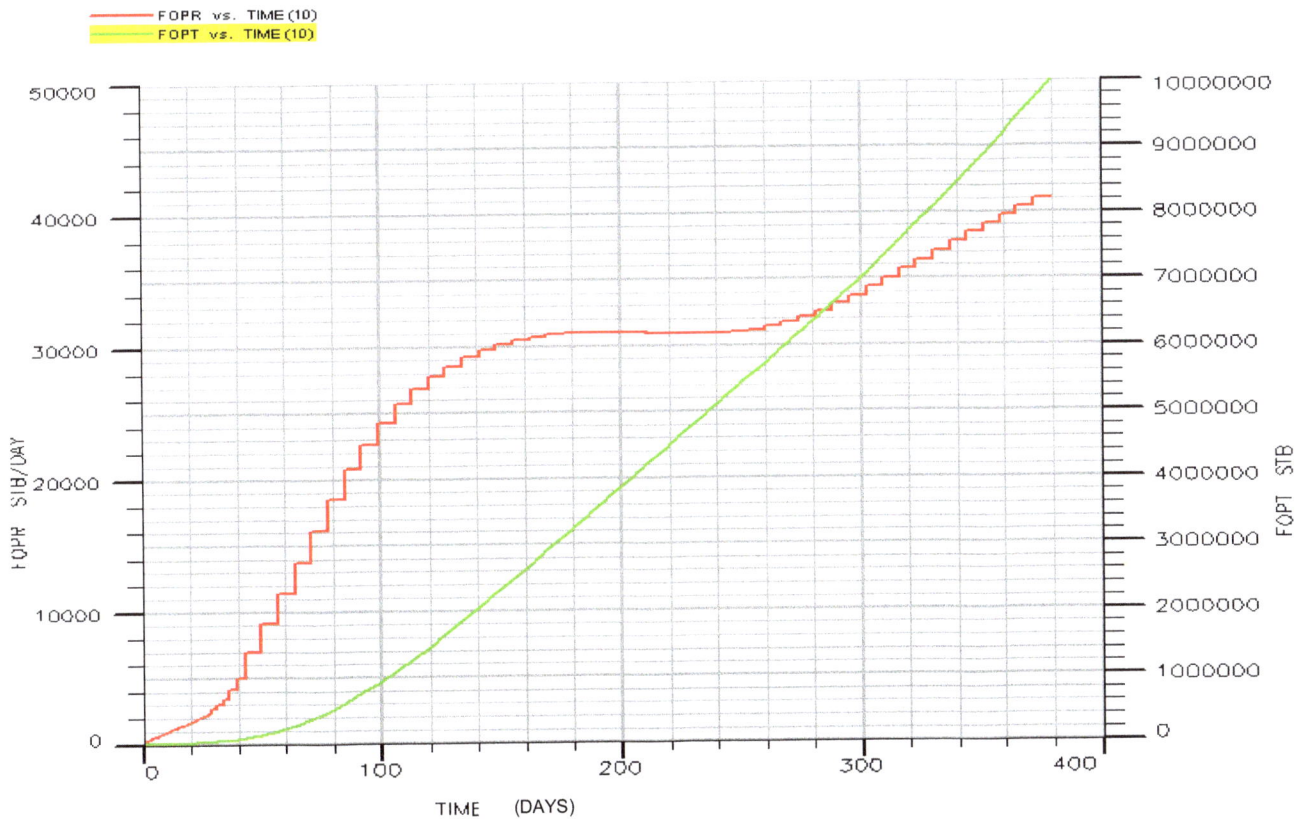

Figure 2. Improved FCAGD process simulation - FOPR, FOIP versus Time Graph.

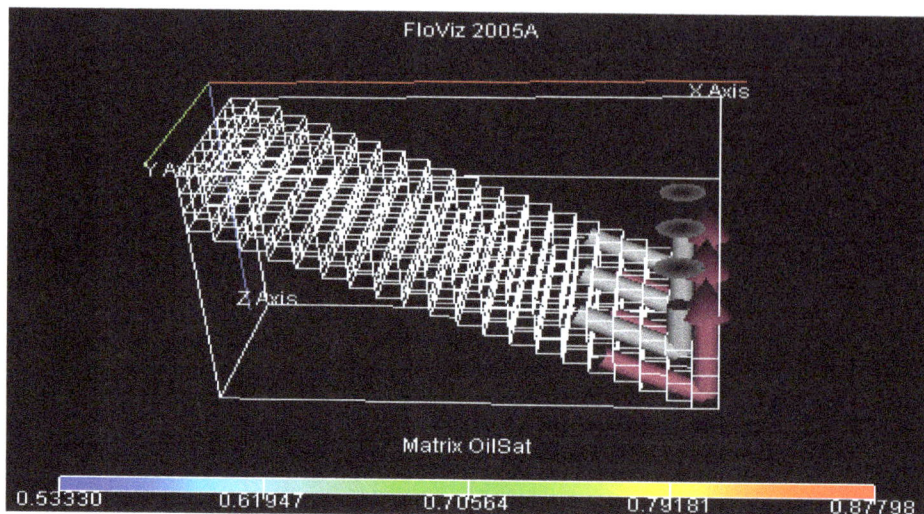

Figure 3. FCAGD process simulation – wells configuration.

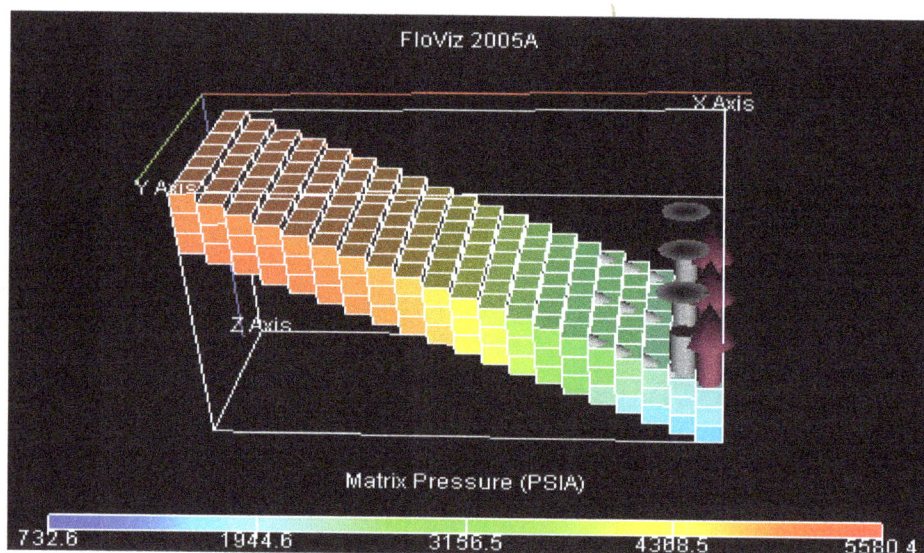

Figure 4. FCAGD process simulation – Matrix pressure (psia).

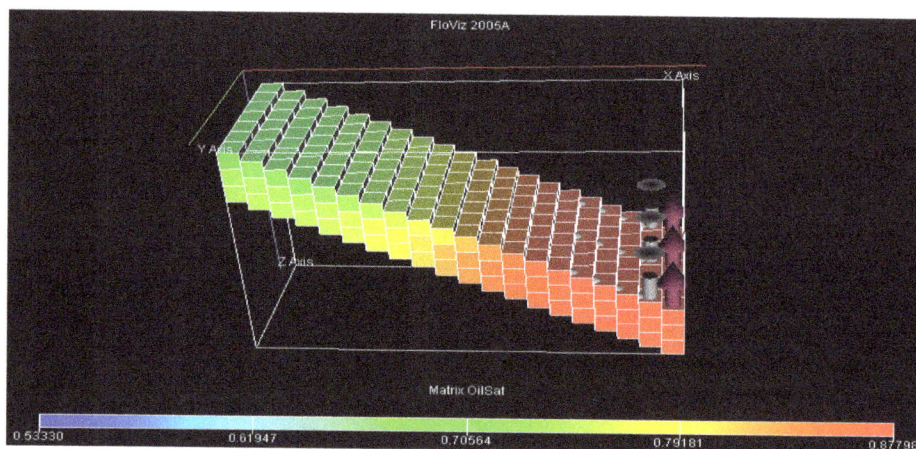

Figure 5. FCAGD process simulation – matrix oil saturation.

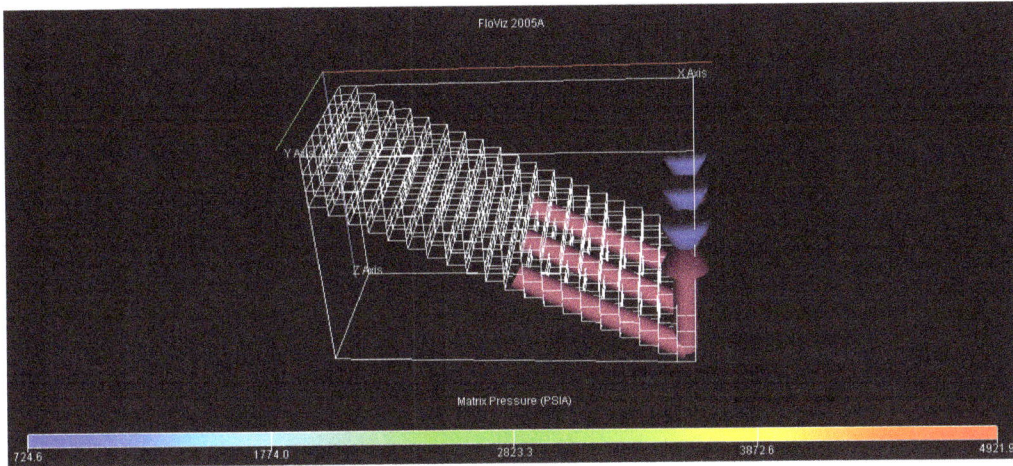

Figure 6. Improved FCAGD process simulation – huff and puff well configuration.

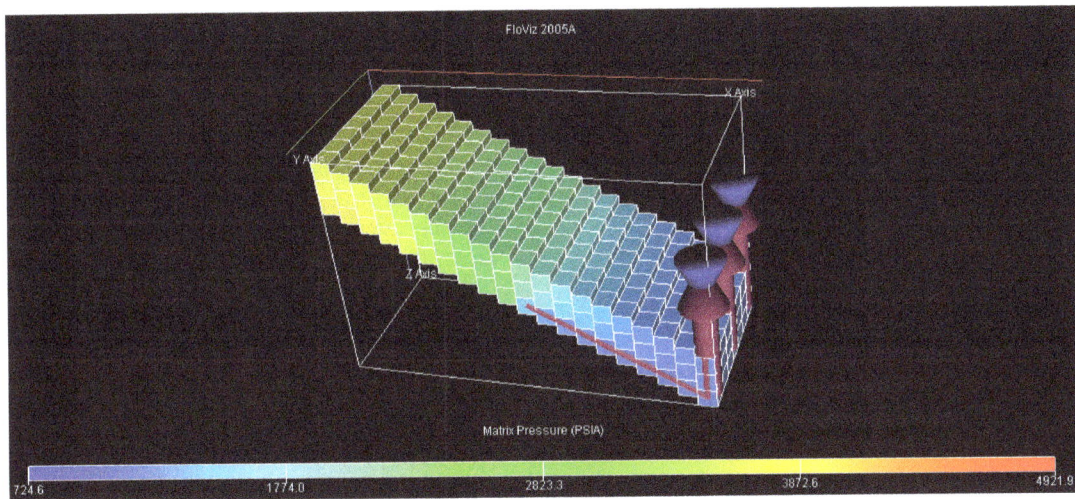

Figure 7. Improved FCAGD process simulation – Matrix pressure (psia).

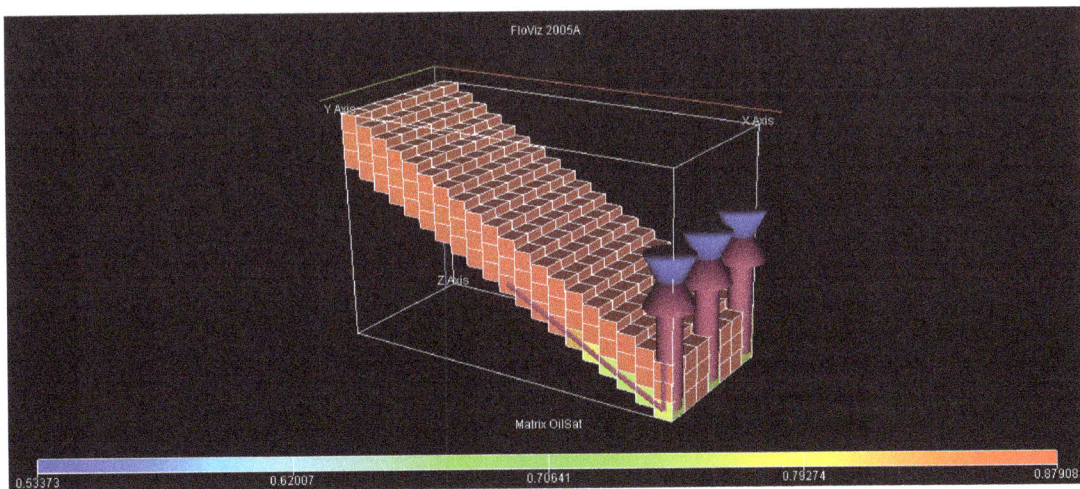

Figure 8. Improved FCAGD process simulation – matrix oil saturation.

improvements such as: increase in wells lengths and diameters and vertical distance, or use of "water alternating oxygen injection" for a better combustion quality through the horizontal wells and etc...) and the improved simulation of the FCAGD process.

As a result and as shown in the related graphs, the oil recovery factor of the simulation of the FCAGD process has been gained to about 48%. But, in comparison to its improved one, it is less efficient because as a result of the improved FCAGD process simulation, the oil recovery factor has been increased to about 52% in the reservoir. Some of the important formulas which could be effective to validate the thermal simulation are expressed as below:

A) Conductive heat transfer:

$$Q = (T_1 - T_2) / (L / KA) \qquad (1)$$

B) Convective heat transfer:

$$Q = (T_{surf} - T_{envr}) / (1 / h_{conv} A_{surf}) \qquad (2)$$

C) SAGD thermal process mechanisms:

$$q = 2L\sqrt{(\beta K g \alpha \phi \Delta S_O h) / (m v_s)} \qquad (3)$$

D) ISC thermal process mechanisms (Fuel (coke) amount):

$$C_m = \frac{V_g}{CV_b}\left[\frac{4C_{N2}C_{O2}}{C_{N2}}\right] - 4C_{O2} + 8C_{O2} + 10C_{CO} \qquad (4)$$

E) ISC thermal process mechanisms (Combustion heat):

$$\Delta H = \frac{174000m}{(m+1)(n+12)} + \frac{52500}{(m+1)(n+12)} + \frac{61500n}{(n+12)} \qquad (5)$$

F) Heat loss for steam injection (sum of heat losses of convection and radiation):

$$Q_{lp} = Q_{lc} + Q_{lr} \qquad (6)$$

G) SAGD process capillary and gravity forces:

$$u = [\Delta \int g(H-z) - p_c] / [(\mu_g / kk_{rg})z + (\mu_o / kk_{ro})(H-z)] \qquad (7)$$

It is shown that the gravity force has a positive impact on production but the capillary and viscous forces have negative impacts on production.

Conclusions

1. In thermal heavy oil recovery enhancement methods, the heat transfer to reservoir heavy oil can result in the reduction of oil viscosity and density and also the increase in its mobility for more recovery factor.
2. The FCAGD process is a novel thermal heavy oil recovery enhancement method which is a specific combination of the two *in-situ* combustion processes (mostly in terms of the process mechanism and the combustion reactions and the injected gas type) and the Steam Assisted Gravity Drainage process (mostly in terms of well configuration). This novel process can be substituted with other thermal heavy oil recovery enhancement methods under proper technical circumstances with a greater recovery factor of oil produced.
3. This novel FCAGD thermal heavy oil recovery enhancement process has been presented by an improved simulation with an Eclipse500 simulator and has been analytically validated with Visual Basic.net code.
4. The recovery factor of the ordinary simulation of the FCAGD process (without the mentioned improvements) has been gained to about 48%. But, in comparison to its improved one, it is less efficient because as a result of the improved FCAGD process simulation, the oil recovery factor has been increased to about 52% in the reservoir.

ACKNOWLEDGEMENTS

It is considerable to thank the Petroleum/Chemical Engineering Departments of Amirkabir University of Technology, Tehran, Iran for backing up the study on this scientific matter.

REFERENCES

Awoleke OG, Castanier LM, Kovscek AR, Stanford University (2010). An Experimental Investigation Of *In-Situ* Combustion In Heterogeneous Media, Canada, SPE Paper No. 137608,

Dang CTQ, Ngoc TBN, Wisup B, Huy XN, Tho NT, Taemoon C, Sejong University (2010). Investigation of SAGD Recovery Process in Complex Reservoir, Sejong University, SPE Paper No. 133849.

Farajzadeh R, Andrianov A, Krastev R, Hirasaki GJ, Rossen WR (2012). Foam-Oil Interaction in Porous Media: Implications for Foam Assisted Enhanced Oil Recovery, Oman, SPE Paper No. 154197.

Kristensen MR, Technical University of Denmark, Gerritsen MG, Stanford University, Thomsen PG, Michelsen ML, Stenby EH, Technical University of Denmark (2008). Impact of Phase Behavior Modeling on *In-Situ* Combustion Process Performance, USA, SPE Paper No. 113947.

Oskouei S, Javad P (2011). Effect of Initial Gas Saturation on Steam Assisted Gravity Drainage Process, Canada, SPE Paper No. 143452.

Parikshit SL (2009). Novel *In-Situ* Combustion Technique Using a Semi-Permeable Igniter Assembly, Delft University of Technology, SPE Paper No. 125583.

Rahnema H, Mamora D (2010). Combustion Assisted Gravity Drainage (CAGD) Appears Promising, Canada, SPE Paper No. 135821.

Tavallali M, Maini B, Harding T (2011). Evaluation of New Well Configurations for SAGD in Athabasca McMurray Formation, Austria.

University of Calgary, Annual Conference and Exhibition, 23-26 May 2011, Vienna, Austria, SPE Paper No. 143487.

Thorne T, Zhao L (2009). The Impact of Pressure Drop on SAGD Process Performance, Canada, SPE Paper No. 09-09-41.

Yucel A, Yannis CY (2005). The Effect of Heterogeneity on *In-Situ* Combustion: Propagation of Combustion Fronts in Layered Porous Media, Canada, SPE Paper No. 75128-PA.

6

Numerical modelling of the fluid dynamics in a bubbling fluidized bed biomass gasifier

Silva J. D.

University of Pernambuco-UPE, Polytechnic School in Recife, Rua Benfica - 455, Environmental and Energetic Technology Laboratory, Madalena, Cep: 50750-470, Recife - PE, Brazil. E-mail: jornandesdias@poli.br.

The fluid dynamics presented in this paper for a fluidized bed gasifier consists of estimating the volume fraction fields of gas and solid phases, as well as the velocity fields of gas and solid phases and the pressure drop inside the fluidized bed. However, understanding of the flow characteristics has an important role in understanding of the operation of the gasifier. The analyzed system consists of the fundamental equations of mass balance for the gas and solid phases, as well as the equations of momentum balance for gas and solid phases and an equation for the pressure given by the sum of the momentum balance equations of the gas and solid phases. The sets of equations developed form a system of one-dimensional partial differential equations (PDEs). The PDEs system has been transformed in a coupled ordinary differential equation (ODEs) system. The ODEs system has been solved using an implementation of the Runge-Kutta Gill method. The objective of this work is to obtain the profiles of state variables such as volume fraction of gas (ε_g) and solid phases (ε_s), the velocity profiles of gas (V_g) and solid phase (V_s) and pressure field (Δp).

Key words: Fluid dynamics, gasifier, fluidized bed, gas-solid, biomass.

INTRODUCTION

Gasification is a conversion of any solid or liquid fuel into an energetic gas through the process of partial oxidation due to elevated temperatures. The process of gasification occurs naturally in four distinct physical-chemical phases with temperatures of different reactions, such as the drying of biomass, devolatization or pyrolisis, reduction and burning. Each of these processes can be analyzed in different equipment, depending on the determined sequence by the characteristics of the project.

The technique of gasification is extremely versatile but there are lots of problems in transforming this theoretical potential into a commercially competitive technology, in spite of already being practical and viable. The difficulties lie not in the basic process of gasification but in the project of a device that transforms the solid fuel into a gas of quality with trusty and security adapted to the particular conditions of the fuel and of the operation (Abdullah et al., 2003; Basu, 1999). In the case of the gases produced being used for electric energy generation, the requirements of cleanness of these gases become of extreme necessity due to the sensitiveness of the gas turbines.

The available gasifiers can be classified in three models such as counter-current flux, for-current flux and fluidized bed. The gasifier of countercurrent flux is a device in which the pyrolized biomass and the air enter into different directions, the gas coming out through the upper part. The tars produced during this phase are dragged by the gases that come out from the gasifier. The biomass is gasified in the reduction zone using the energy generated in the chemical reactions that occur in the burning zone (Gabra et al., 2001a, b, c).

The co-current gasifier is characterized by the presentation of the feeding of biomass and air for the burning through the upper ending and producing gases almost tar less because the products of pyrolisis are forced through a burning zone where the biomass is found incandescent thermally destroying the tars formed resulting in clean gases (Graham and Walsh, 1996).

The fluidized gasifier is characterized by the formation of a biomass bed in suspension produced by the effect of the flux of forced air through a distributor. The fuel particles are maintained suspended into an inert particles bed (sand, ashes, alumina, etc.), fluidized by the flux of air. The biomass is fed in reduced dimensions to allow the fluidization (Heitor and Whitelaw, 1986).

A limiting factor for this kind of equipment is the level of humidity acceptable in the biomass for the process whose limit is around 30% due to the provoked instability by the water vapor in the burning zone. This way, it becomes extremely necessary an operation of pre-drying the biomass that presents humidity superior to 30%. For small plants, this pre-drying does not show major technical or economic problems. However, for the installations of big plants that demands the handling and the storage of thousands of tons monthly and this step must be considered as an integrated part of the gasification process (Kunii and Levenspiel, 1991; Levenspiel, 2002).

Two phases can be identified in the transversal section of the bed: (i) the emulsion; (ii) the bubbles. The emulsion contains the solid particles and the gas that move (process of the filtering of the gas) through them. The flux of gas in the emulsion is limited by the minimum velocity of fluidization (Larson and Williams, 1990). Any greater amount of gas passes through the bed as bubbles. The bubbles are practically free of solid particles but in its passage through the bed, some particles are dragged by them.

The objective of this paper is to obtain the profiles of the state variables such as the volumetrical fractions of the gas (ε_g) and solid (ε_s) phases, velocity of the gas phase (V_g), velocity of the solid phase (V_s) and the field of pressure (Δp) of a fluidized bed gasifier.

DESCRIPTION OF THE MATHEMATICAL MODEL

The applicability of the moment balance equations, energy and mass focus the sizing of gasifiers. From these balance equations, it is possible to obtain parameters that serve to characterize the process that occurs in the gasifiers. The knowledge of these optimized parameters makes the project of the equipment technically viable for an industrial scale. In the development of the model, some hypothesis were considered: (i) the fluxes are one-dimensional, (ii) the fluid phase is compressible, (iii) all the particles have the same dimension; (iv) the irregular movement and the colliding of the particles are ignored; (v) the friction on the wall is ignored. Having said that, based on these hypothesis it was developed the following fluid dynamic model for the balances of mass and moment:

Mass balance for the gas phase:

$$\frac{\partial \varepsilon_g}{\partial t} + V_g \frac{\partial \varepsilon_g}{\partial z} = \left(1 - \varepsilon_s\right) \frac{g}{V_g} \tag{1}$$

Mass balance for the solid phase:

$$\frac{\partial \varepsilon_s}{\partial t} + V_s \frac{\partial \varepsilon_s}{\partial z} = \left(1 - \varepsilon_g\right) \frac{g}{V_s} \tag{2}$$

Moment balance for the gas phase:

$$\rho_g \left(\varepsilon_g \frac{\partial V_g}{\partial t} + 2 \varepsilon_g V_g \frac{\partial V_g}{\partial z} \right) = -\rho_g \left(V_g \frac{\partial \varepsilon_g}{\partial t} + V_g^2 \frac{\partial \varepsilon_g}{\partial z} \right) - \left(1 - \varepsilon_s\right) \frac{\partial P}{\partial z} - \varepsilon_g \rho_g g - F_s \tag{3}$$

Moment balance for the solid phase:

$$\rho_s \left(\varepsilon_s \frac{\partial V_s}{\partial t} + 2 \varepsilon_s V_s \frac{\partial V_s}{\partial z} \right) = -\rho_s \left(V_s \frac{\partial \varepsilon_s}{\partial t} + V_s^2 \frac{\partial \varepsilon_s}{\partial z} \right) - \left(1 - \varepsilon_g\right) \frac{\partial P}{\partial z} - \varepsilon_s \rho_s g - F_s \tag{4}$$

The addition of Equations (3) and (4) results in an only equation for the pressure drop, being expressed as:

$$-\frac{\partial P}{\partial z} = \rho_s \left[\left(\varepsilon_s \frac{\partial V_s}{\partial t} + V_s \frac{\partial \varepsilon_s}{\partial z} \right) + \left(\frac{\partial V_s}{\partial z} 2 \varepsilon_s V_s + V_s^2 \frac{\partial \varepsilon_s}{\partial z} \right) \right] + \rho_g \left[\left(\varepsilon_g \frac{\partial V_g}{\partial t} + V_g \frac{\partial \varepsilon_g}{\partial z} \right) + \left(\frac{\partial V_g}{\partial z} 2 \varepsilon_g V_g + V_g^2 \frac{\partial \varepsilon_g}{\partial z} \right) \right]$$
$$+ \left(\rho_g \varepsilon_g + \rho_s \varepsilon_s \right) g \tag{5}$$

in which:

$$F_s = \frac{(\rho_s - \rho_g)\,g\,\varepsilon_g(1-\varepsilon_s)^{(1-n)}(v_g - v_s)}{v_t} \; ; \; v_t = \left(\frac{\rho_s - \rho_g}{\rho_s}\right)g\,\tau \quad \text{and}$$

$$\tau = \frac{d_p^2 \rho_s}{k\,\mu_g}$$

The initial and boundary conditions of equations (1) to (5) are given in Table 1.

SOLUTION OF THE SIMULATION MODEL

Equations of the model together with the initial and boundary conditions form a coupled partial differential equation system (PDEs), which characterizes a problem of initial values and boundary. These equations of model are very complex to be solved analytically. Having said that, it was used the methods of the lines (ML) to solve the system of PDEs. This method discretizes the partial derivades for method such as finite differences orthogonal collocation etc, forming a system of ODEs.

The system of ODEs will be solved with the implementation of Runge-Kutta Gill's method (Rice and Duong, 1995). In this present paper, this methodology was used:

Mass balance for the gas phase, discretized:

$$\frac{d\varepsilon_g}{dt} = \left[1 - \left(\varepsilon_s\right)_j^{(k)}\right]\frac{g}{\left(V_g\right)_j^{(k)}} - \frac{2\left(V_g\right)_j^{(k)}}{\Delta z}\left[\varepsilon_{g,0} - \left(\varepsilon_g\right)_j^{(k)}\right]$$

(6)

Mass balance for the solid phase, discretized:

$$\frac{d\varepsilon_s}{dt} = \left[1 - \left(\varepsilon_g\right)_j^{(k)}\right]\frac{g}{\left(V_s\right)_j^{(k)}} - \frac{2\left(V_s\right)_j^{(k)}}{\Delta z}\left[\varepsilon_{s,0} - \left(\varepsilon_s\right)_j^{(k)}\right] \quad (7)$$

Moment balance for the gas phase, discretized:

$$-\frac{dV_g}{dt} = \frac{\left[1 - \left(\varepsilon_s\right)_j^{(k)}\right]}{\left(\varepsilon_g\right)_j^{(k)}}\left\{g + \frac{2}{\rho_g \Delta z}\left[P_0 - \left(P\right)_j^{(k)}\right]\right\}\frac{4\left(V_g\right)_j^{(k)}}{\Delta z}\left[V_{g,0} - \left(V_g\right)_j^{(k)}\right] + \frac{\left(F_s\right)_j^{(k)}}{\rho_g\left(\xi_g\right)_j^{(k)}} \quad (8)$$

Moment balance for the solid phase, discretized:

$$-\frac{dV_s}{dt} = \frac{\left[1 - \left(\varepsilon_g\right)_j^{(k)}\right]}{\left(\varepsilon_s\right)_j^{(k)}}\left[g + \frac{2}{\rho_s \Delta z}\left[P_0 - \left(P\right)_j^{(k)}\right]\right] + \frac{4\left(V_s\right)_j^{(k)}}{\Delta z}\left[V_{s,0} - \left(V_s\right)_j^{(k)}\right] + \frac{\left(F_s\right)_j^{(k)}}{\rho_s\left(\xi_s\right)_j^{(k)}} + g \quad (9)$$

Pressure drop, discretized

$$-\frac{dP}{dz} = \rho_s\left\{ \begin{array}{l} -g\left[1-\left(\varepsilon_g\right)_j^{(k)}\right] - \frac{2\left[1-\left(\varepsilon_g\right)_j^{(k)}\right]}{\rho_s \Delta z}\left[P_0 - \left(P\right)_j^{(k)}\right] \\ -\frac{\left(F_s\right)_j^{(k)}}{\rho_s} + \frac{2\left(V_s\right)_j^{(k)}\left[\xi_{s,0} - \left(\varepsilon_s\right)_j^{(k)}\right]}{\Delta z}\left[1+\left(V_s\right)_j^{(k)}\right] \end{array} \right\} + \rho_g\left\{ \begin{array}{l} -g\left[1-\left(\varepsilon_s\right)_j^{(k)}\right] - \frac{2\left[1-\left(\varepsilon_s\right)_j^{(k)}\right]}{\rho_g \Delta z}\left[P_0 - \left(P\right)_j^{(k)}\right] \\ -\frac{\left(F_s\right)_j^{(k)}}{\rho_g} + \frac{2\left(V_g\right)_j^{(k)}\left[\varepsilon_{g,0} - \left(\varepsilon_g\right)_j^{(k)}\right]}{\Delta z}\left[1+\left(V_g\right)_j^{(k)}\right] \end{array} \right\} \quad (10)$$

Where:
$$F_s = \frac{(\rho_s - \rho_g)\,g\left(\varepsilon_g\right)_j^k\left[1-\left(\varepsilon_s\right)_j^k\right]^{(1-n)}\left[\left(v_g\right)_j^k - \left(v_s\right)_j^k\right]}{v_t}$$

Table 1. Initial and boundary conditions.

Initial conditions	Boundary conditions
$\varepsilon_g\big|_{t=0} = 0$	$\varepsilon_g\big|_{z=0^+} = \varepsilon_{g,0}$
$\varepsilon_s\big|_{t=0} = 0$	$\varepsilon_s\big|_{z=0^+} = \varepsilon_{s,0}$
$V_g\big|_{t=0} = 0$	$V_g\big|_{z=0^+} = V_{g,0}$
$V_s\big|_{t=0} = 0$	$V_s\big|_{z=0^+} = V_{s,0}$
——	$P\big|_{z=0^+} = P_0$

Table 2. Parameters used in the simulation.

Parameter	Unit
$\rho_s = 1{,}21$	$kg.m^{-3}$
$\rho_g = 1150$	$kg.m^{-3}$
$\mu_g = 1{,}8x10^{-5}$	$Pa.s$
$g = 9{,}8$	$m.s^{-2}$
$d_g = 500$	μm
$n = 1{,}37$	-
$k = 0{,}8$	-
$\Delta t = 10^{-2}$	s

RESULTS AND DISCUSSIONS

The solution of the developed model provided the behavior of the volumetric fraction of the gas phase (ε_g), volumetric fraction of the solid phase (ε_s), velocity of the gas phase (V_g), velocity of the gas phase (V_s) and pressure drop (ΔP) versus the time in the exit of the burning zone of the fluidized bed gasifier. The data used for the feeding of the computer code developed are presented in Table 2.

A model validation procedure was established by comparing between the model for the superficial velocity of the gas phase and Sirinivasan and Angirasa (1988) according to Figure 1. Figures 2 and 3 show a dynamic evolution of the volumetric fractions (eg and es) at z = H. It is observed that there was an increase of these fractions with the increase of the parameter eg, 0 at the inlet of the system. The volumetric fractions will encounter an invariable state around t = 100. The two fractions presented the very similar results. Figures 4 and 5 show the dynamic evolution of the velocities in the gas and

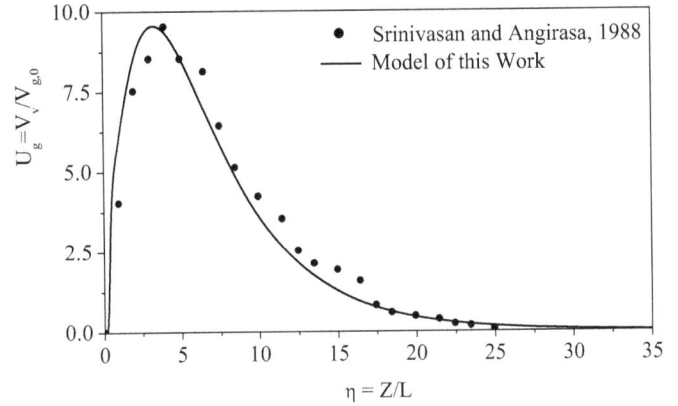

Figure 1. Comparisons of dimensionless curves for the solution of the model for the superficial velocity of the gas phase and Sirinivasan and Angirasa (1988).

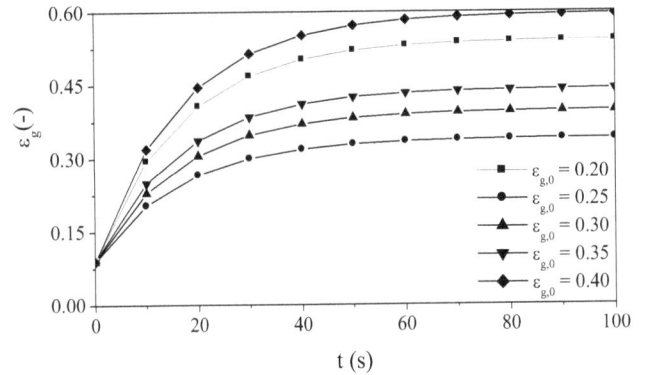

Figure 2. Profiles of the volumetric fraction of the gas phase in the exit of the fluidized bed gasifier countercurrent with the length at 1 m.

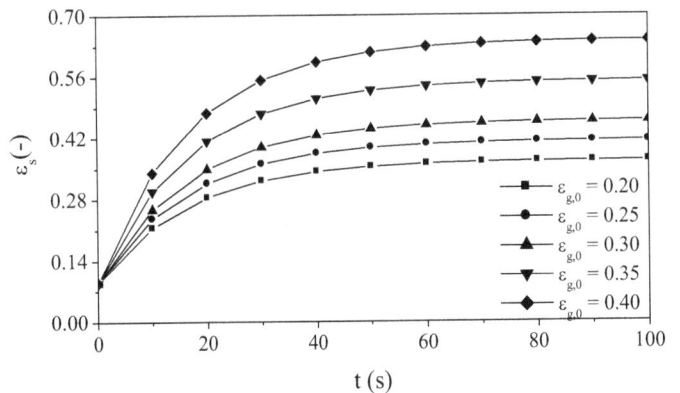

Figure 3. Profiles of volumetric fraction in the solid phase in the exit of the fluidized bed gasifier countercurrent with the length at 1 m.

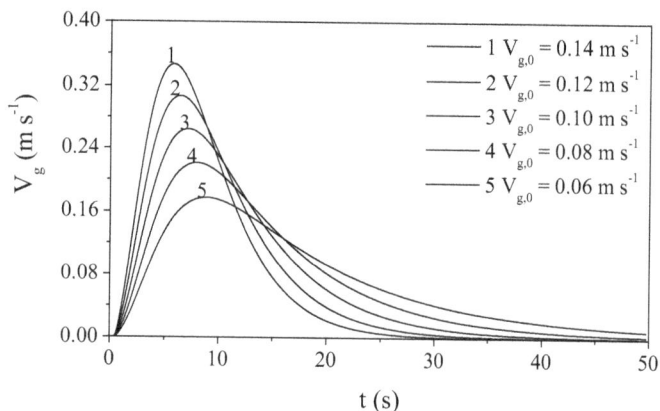

Figure 4. Profiles of the superficial velocity of the gas phase in the exit of the bubbling fluidized bed gasifier countercurrent with the length at 1 m.

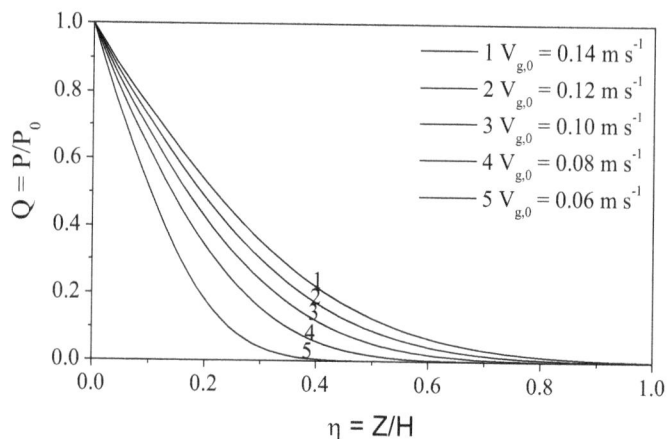

Figure 6. Profiles of the pressure drop in the bubbling fluidized bed gasifier countercurrent with the length at 1 m.

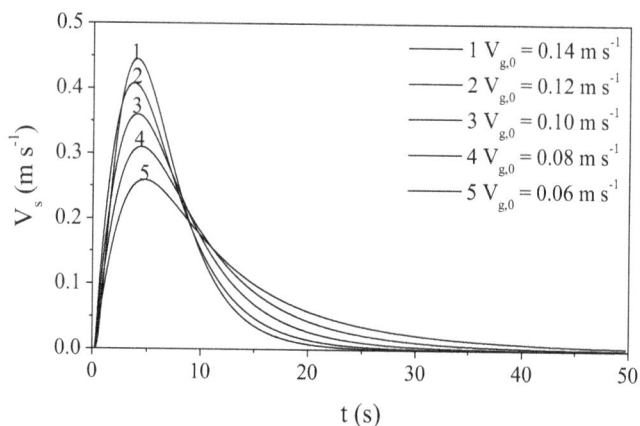

Figure 5. Profiles of the superficial velocity of the solid phase in the exit of the bubbling fluidized bed gasifier countercurrent with the length at 1 m.

solid phases (Vg and Vs) at z = H. The graphs show that such velocities also suffered substantial increase with the rising of variable Vg,0 at the inlet but from 30 to 50 s in the velocity Vg and from 25 to 50 s in the velocity Vs there was a fall until zero. It could also be observed that with the rising at the inlet variable Vg,0 the fall of the velocities (Vg and Vs) were quicker.

Figure 6 shows a pressure field along the region of fluidization. It is observed the effect of velocity Vg,0 at the inlet under the pressure inside the fluidized bed showing in the graph that the velocity was proportional to the rising of the pressure until zero.

Conclusion

The modelling of the gas-solid process of the fluidized

bed was developed in relation to the volumetric fractions (ε_g and ε_s) and the velocities (Vg and Vs) as well as an equation for the pressure drop resulting from the adding of equations V_g and V_s. The simulations of this model led us to the following conclusions:

i. The validation portrayed a satisfactory prognostication for the numerical experiment conducted for the variable V_g;

ii. It was allowed by the developed model to analyze the sensitivity of the variables ε_g and ε_s with the variable $\varepsilon_{g,0}$, at the inlet, as well as the variables V_g, V_s and p by using the variable at the inlet $V_{g,0}$;

iii. It was shown by the variables $\varepsilon_{g,0}$ and $V_{g,0}$ at the inlet, great influence on the behavior of the fluidynamic variables of the gas-solid process concerning the control of the process in this equipment;

iv. In this paper, it was performed using simulations only with $\varepsilon_{g,0}$ and $V_{g,0}$ of feeding. However, other variables at the inlet can be tested.

ACKNOWLEDGEMENT

The author would like to thank CNPQ (National Council of Scientific and Technological Development) for the financial support given (Process 48354 / 2007 / Project / Title: Computer Modeling and Simulation for the Development of the technology of a Gasification Plant in Combined Cycle for the Generation of Electric Energy./Edict CNPq 15/2007 – Universal).

Nomenclature

Dp: Diameter of the particles, m
Fs: Force of interaction between the gas and solid phases per volume unit, $kg\ s^{-2}\ m^{-2}$

G: Gravity acceleration, m s^{-2}
H: Length of the fluidization zone, m
K: Constant of equation (9)
N: Constant of equation (10)
P: Pressure, Pa
P_0: Initial pressure, Pa
Q: Dimensionless pressure
T: Temporal coordinate, s
Vg: Velocity of the gas phase, m s^{-1}
$V_{g,0}$: Velocity of the initial gas phase, m s^{-1}
Vs: Velocity of the solid phase, m s^{-1}
Vt: Ending velocity of the liquid particle, m s^{-1}
U: Dimensionless Velocity
Z: Special coordinate, m

Greek letters

ε_g: Volumetric fraction of the gas phase
$\varepsilon_{g,0}$: Volumetric fraction of the initial gas phase
ε_s: Volumetric fraction of the solid phase
μ_g: Coefficient of viscosity of the gas phase, Pa s
ρ_g: Density in the gas phase, kg m^{-3}
ρ_s: Density in the solid phase, kg m^{-3}
τ : Factor of relaxation ($0 \leq \tau \leq 1$) for the gas phase, s
Δt: Time pace, s

REFERENCES

Abdullah MZ, Husam Z, Yin PSL (2003). "Analysis of coal flow fluidization test results for various biomass fuels, Biomass Bioenergy, 24: 487-494.

Basu P (1999). "Combustion of coal circulating fluidized-bed boilers: a review." Chem. Eng. Sci, 54: 5545-555.

Gabra M, Nordin A, Ohman M, Kjellstrom B (2001c) "Alkali retention/separation during bagasse gasification: a comparison between a fluidized bed and a cyclone gasifier". Biomass Bioenergy, 21: 461-476.

Gabra M, Pettersson E, Backman R, Kjellstrom B (2001a) "Evaluation of cyclone gasifier performance for gasification of sugar cane residue-Part I: gasification of bagasse". Biomass Bioenergy, 21: 351-369.

Gabra M, Pettersson E, Backman R, Kjellstrom B (2001b). "Evaluation of cyclone gasifier performance for gasification of sugar cane residue-Part I: gasification of cane trash". Biomass Bioenergy, 21: 371-380.

Graham RL, Walsh ME (1996). "Evaluating the economic costs, benefits and tradeoffs of dedicated biomass energy systems: the importance of scale", In: Second Biomass Conference of the Americas: Energy, Environment, Agriculture, and Industry, Portland, Oregon, pp. 207-215.

Heitor MV, Whitelaw JH (1986). "Velocity temperature and species characteristics of the on in a gas turbine combustor", Combust. Flame, 1: 64-76.

Kunii D, Levenspiel O (1991). "Fluidization Engineering", second ed., Butterworth- Heinemann, Boston, pp. 137-164.

Larson ED, Williams RH (1990). "Biomass-Gasifier Steam-Injected Gas Turbine Cogeneration". J. Engine. Gas Turbines Power, 112: 157-163.

Levenspiel O (2002). "Modeling in chemical engineering". Chem. Eng. Sci., 57: 4691-4696.

Rice RG, Duong DD (1995). "Applied mathematics and modeling for chemical engineers, John Wiley & Sons, Inc, United States, pp. 706.

Srinivasan J, Angirasa D (1988). "Numerical study of double-diffusive free convection from a vertical surface". Int. J. Heat, Mass Transfer, 31: 2033-2038.

7

Statistical analysis and evaluation of lithofacies from wireline logs over 'Beleema' field, Niger Delta, Nigeria

Enikanselu P. A* and Ojo A. O

Department of Applied Geophysics, Federal University of Technology, P. M. B. 704, Akure, Ondo state, Nigeria.

A statistical analysis of well-log data for the purpose of estimating and evaluating lithofacies with depth, around 'Beleema' field, Niger Delta, was carried out. The principal component analysis (PCA) technique and the bulk volume water and the grain size relationships formed the main principles of analysis. The PCA was applied to selected lithology-sensitive logs, each serving as a variable within the trivariate statistical system. It involved determining the first principal component (PC1) in each well, plotting them against depth and segmenting into intervals with similar statistical characteristics. The gamma ray log was used as control for the segmentation. A total of forty-three (43) electrofacie-blocked units, grouped into two major facies - sand and shale (major) - as well as two shaly-sand and sandy-shale (minor), were identified within the wells studied. Also, the bulk volume water (BVW) was observed to vary from 0.0224 in coarse grained sand facies to 0.0892 in silt grained sand facies. The grain size values obtained varied from about 0.0625 mm in silt grained facies to a range of 0.5 to 1.0 mm in coarse grained facies. The computed BVW curves closely mimic the field gamma ray traces; such that the former could be confidently employed where the latter is unavailable.

Key words: Lithofacies, principal component analysis (PCA), bulk volume water (BVW), irreducible water saturation and producibility.

INTRODUCTION

Lithofacies identification is important for many geological and engineering disciplines. Lithofacies, rock or sediment units, characterized by texture or other features can be used to correlate and predict important reservoir characteristics such as permeability and porosity (Chikhi et al., 2005). Identifying various lithofacies of the reservoir rocks is a primary task for petroleum reservoir characterization. Traditionally, lithofacies are identified from cores. However, while core data provide direct observations of lithofacies, they are costly to acquire and recovery is often less than 100% as they seldom encompass the entire stratigraphic interval of interest (Chang et al., 2000). Also, core description can be time consuming and dependent on geologists' extensive wealth of experience. Therefore, an alternative lower-cost method yet providing similar or improved accuracy and resolution is desirable.

A number of approaches have been applied in the estimation and evaluation of lithofacies from well log data; all of which involve intense computer analysis and programming. Some of these methods include the hybrid neural networks which combine probabilistic neural method with radial-bias function (Chikhi et al., 2005), clustering method, litho quick-look approach and the multidimensional histogram approach (Frew, 2004) and cluster analysis technique (Saggaf and Nebrija, 2000). Others include Doveton (1986) who provided a clue to mathematical analysis of log trends and patterns and Elek (1988) who showed how principal-component analysis could be applied to zonation and well-log correlation. Asquith (2004) related the bulk volume water (BVW) with the grain sizes of facies with depth.

In this paper, we have subjected a suite of well-logs obtained from the "Beleema field" of the Niger Delta to the principal component analysis (PCA). The PCA technique is a procedure that transforms a number of

*Corresponding author. E-mail: pa.enikan@yahoo.com.

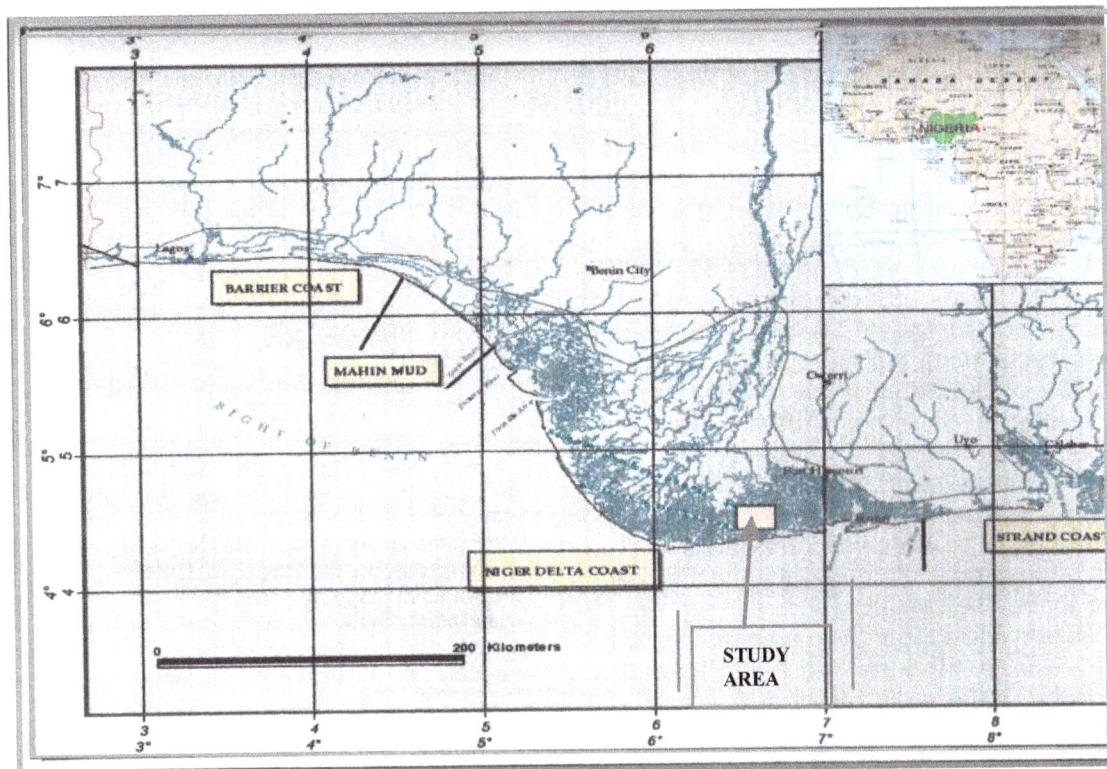

Figure 1. Map of the Nigerian Coastline (Awosika and Folorunso, 2002) showing the study area.

possibly correlated variables into a smaller number of uncorrelated variables called principal components. The first principal component accounts for the variability in the data while the succeeding components account for as much of the remaining variability. In doing this, the mean, standard deviation, variance and covariance were computed. The covariance matrix obtained is resolved to obtain the eigenvectors and the eigenvalues of the data set. The variable that accounts for the most variability is usually depicted by the possession of the highest eigenvalues and hence represents the first principal component.

In the study, the PCA method and the grain size technique were integrated to proffer a statistical approach to the estimation of the facies and evaluate their producibility. Not only does the PCA assist in the estimation of lithofacies, it is useful in the reduction of the error thus ensuring accuracy of prediction. The method is amenable to direct digitized numerical well log data values. The results can then be used to predict lithofacies in non-cored wells (or un-cored intervals in cored wells) or more especially in wells that do not have useful lithofacies identification logs (Chang et al., 2000).

The use of the gamma ray log as control in the estimation is to provide information on the subsurface geology which is useful in bridging the gap between the geologists and the engineers. It also helps to integrate the statistical analysis with the geology of the environment with a view to maximize the accuracy of the estimation.

Location and geology of the study area

The Niger Delta province is situated in the Southern part of Nigeria between latitudes 4 and 6°N and Longitudes 3 and 9°E (Nwachukwu and Chukwura, 1986). The location of 'Beleema' Field (Figure 1) is situated in the Gulf of Guinea which is part of the Niger Delta Province. From the Eocene to the present, the delta has prograded south-westward, forming depobelts that represent the most active portion of the delta at each stage of its development (Doust and Omatsola, 1990).

These depobelts form one of the largest regressive deltas in the world with an area of about 300,000 km^2 (Kulke, 1995), a sediment volume of 500,000 km^3 and a sediment thickness of over 10 km in the basin depocenter. The Niger Delta Province contains only one identified petroleum system (Kulke, 1995; Ekweozor and Daukoru, 1994). This system is referred to as the Tertiary Niger Delta (Akata–Agbada) Petroleum System

Lithostratigraphy

The regional lithostratigraphy of the Niger Delta reveals that it consists of three broad facies units or formation.

These are the continental top facies (Benin Formation), the paralic delta front facies (Agbada Formation) and the Akata Formation which is the pro-delta facies (Short and Stäuble, 1967). The Benin Formation is the shallowest unit and occurs throughout the entire onshore and part of the offshore Niger delta and very little hydrocarbon accumulation has been associated with its highly porous and generally fresh water-bearing sand (Short and Stauble, 1967). It consists predominantly of fresh water-bearing continental sands and gravels deposited in an upper deltaic plain environment. The overall thickness of the Formation varies from 1000 feet(305 m) in the off-shore to 10,000 feet (3050 m) onshore. Various structural units are identifiable within the formation: point bars, channel fills, and natural levees. The oldest known age of the Benin Formation at the surface is Miocene while at the subsurface, Oligocene.

The Agbada Formation underlies the Benin Formation and consists of alternation of coastal fluviomarine sand, siltstone and shale. The formation occurs in the subsurface throughout the entire Niger Delta area with thickness ranging from 300 to 4500 m. Most structural traps observed in the Niger delta developed during syn-sedimentary deformation of the Agbada paralic sequence (Evamy et al., 1978). The primary seal rock is the interbedded shale within the Formation.

The Akata Formation is the basal sedimentary unit and it is characterized by uniform dark grey marine shale, deposited as the high speed delta advance into deep water. It consists of sandstone, siltstone and plant re-mains in the upper parts - rich in fauna of both planktonic and benthonic type. The Akata shales are typically under-compacted and over-pressured and form diapiric structures including shale swells and ridges which often intrude into overlying Agbada Formation. The Akata Formation is thought to be the main source rock of liquid hydrocarbon in the Niger Delta complex.

MATERIALS AND METHODS

The materials used for this study include: bulk density (RHOB), gamma ray (GR), neutron porosity (NPHI), sonic (Δt), deep resistivity (RES), and water saturation (S_w) logs and the base map of the study area. These lithology-sensitive logs were selected from four of the six wells in the field; the two others were used for bulk volume water estimation and grain size analysis. In each well, digitized log data from three lithology identification well-logs were imported from their 'Excel' or 'ASCI' formats unto the 'MATLAB' scientific package. The three selected logs were turned into a population of numerical data sets having zero mean and unit variance. They were then subjected to principal component analysis. The well-log in each well whose eigenvector has the highest eigenvalues becomes the first principal component (PC1) that is, it contains the highest degree of variability of the log data which are representative of lithofacies.

These first principal component logs were selected and segmented (blocked) into intervals of similar statistical behavior. The bulk volume water (BVW), that is, the product of the average water saturation of each facies and the corresponding effective porosity, as computed from either the density or neutron logs, was estimated for the interval under consideration. The BVW values

were subsequently used to classify the facies into grain size ranges (Asquith, 2004).

Both the bulk volume water and the grain sizes were used to predict the producibility of the entire interval; since at irreducible water saturation, it is expected that the facies would produce water–free hydrocarbon. The producible reservoir facies identified in the selected wells were correlated with their respective grain sizes using the gamma ray (GR) log. Correlations of the first principal components were also carried out with the predicted facies.

Theoretical background

The principal component analysis is a classical statistical method of analyzing, discriminating and reducing high dimension data. It involves the conversion of a multivariate set of closely correlated variables into a set of values of uncorrelated variables called principal components, via the eigenvector-eigenvalue based statistical approach. The number of the principal components may be less than or equal to the number of the original variables.

The procedure involves calculating the variance of the entire multivariable system; comparing the variances to obtain a covariance matrix, from where the eigenvalues as well as the eigenvector are deduced. The variance and eigenvalue of the variables are a function of the degree of variability in the data. Hence, the first principal component (PC1) is significant and important in that it not only accounts for as much of the variability in the data set compared to all other principal components, but also reveals the variable having the strongest pattern of distribution among the variables.

The major advantage of the PCA is that it helps in identifying patterns in data, thus expressing the data in such a way as to highlight their similarities and differences. Since patterns in data can be hard to find in data of high dimension, where the luxury of graphical representation is not available, PCA is a powerful tool for analyzing data. The other advantage of PCA is that, once these patterns are found in the data, it is possible to compress or reduce the number of dimensions, without much loss of information.

RESULTS AND DISCUSSION

In all, wireline logs from four of the six wells procured from the study area were used for the estimation and evaluation of the lithofacies. Wells 2, 4, 5 and 6 had gamma ray, neutron, density, resistivity logs (the dual laterolog) and sonic logs, for the PC analysis. Wells 1 and 3 had effective porosity (E-PHI), V_{shale}, and caliper logs, for BVW estimation and grain size analysis. Where these logs were unavailable, the S_w and Φ_e were computed using the Archie (1942) formulation. The gamma ray log (being a primary lithology log) was used as control log in the blocking (or segmentation) of the first principal component traces. The generated bulk volume water (BVW) traces were used to predict the grain sizes of the respective facies which were in turn related to the irreducible water saturation (S_{wir}) from which the producibility of such facies was deduced. The details are presented in the following.

Principal component analysis (PCA)

Table 1 shows the results of the PCA obtained from wells

Table 1. Results of principal component analysis in wells 2, 4, 5 and 6.

Wells	Well 2			Well 4			Well 5			Well 6		
Variable	ΔT	GR	RES	NPHI	LLD	RHOB	RHOB	LLD	NPHI	RES	RHOB	NPHI
Mean	124.9	28.9	147.7	0.345	5.49	2.085	2.18	106.2	0.345	32.6	2.1340	0.340
Standard deviation	2.85	14.9	83.73	0.082	2.25	0.239	0.096	56.6	0.065	43.4	0.0574	0.073
Variance	8.14	222.5	7011.1	0.007	5.068	0.057	0.008	3206	0.004	188	0.0033	0.005
Covariance	8.14	13.44	-81.8	0.006	-0.08	-0.01	-0.004	-0.08	0.004	188	-0.143	-0.67
	13.4	222.5	-858	-0.11	0.245	0.057	0.009	1.61	-.004	-0.1	0.0033	-0.00
	-81.8	-858	7011	-0.08	5.096	0.245	1.6147	3205.	-0.09	-0.7	-0.001	0.005
Eigenvectors	0.99	0.03	-0.011	0.973	-0.02	-0.231	0.5047	-0.00	0.863	0.000	0.000	0.999
	-0.03	0.99	-0.123	0.232	0.04	0.972	-0.863	0.00	0.504	0.896	-0.45	8.461
	0.01	0.12	0.992	0.004	0.99	-0.051	0.0006	0.99	-2.76	0.444	0.896	0.001
Eigenvalues	7.08	0.00	0.00	0.003	0.00	0.00	0.0000	0.00	0.002	0.01	0.000	0.000
	0.00	116	0.00	0.000	5.11	0.00	0.0097	0.00	0.000	0.00	0.006	0.000
	0.00	0.00	7119	0.000	0.00	0.05	0.0000	3206	0.000	0.00	0.000	188.1
Percentage	0.009	1.59	98.30	0.012	99.0	0.93	0.0003	99.9	0.001	0.01	0.003	99.95

2, 4, 5 and 6. In well 2, the selected logs were the sonic, gamma ray and resistivity logs and the computed means respectively. The process revealed the highest value of standard deviation in the resistivity log with a value of 83.73 Ω-m. while the lowest value of 2.85 μs/feet was recorded for the sonic log. The variances of each data set also showed the highest value in the resistivity log while the lowest was in the sonic log.

An estimation of the covariance revealed a three by three matrix; resolved to produce the eigenvalues for the trivariate system, a diagonal matrix having values of 7.08 μs/feet, 116 API and 7119 Ω-m in sonic, gamma ray and resistivity logs respectively. This result showed that most of the variability was contained in the resistivity log; thus constituting the first principal component (PC1) in the well. The first principal component was selected and transformed into its principal component trace and segmented into eleven (11) electrofacies zones while the gamma ray log served as control. Similarly, wells 4, 5 and 6 had 11, 11, and 10 electrofacies zones respectively (Figure 2), totaling forty three (43). were 124.9 μs/feet, 28.9 API and 147.73 Ω-m

Figure 2 shows the observed inverse relationship between resistivity and the gamma ray response for the wells studied. The 'highs' on the gamma ray log coincide with the 'lows' on the resistivity logs and vice versa. Such areas are C2, C4, C6, C8 etc. This is characteristic of a facies unit dominated and influenced by quartz (Barrash et al., 1997). We adduce the fairly low gamma ray log response observed to the presence of scanty radioactive materials in sand while the relatively high resistivity values are a reflection of the low conductivity of the sandy materials. Facies recognized with these characteristics therefore reflect that of a shaly-sand. The sand-bearing facies are characterized by very low gamma ray responses and high resistivity; for depth intervals of 4,650 to 4,900 feet (1417.32 to 1493.52 m) and 4,910 to 5,090 feet (1496.57 to 1551.43 m); notwithstanding observed occasional thin shale lenses.

In all, a total of 11 electrofacies zones were delineated into shaly-sand, sandy-shale, sand and shale. These are identified as C1 through C11 (Figure 2).

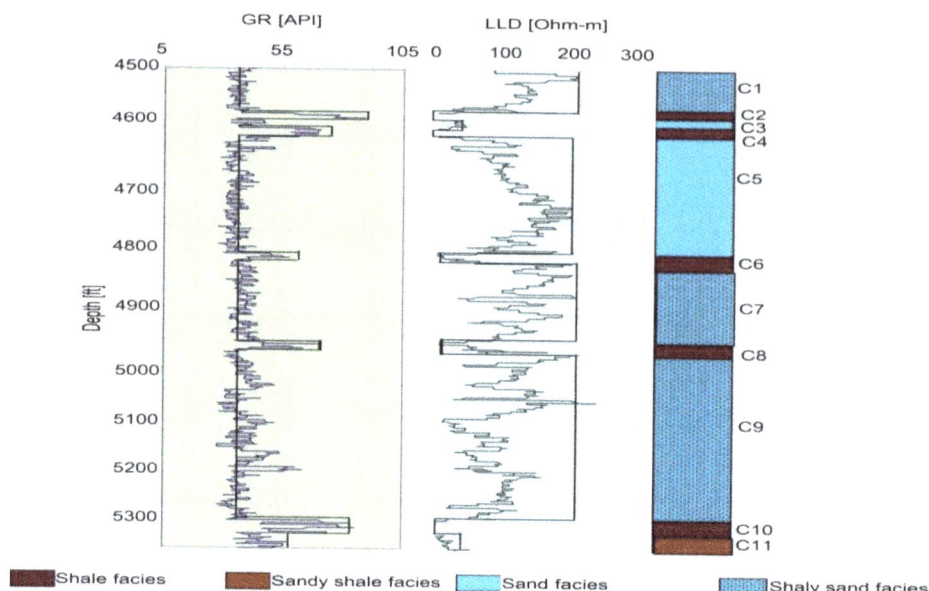

Figure 2. The segmented gamma ray (blue), PC1 (black) and the interpreted facies in well 4.

Similarly, in well 4, the neutron, deep laterolog and the density logs were analysed and the estimated mean for the three logs were found to be 0.345, 5.49 Ω -m and 2.085 g/cc respectively (Table 1). The standard deviation and the variance showed the highest value in the deep resistivity (laterolog) with values of 2.25 and 3.0956 Ω-m respectively. This is in contrast to the low values recorded for the bulk density and the neutron logs observed to be 0.239 and 0.057 g/cc in the density log and 0.082 and 0.007 in the neutron log.

The covariance matrix obtained for the three well logs, when resolved statistically, produced a group of eigenvectors whose eigenvalues were computed to be 0.003, 5.11 Ω-m and 0.05 g/cc for the neutron, deep laterolog and the density logs respectively. From the analysis, most of the variability, therefore resided in the deep laterolog; hence the first principal component.

Transformation of the principal component and segmentation produced eleven (11) blocked electrofacies zones (Figure 2) grouped into shaly-sand, sandy-shale, sand and shale. The shaly sand facies occur at a depth of 4,500 to 4,570 feet (1371.6 to 1392.94 m) and 4,800 to 4,820 feet (1463.04 to 1469.13 m) and are designated C1 and C6. The sand bearing facies occurs at depths of 4,610 to 4,800 feet (1405.13 to 1463.04 m), 4.810 to 4,940 feet (1466.09 to 1505.71 m) and 4,970 to 5,250 feet (1514.86 to 1600.20 m) and has been designated C5, C7 and C9.

In well 5, mean values were 106.2 Ω-m, 2.18 g/cc and 0.345 for the deep laterolog, density and the neutron logs respectively, and standard deviation values were 56.6 Ω-m, 0.096 g/cc and 0.065 respectively. The variance is highest in the deep laterolog with values of 3206 Ω-m, and lowest in the neutron log with values of 0.004;

density values being 0.008 g/cc. The first principal component was the deep laterolog and the gamma ray served as control for the transformation and segmention to produce eleven (11) electrofacies-blocked zones (not shown) grouped into four facies viz: sand, shaly sand, sandy shale and shale bodies.

The sand bearing facies at depths of 4,675 to 5,075 feet (1424.94 to 1546.86 m) shows a characteristic increase in the resistivity response but a lower gamma ray response due to low radioactive materials present in them. However, a shaly-sand facies, characterized by a sharp increase in the resistivity response and a sudden drop in the resistivity response, interrupted the sequence at a depth of 4,800 feet. This suggests the possible presence of a gaseous hydrocarbon. Shaly sand facies occurs at a depth interval of 4,500 to 4,570 feet (1371.6 to 1392.94 m) while a sandy shale occurs at a depth interval of 5,080 to 5,120 feet (1548.38 to 1560.58 m).

In well 6, the mean values of 32.6 Ω-m, 2.1340g/cc and 0.341 were recorded for the resistivity, density and the neutron logs respectively. The standard deviation also increased in that order with resistivity log having the highest value of 43.4 Ω-m and the density and neutron logs with values of 0.0574g/cc and 0.073 respectively. However, the variance was highest in the resistivity log (1883 Ω-m) but lowest in the neutron log (0.0033). It was found to be moderate in the density log, (0.005 g/cc). A resolution of the covariance matrix obtained from the analysis revealed eigenvectors having eigenvalues of0.002 Ω-m, 0.0056g/cc and 188.1 in the resistivity, density and neutron logs respectively. The neutron log was therefore identified as the PC1.

The first principal component in the well, Neutron log and the gamma ray log, when transformed and segmented

Table 2. Computed bulk volume water and grain size analysis of delineated facies.

	Well 4					Well 5			
DI [m]	FU	BVW	GSR[mm]	DF	DI [m]	FU	BVW	GSR [mm]	DF
1372-1389	C1	0.0347	0.025 - 0.035	Medium grain shaly sand	1372-1389	C1	0.0263	0.025-0.035	Medium grain shaly sand
1389.8-1395.5	C2	0.0693	0.05- 0.07	Very fine grain shale	1389.6-1402.0	C2	0.0681	0.05-0.07	Very fine grain shale
1396.0-1402.1	C3	0.0752	0.07 - 0.09	Silty sand	1402.1- 1414.3	C3	0.0439	0.035-0.05	Fine grain sand
1402.1-1408.2	C4	0.0653	0.05- 0.07	Very fine grain slide	1414.3-1456.9	C4	0.0519	0.05-0.07	Very fine grain shale
1408.2-1463.0	C5	0.0403	0.035-0.05	Fine grain sand	1426.5-1456.9	C5	0.0416	0.035-0.05	Fine grain sand
1463.0-1471.0	C6	0.0520	0.05-0.07	Very fine grain shaly sand	1456.9-1463.0	C6	0.0573	0. .05-0.07	Very fine grain shaly sand
1470.9-1508.8	C7	0.0249	0.02- 0.025	Coarse grain sand	1463.0-1553.9	C7	0.0224	0.02-0.025	Coarse grain sand
1509.1-1514.9	C8	0.0555	0.05-0.070	Very fine grained shale	1553.9-1562.1	C8	0.0531	0.05-0.07	Very fine grain shale
1515.2-1600.0	C9	0.0340	0.025-0.035	Medium grained sand	1562.1-1571.2	C9	0.0872	0.07-0.09	Silty grain sand
1600.2-1607.8	C10	0.0576	0.05-0.070	Very fine grained shale	1571.2-1581.9	C10	0.0536	0.05-0.07	Very fine grain shale
1607.8-1615.0	C11	0.0892	0.07 - 0.09	Silty sand	1581.9-1615.4	C11	0.04201	0.035-0.05	Fine grain sand

	Well 6			
DI [m]	FU	BVW	GSR[mm]	DF
1372 -4630	C1	0.0333	0.025-0.035	Medium grain shaly sand
1411.2-1414.3	C2	0.0503	0.05 - 0.07	Very fine grain shale
1414.3- 1417.3	C3	0.0467	0.035- 0.05	Fine grain sand
1417.3-1420.4	C4	0.0564	0.05 - 0.07	Very fine-grain shale
1420.4-1447.8	C5	0.0341	0.025 - 0.035	Medium grained sand
1447.8-1450.8	C6	0.0512	0.05 - 0.07	Very fine grain shaly sand
1450.8-1478.3	C7	0.0243	0.02 - 0.025	Coarse grain sand
1478.3-1505.7	C8	0.0625	0.05 - 0.07	Fine grain shaly sand
1505.7-1600.2	C9	0.0302	0.025 - 0.035	Medium grain sand
1600.2-1615.4	C10	0.0848	0.07 - 0.09	Silty shale

DI: delineated interval; FU: facies unit; BVW: bulk volume water; GSR [mm]: grain size ranges; DF: delineated facies.

produced ten (10) electrofacie- blocked zones comprising shaly-sand, sand and the shale bodies.

The shaly-sand occurs between the depths of4,500 to 4,600 feet (1371.6 to 1402.08 m), 4,850 to 4,950 feet (1478.28 to 1508.76 m) and 4,730 to 4,760 feet (1441.7 0 to 1450.85 m). They are characterized by a generally low density response but having intermittent increase in dense shale material in them. Sand bearing facies, however occurs at depths of 4,660 to 4,720 feet (1420.37 to 1438.66 m), 4,760 to 4,850 feet (1450.848 to 1478.28 m) and 4,940 to 5,260 feet (1505.71 to 1603.25 m). They are generally characterized by low bulk density responses.

Bulk volume water estimation and grain size analysis

The bulk volume water was estimated for the

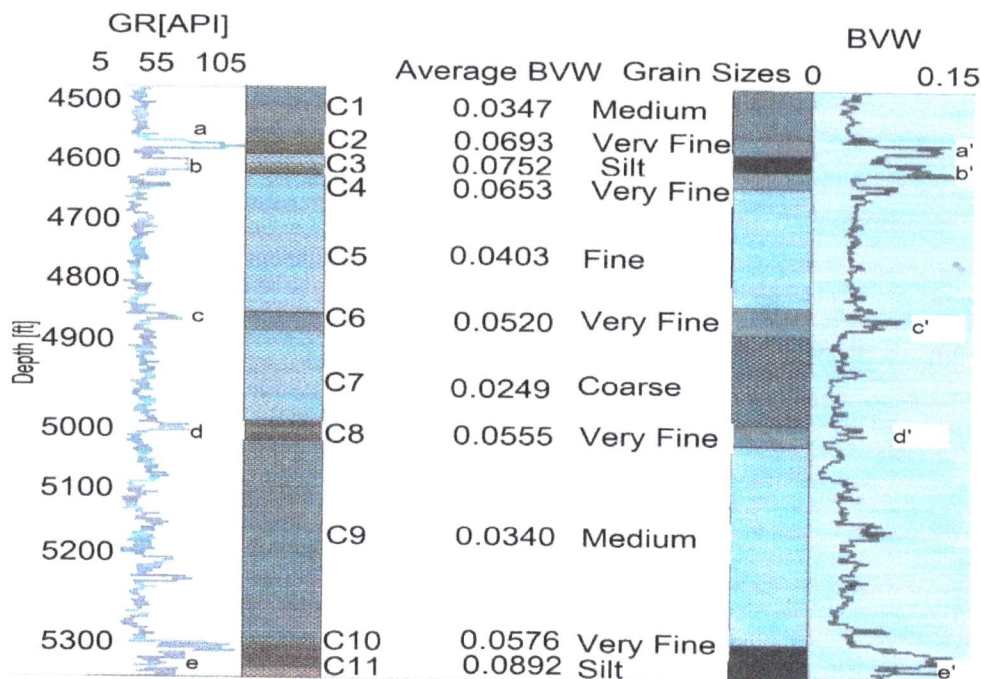

Figure 3. Bulk volume water at Irreducible water saturation correlated with gamma ray log in well 4.

entire depth interval that is, 4500 to 5300 feet (1371.6 to 1615.4 m). Table 2 shows the results as computed in wells 4, 5, and 6. The grain sizes (Asquith, 2004) were also deduced. Usually, the lower the values of the bulk volume water, the lower the irreducible water saturation, hence the higher the hydrocarbon saturation. At this point, the facies materials usually possess coarser grain texture. Conversely, the finer the facies material, the higher the bulk volume water value; and so the higher the irreducible water saturation (S_{wir}). At this point, the reservoir facies becomes less productive; producing a great deal of water with little or no hydrocarbon.

Well 4

Table 2 shows the results of the BVW and the grain sizes of the delineated facies in well 4 for the depth interval of interest. Further, Figure 3 shows the correlation of the computed bulk volume water at irreducible water saturation with the field gamma ray log in the well. The BVW values ranged from as low as 0.00249 in medium grained sand facies to as high as 0.0892 in fine grained silt. The low values were encountered in facies C1, C5, C7 and C9, the lowest being at C7, C1 and C9. Hence these facies would be at relatively low irreducible water saturation, hence their producibility increases from C9 to C1 and finally to C7 (Figure 3).

In Figure 3, a strong correlation was observed to occur between the gamma ray trace and the bulk volume water trace in the well. Increase in the bulk volume water value showed a corresponding increase in the gamma ray

values (points a/a′, b/b′ to........... e/e′ etc., Figure 3). This can be explained by the fact that very fine grained materials in shale contain more pore spaces that hold back more amount of water in its matrix, hence leading to higher bulk volume water values.

The bulk volume water estimate increases from as low as 0.0263 in a medium grained shaly sand facies to as high as 0.0872 in the silt in the well. From the grain size analysis, a coarse grained sand facies was encountered at a depth of 4,820 to 5,080 feet (1469.14 to 1548.38 m) while a medium grained shaly-sand facies exists at a depth of 4,500 to 4,550 feet (1371.6 to 1386.84 m). Fine grained sand facies are encountered at depths of 4,680 to 4,795 feet (1426.46 to 1461.52 m), 5,250 to 5,300 feet (1600.2 to 1615.44 m) and 4,600 to 4, 640 feet (1402.08 to 1414.27 m) (Table 2).

From the analysis, the bulk volume water (BVW) estimate showed that there exist low values at facies C7, C1. These derived values possibly implied that these two facies are at irreducible water saturation and the grain sizes are larger, with C7 having a coarser grain size while C1 has a medium grained size. Hence they are more producible than any other facies in the well (Figure 3).

Well 6

In well 6, the bulk volume water increases from as low as 0.0243 in the sand facies in C7 to as high as 0.0848 in silt. Grain size analysis revealed a medium grained shaly-sand at depths of 4,500 to 4,620 feet (1371.6 to 1408.18 m) and medium grained sand at 4,940 to 5,500 feet

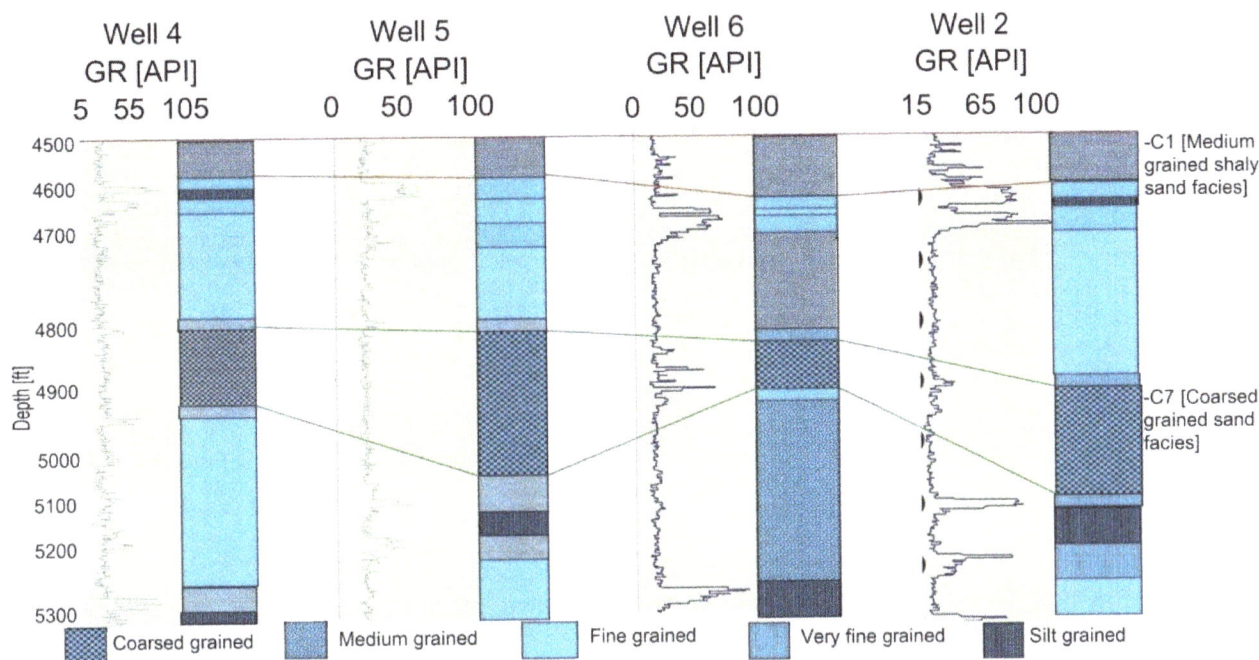

Figure 4. Reservoir facies correlation of gamma ray log with grain sizes in wells 4, 5, 6 and 2.

(1505.71 to 1676.4 m). Coarse grained sand was encountered at a depth of 4,760 to 4,850 feet (1450.85 to 1478.28 m).

The decrease in the bulk volume water in facies C1, C5, C7 and C9 possibly indicate that these facies are at irreducible water saturation, and hence a larger grain sizes than the others in the well. They all have higher porosity and permeability and hence higher producibility.

Reservoir facies correlation

Having identified the producible facies units in each well using the bulk volume water and the grain sizes, a relationship between the grain sizes of the entire interval under consideration is established with that of the gamma ray response by correlation in order to predict the reservoir facies in the entire interval. The result is as shown in Figure 4.

The reservoir facies correlation indicates that producible facies that may be of interest are those of facies C1 and C7 due to their low bulk volume water (0.0347 and 0.0249) respectively (Figure 3). Thus, this indicates low irreducible water saturation. Also, their large grain sizes are indicative of high porosity and permeability

Conclusion

The principal component analysis technique has been effectively applied to well-log data obtained from the Beleema field in Niger Delta to evaluate lithofacies of similar statistical characteristics and their behavior. The variables have been selected in four wells and segmented into depth intervals of similar statistical behavior using gamma ray log as control. A resultant forty-three (43) electrofacie-blocked units (Figure 4) were delineated in the four of the six wells studied and were grouped into major and minor facies.

The result of the BVW analysis revealed varied bulk volume water values from as low as 0.0224 in coarse grained sand to 0.0892 in silt grained sand facies. From these, the grain sizes of the facies units as well as their respective producibility were deduced. The facie grain sizes varied from as high as 1.0 to 0.5 mm and 0.5 to 0.25 mm in coarse sand facies and medium grained shaly-sand facies respectively. Grain size values less than 0.0625 mm were associated with silt in the shale body. The producible coarse grained sand facies were found to possess very low bulk volume water of 0.0249, 0.0224 and 0.0243 in wells 4, 5 and 6 respectively. Also, the producible medium-grained sand facies were found to have bulk volume water of 0.0347, 0.0263 and 0.03333 in wells 4, 5, and 6 respectively. These two producible facies were at low irreducible water saturation and of higher grain sizes. Increase in the bulk volume water value showed a corresponding increase in the gamma ray values. This could be due to the fact that very fine-grained materials in shale contained more pore spaces that hold back much water in its matrix, hence leading to higher bulk volume water.

From the analysis, a strong visual correlation between the gamma ray and the bulk volume water traces was observed in the wells; indicating the possibility of employing the bulk volume water trace as a surrogate for facies estimation in the absence of gamma ray log.

Based on the results, coupled with the grain sizes and information from predicted irreducible water saturation, two viable and producible reservoir facies were mapped and correlated within the depth interval of study. These were the medium grained shaly-sand facies at a depth of 4500 to 4580 feet (1371.6 to 1395.98 m) and the coarse grained sand facies at a depth of 4900 to 4980 feet (1493.52 to 1517.90 m). The observed good correlation between the gamma ray and the bulk volume water (BVW) traces suggests that the latter could be employed in areas where the former is unavailable.

ACKNOWLEDGEMENTS

The authors are grateful to the Chevron Nigeria limited for providing us the geophysical data with which the work was carried out. Also, the Department of Applied Geophysics, Federal University of Technology, Akure, Nigeria, permitted us the use of their computer facilities for the computations.

REFERENCES

Archie GE (1942). The electrical resistivity log as an aid in determining some reservoir characteristics. J. Petroleum Technol., 5: 54 – 62.

Asquith GB (2004). Basic well log analysis for Geologists. American Association of Petroleum Geologists ser Tulsa, pp. 216

Awosika LF, Folorunsho R (2002). An Outline of the Different Geomorphology of the Niger Delta in Shelf Circulation Patterns Observed from Davies Drifter Off the Eastern Niger Delta in the Gulf of Guinea. Nigeria Institute for Oceanography and Marine Research Report; Victoria Island, Lagos.

Chang CH, Kopaska-Merkel DC, Chen HC, Duran R (2000). Lithofacies Identification Using Multiple Adaptive Resonance Theory of Neural Networks and Group Decision Expert System. Comput. Geosci., 26: 591- 601.

Chang HC, Kopaska-Merkel DC, Chen HC (2002). Identification of lithofacies using Kohonen self-organizing maps. Comput. Geosci., 28: 223–229.

Chikhi S, Batouche M, Shout H (2005). Hybrid Neural Network Methods for Lithology Identification in the Algerian Sahara. Int. J. Comput. Intell., 1: 6-12.

Doust H, Omatsola E (1990). Niger Delta, in: Edwards, J. D., and Santogrossi, P.A. (eds.) Divergent/passive Margin Basins. American Association of Petroleum Geologists, Memoir 48, Tulsa, pp. 239-248.

Doveton JH (1986). Log analysis of subsurface geology - Concepts and Computer methods, third ed. John Wiley & Sons, New York.

Ekweozor CM, Daukoru EM (1994). Northern delta depobelt portion of the Akata-Agbada Petroleum system, Niger Delta, Nigeria, in: Magoon, L.B., and Dow, W.G. (eds.), The Petroleum System — From Source to Trap, American Association of Petroleum Geologists Memoir, 60, Tulsa, pp. 599-614.

Elek I (1988). Some applications of principal component analysis; well-to-well correlation, zonation. Geobyte, 3:(20): 46-55.

Evamy BD, Haremboure J, Kamerling PWA, Molloy FA, Rowland PH (1978). Hydrocarbon habitat of tertiary Niger delta. Am. Assoc. of Petroleum Geologists Bull., 62: 277-298.

Frew K (2004). Lithofacies estimation tool with the most comprehensive kit. Schlumberger Softw. Solut., 1: 1-6.

Kulke H (1995). Nigeria, in: Kulke, H. (eds.), Regional Petroleum Geology of the World. Part II: Africa, America, Australia and Antarctica: Berlin, Gebrüder Borntraeger, pp. 143-172.

Nwachukwu JI, Chukwurah PI (1986). Organic matter of Agbada Formation, Niger Delta, Nigeria. American Association of Petroleum Geologists Bulletin, 70: 48-55.

Saggaf M, Nebrija L (2000). Estimation of Lithology and Depositional Facies from Wire-Line Logs. Am. Assoc. Petroleum Geologists Bull., 84: 1633-1646.

Short KC, Stauble AJ (1967). Outline of the geology of the Niger delta. Am. Assoc. Petroleum Geologists Bull., 51: 761-779.

8

Reservoir characterization and evaluation of depositional trend of the Gombe sandstone, southern Chad basin Nigeria

Adepelumi, A. A*, Alao, O. A and Kutemi, T. F

Department of Geology, Obafemi Awolowo University, Ile-Ife, Osun State, Nigeria.

This study attempts a reservoir characterization and evaluation of depositional trend of sands to assess the petrophysical qualities of Gombe sandstone as a potential reservoir unit for hydrocarbon accumulation in the Chad basin. The economic viability in parts of the Chad basin outside Nigeria and other structurally related contiguous basins such as Doba, Doseo, Bongor fields, and Termit-Agadem basin in the Niger Republic have been investigated. The investigations revealed commercial petroleum accumulations which necessitated the need to assess the petroleum potentials of the Nigerian portion of the Chad basin. Four sand units within Gombe formation penetrated by five wells (Ngammaeast 01, Ngornorth 01, Kinasar 01, Ziye 01 and Murshe 01) were delineated, correlated and their continuity estimated across the studied wells. Petrophysical parameters of these sand units such as porosity, permeability, water saturation, hydrocarbon saturation, bulk water volume, etc were computed and interpreted. Net-to-gross (NGR) values of these sand units were also calculated, and NGR maps were contoured for each sand unit. The direction of deposition of the sands was thus inferred to be east-west. The interpretation suggests that Gombe sandstone in Chad basin is a potential reservoir for hydrocarbon accumulation.

Key words: Petrophysical, sandstone, net-to-gross (NGR), well, Nigeria.

INTRODUCTION

Reservoir rocks are mainly sedimentary rocks which are products of recycling of rock debris from weathering of pre-existing rocks, for example, sandstone, limestone, dolomite, and shale. Essentially, the pore spaces of the reservoir rocks are interconnected such that petroleum is able to migrate and accumulate in a trap. This intercomnected nature of pore spaces is described as permeability which is a measure of the ease of flow of hydrocarbon in the reservoir. A good reservoir system comprises the reservoir, the trap and an impervious stratum overlying the reservoir usually referred to as caprock or seal for example, shale. Within the Chad basin, the source rocks could also provide suitable seals (Avbovbo et al., 1986).

In the Chad basin, source rocks are mainly in the Gongila formation (Olugbemiro et al., 1997; Obaje et al.,

2006; Adepelumi et al., 2010) and in the Fika shale (Petters and Ekweozor, 1982). Likely reservoirs are the sandstone facies in the Gongila and Fika formations and Gombe sandstone. Obaje et al. (2004) carried out geo-chemical studies to assess the qualities of source rocks penetrated by four wells (Kemar 01, Murshe 01, Tuma 01 and Ziye 01) in the Nigerian sector of the Chad basin. He concluded that fair to poor source quality entirely gas-prone source rocks are inherent in the sequences pene-trated by the studied wells. Also, from biomarker chromatograms and extract vs. TOC plots, the presence of oil shows in Ziye 01 well at a depth of 1210 m was indi-cated, he thus concluded that generated hydrocarbons would be overwhelmingly gaseous.

Gombe sandstone within the Chad basin directly overlies the Fika shale which is considered a potential source rock within the basin. Gombe sandstone is essen-tially a sequence of estuarine to deltaic sandstone. It is also discovered that the Gombe sandstone is restricted to the western part of the basin (Ali and Orazulike, 2010).

*Corresponding author. E-mail: adepelumi@gmail.com.

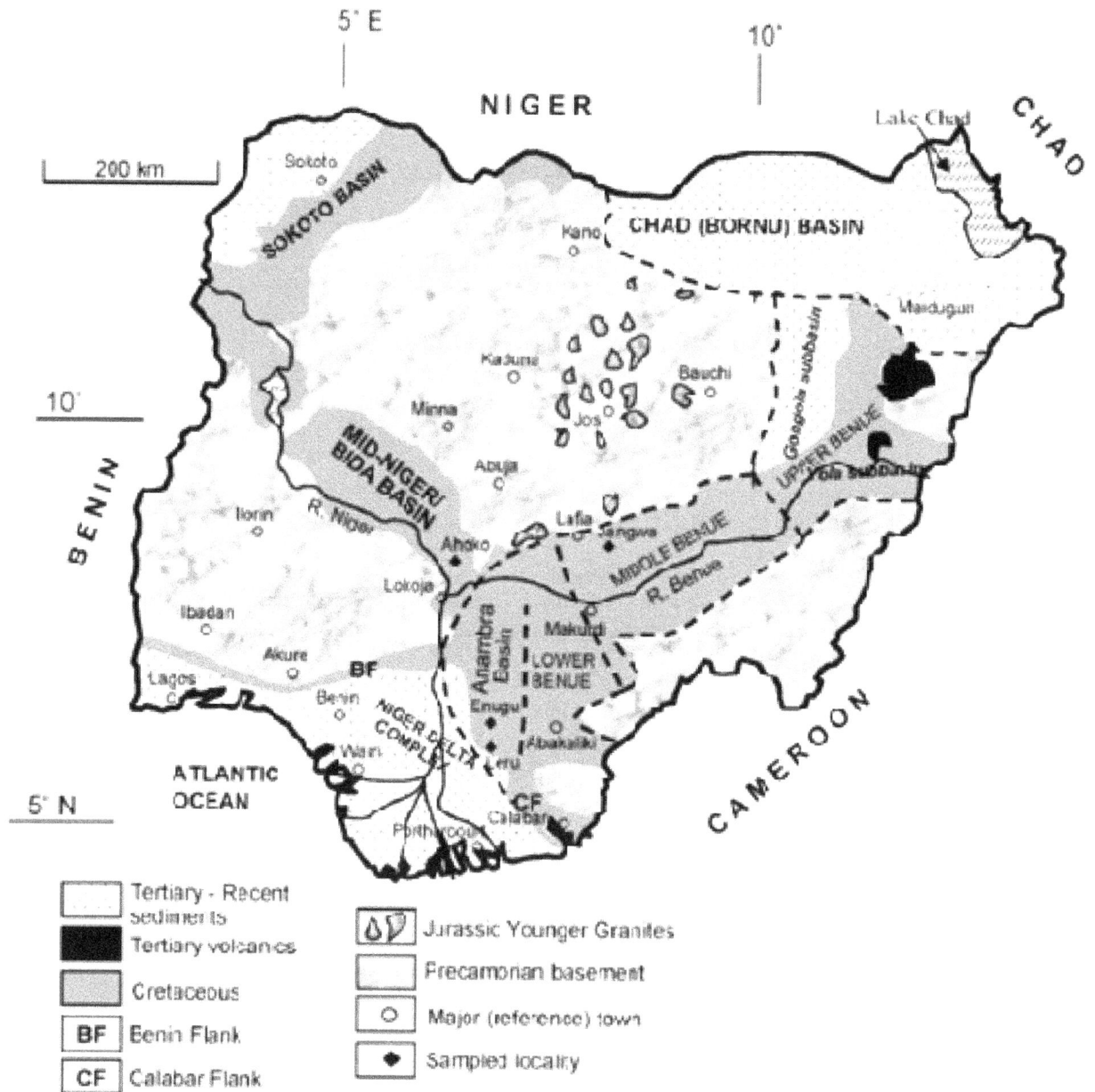

Figure 1a. Geologic map of Nigeria showing the inland basins and the Nigerian sector of Chad basin (Obaje et al., 2006).

The economic viability of parts of the Chad basin outside Nigeria and other structurally related contiguous basins such as Doba, Doseo, Bongor fields, and Termit-Agadem basin in the Niger Republic have been investigated. The investigations revealed commercial petroleum accumulations and this necessitate the need to assess the petroleum potentials of the Nigerian portion of the Chad Basin. Thus, through this project, we aimed to carry out petrophysical evaluation of Gombe sandstone using lithologic, resistivity and porosity logs, and also to confirm the results of this evaluation from various cross-plots in order to ascertain the potential and producibility of the formation as a suitable reservoir rock.

GEOLOGY OF CHAD BASIN

The study area is the Chad basin (Figure 1). The Chad basin in Nigeria is a broad sediment-filled depression stranding Northeastern Nigeria and adjourning parts of Chad Republic. It is separated from the upper Benue basin by the Zambuk ridge (Adepelumi et al., 2010). By far the greater part of the Chad basin is located to the north outside Nigeria in the Republic of Chad. It occupies an area of about 2,500,000 km², extending over parts of Algeria, the Niger, Chad, and Sudan Republics as well as the northern parts of Cameroon and Nigeria. The study area is located within longitude 11°45`E and 14°45`E

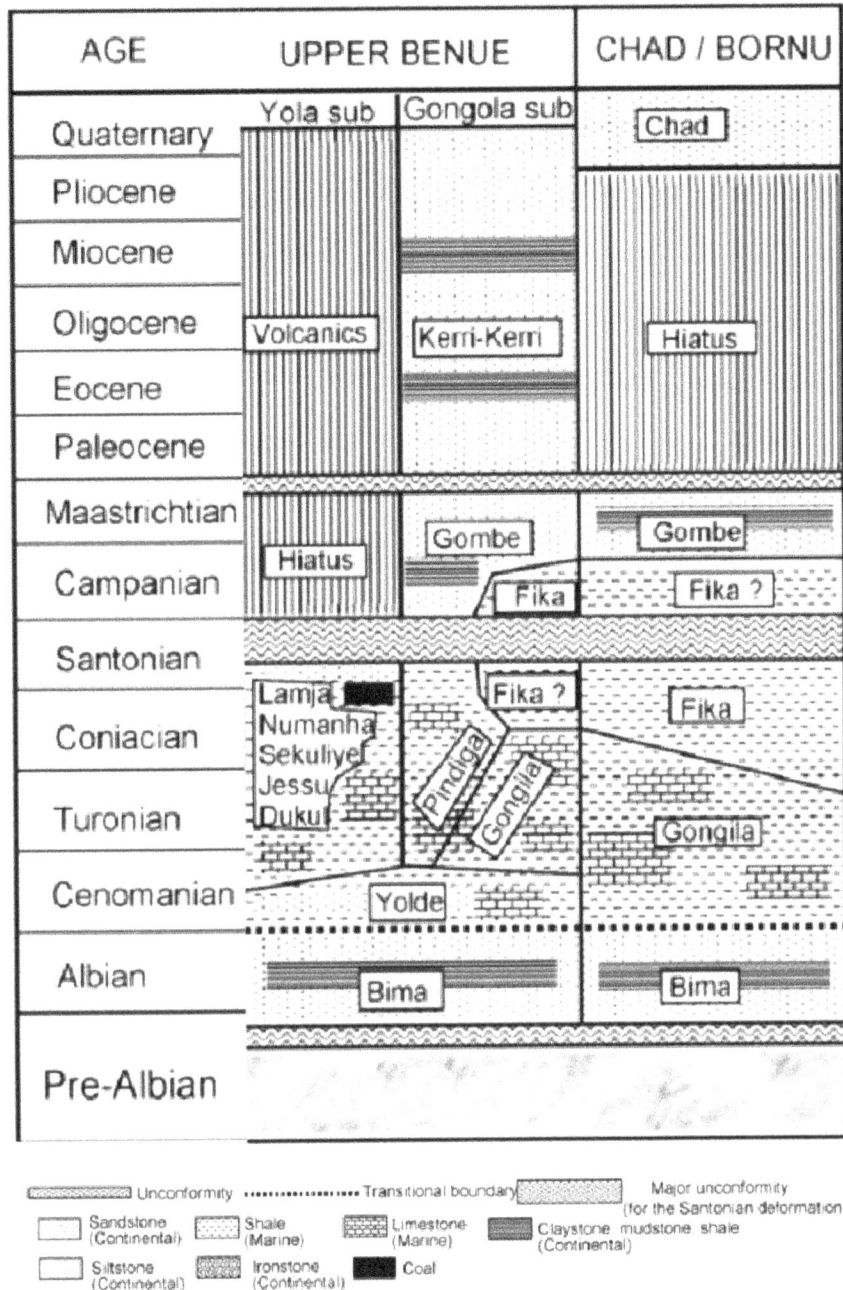

Figure 1b. Stratigraphic successions in the Nigerian sector of the Chad basin in relation to the Benue Trough (Obaje et al., 2006).

and latitude 9°30`N and 13°40`N.

The origin of the Chad basin has been generally attributed to the rift system that developed in the early Cretaceous when the African and South American lithospheric plates separated and the Atlantic opened. Pre-Santonian Cretaceous sediments were deposited within the rift system (Obaje et al., 2004, 2006). The Basin has been developed at the intersection of many rifts, mainly in an extension of the Benue Trough. Major grabens then developed and sedimentation started.

Sedimentary sequences were deposited from the Paleozoic to Recent, accompanied by a number of stratigraphic gaps. Sediments are mainly continental, sparsely fossiliferous, poorly sorted, and medium- to coarse-grained, feldspathic sandstones called the Bima Sandstone. A transitional calcareous deposit - Gongila formation - that accompanied the onset of marine incursions into the basin, overlies the Bima Sandstones. These are overlain by graptolitic shale (Okosun, 1995). The oldest rocks in the Chad basin belong to Bima Sandstone and the youngest to the Chad Formation as shown in the stratigraphic column of the study area (Figure 1b).

Bima sandstone

This formation has essentially the same lithology as in the upper Benue Basin. It is largely constituted of coarse feldspathic and cross-bedded sandstones. It is, however, thinner in the Chad Basin. It has been dated Albian.

Gongila formation

This unit consists of sequence of sandstones, clays, shales and limestone layers. It varies laterally into massive grey limestone overlain by sandstone, siltstones, thin limestone and shales with shally limestone. To the south at Kupto, however, a thick limestone is overlain by sandstones, mudstones, and shales with lime stones (Carter et al., 1963).

The limestone horizons are richly fossiliferous with abundant ammonites, pelecypods and echinoid remains, on the basis of these, Carter et al. (1963) assigned an Early Turonian age to the formation.

Fika shale

This formation consists of blue-grey shale, at times gypsiferous; with one or two non-persistent limestones horizons. A maximum thickness of 430 m has been penetrated in by boreholes near Maiduguri. Fossils of the Fika shale consist mainly of fish remains and fragments of reptiles suggesting a Cenomanian to Maastrichtian age (Dessauvagie, 1975). However, Dessauvagie (1975) suggests a pre-Santonian upper age limit for the formation based on stratigraphic evidence.

Gombe sandstone

This unit is a sequence of estuarine and deltaic sandstone, siltstone and subordinate shale. Thin coal seams are locally present. In outcrop many of the sandstones and siltstones are ferruginised forming low-grade ironstones. The macrofauna is limited and consists of a few indeterminate lamellibranchs (Carter et al., 1963). Shell-BP palynologists dated the coal late Senonian - Maastrichtian.

Kerri-Kerri formation

This consists of loosely cemented, coarse to fine-grained sandstone, massive claystone and siltstone; bands of ironstone and conglomerate occur locally. The sandstone is often cross bedded. The sediments are lacustrine and deltaic in origin and have a maximum thickness of over 200 m (Du Preez and Barber, 1965). The coal in the formation has yielded palynomorphs on the basis of which Shell-BP palynologists dated it Paleocene and later by Adegoke et al. (1986).

Chad formation

This formation is a succession of yellow and grey clay, fine- to coarse-grained sand with intercalations of sandy clay and diatomites. Its thickness considerably varies. It is estimated to be about 800 m thick on the western shore of Lake Chad. Vertebrate remains (*Hippopotamus imaguncula*) and diatoms collected from it indicate an Early Pleistocene (Villafranchian) age. However, its age is considered to range from Pliocene to Pleistocene. The Chad basin is capped by Tertiary volcanic rocks. The Biu Plateau Basalts underlie the Pleistocene diatomite deposits near Bulbaba but overly Cretaceous rocks (Carter et al., 1963). They are thus most probably of Tertiary age. The basalts consist of fine-grained, dense olivine-bearing varieties.

METHODOLOGY

The well log data used in this study were recorded as part of the Chad basin petroleum drilling program of the Nigerian National Petroleum Corporation (NNPC) database available in the Department of Geology, Obafemi Awolowo University, Ile-Ife, Nigeria. The data consist of a suite of well log readings from twenty-three (23) wells drilled in the Chad basin. Five (5) wells were selected for the purpose of this work. The analyses of the well log data began with arrangement of log data in a readable format using Excel Spreadsheet. Some parameters such as sonic-derived porosity and gamma ray index computed on the Excel Spreadsheet were converted into LAS file and imported along with the data already in LAS format into the RokDoc 5.4.4 Software (powered by Ikon Science). These imported data were then plotted as logs. These logs were used in the calculation of the petrophysical parameters which were consequently exported as Excel files and were finally tabulated.

The logs include lithologic logs (spontaneous potential and gamma ray), resistivity logs (induction log deep, short normal, micro-spherically focused log), porosity logs (bulk density, sonic) and caliper logs. The logs include useful information found on the header which includes the well name, depth of drilling, longitude and latitude, surface X and Y coordinate.

The wells used for this study were plotted as they appeared on the base map, that is, from west to east. The plot revealed that the sands reduced in thickness going from east to west. The direction of deposition of the sands was thus inferred to go from proximal (east) to distal (west).

The well logs were carefully studied. The caliper logs which measure the borehole size gives an indication of caving where the readings are inconsistent; the gamma ray logs were used to identify lithologies, that is, sand versus shale; the resistivity logs are also studied carefully in order to delineate hydrocarbon bearing zones in each well, these zones are indicated by high deep resistivity readings. However, high resistivity zones may indicate the presence of fresh water.

Sand units of interest were carefully picked and correlated across the wells to give an idea of the continuity of the reservoirs at different depths across the whole survey area. Petrophysical parameters obtained for reservoir evaluation in the course of this study includes:

1. Volume of shale: To derive V_{sh} from gamma ray, it is imperative that the gamma ray index, that is, I_{GR} is first determined. Using equation 3.1 of Schlumberger (1974):

$$I_{GR} = (GR_{log} - GR_{min}) / (GR_{max} - GR_{min}) \qquad (1)$$

Reservoir characterization and evaluation of depositional trend of the Gombe sandstone, southern Chad...

57

Where I_{GR} = gamma ray index; GR_{log} = gamma ray reading of formation; GR_{min} = minimum gamma ray reading (clean sand or carbonate); GR_{max} = maximum Gamma ray reading (shale).

For the purpose of this work, formula of Larionov (1969) for older rocks was used. Larionov (1969) formula for Tertiary rocks:

$$V_{sh} = 0.083(2^{3.7*IGR} - 1) \qquad (2)$$

On the basis of the following V_{sh} cut-offs, formations are regarded as clean, shaly, or shale zones:

i. V_{sh} < 10% implies a clean sand (Hilchie, 1978)
ii. V_{sh} 10-35% implies a shaly sand
iii. V_{sh} >35% implies a shale zone (Ghorab et al., 2008)

2. Porosity (sonic-derived porosity): Raymer-Hunt Gardner (RHG) equation (Raymer et al., 1980):

$$\Phi_S = 5/8*(\Delta t_{log} - \Delta t_{ma}) / \Delta t_{log} \qquad (3)$$

Wyllie time-average equation (Wyllie et al., 1958):

$$\Phi_S = (\Delta t_{log} - \Delta t_{ma}) / (\Delta t_{fl} - \Delta t_{ma}) \qquad (4)$$

Δt is increased due to the presence of hydrocarbon, to correct for hydrocarbon effect, Hilchie (1978) suggested the following empirical corrections:

$$\Phi = \Phi_S * 0.7 \text{ (gas)} \qquad (5)$$

$$\Phi = \Phi_S * 0.9 \text{ (oil)} \qquad (6)$$

In order to correct for the overestimation of sonic-derived porosity resulting from the effect of shale within formations, the following equation is used:

$$\Phi_{Scor} = \Phi_S - (V_{sh} - \Phi_{Sch}) \qquad (7)$$

Where Φ_S = sonic derived porosity; Δt_{log} = Interval transit time in the formation; Δt_{ma} = Interval transit time in the matrix; Δt_{fl} = Interval transit time in the fluid in the formation; Φ_{Sch} = Apparent porosity of the shale point; Φ_{Scor} = corrected sonic porosity.

3. Effective porosity: This can be calculated from the combination of density and neutron logs but the entire well log data used for the purpose of this study lack neutron log. Hence, the effective porosity was simulated from the total porosity using the RokDoc 5.4.4 software.

4. Net-to-gross ratio (NGR): This is a measure of the proportion of clean sand within a reservoir unit. The gross sand is taken to be the whole thickness; the non-net sand refers to the shaly sequences within the gross sands that tend to divide it into flow units; the net sand is thus obtained by subtracting the non-net sand from the gross sand. The NGR reflects the quality of the sands as potential reservoirs. The higher the NGR value, the better the quality of the sand.

For predictive purposes, a NGR map is produced for the area of study. These maps can be used to estimate the viability of a test well in the exploration stage:

$$NGR = \text{Net sand} / \text{Gross sand} \qquad (8)$$

Where Net sand = gross sand − non-net sand $\qquad (9)$

5. Water and hydrocarbon saturation: For the uninvaded zone, according to Archie (1942):

$$S_w = [(a * R_w) / (R_t * \Phi^m)]^{1/n} \qquad (10)$$

$$S_h = 1 - S_w \qquad (11)$$

Where S_w = water saturation; R_o = resistivity of a water filled formation; R_t = true formation resistivity (that is, deep induction); R_w = resistivity of formation water at formation; Φ = porosity; n = saturation exponent usually taken as 2.0; m = cementation factor; a = tortousity factor

6. Irreducible water saturation: This describes the water saturation at which all the water is adsorbed on the grains in a rock or is held in capillaries by capillary pressure. Because production of water in a well can affect a prospect's economics, it is important to know the bulk volume water and whether the formation is at irreducible water saturation (S_{wirr}). At irreducible water saturation, water does not move and the relative permeability to water is zero. Hence, water saturation varies from 100% to a small value but never goes to zero because some water held in capillaries cannot be displaced:

$$S_{wirr} = (F/2000)^{1/2} \text{ (Asquith and Krygowski, 2004)} \qquad (12)$$

Where F = formation factor

7. Bulk volume water (BVW): This is the product of water saturation and porosity corrected for shale:

$$BVW = S_w * \Phi_e \text{ (Asquith and Krygowski, 2004)} \qquad (13)$$

Where BVW = bulk volume water; S_w = water saturation; Φ_e = effective porosity

If values for BVW calculated at several depths within a formation are consistent, then the zone is considered to be homogeneous and at irreducible water saturation. Therefore, hydrocarbon production from such zone should be water free (Morris and Biggs, 1967).

8. Determination of permeability: Permeability is a measure of the ease with which fluids are transmitted within a rock body. It is related to porosity but not always dependent upon it. From Timur (1968):

$$K = [(100 * \Phi^{2.25}) / S_{wirr}]^{\wedge} 0.5 \qquad (14)$$

Where K = permeability (millidarcy; Φ = porosity; S_{wirr} = irreducible water saturation)

9. Identification of fluid type: Usually, a definite identification of fluid type contained within the pore spaces of formation is achieved by the observed relationship between the Neutron and Density logs. Presence of hydrocarbon is indicated by increased Density log reading which allows for a cross-over. Gas is present if the magnitude of cross-over, that is, the separation between the two curves is pronounced while oil is inferred where the magnitude of cross-over is low (Asquith and Krygowski, 2004).

However, the well log data used for this study lack Neutron compensated logs thereby making the identification of fluid type unfeasible. Hence, a general indication of fluid type, that is, the presence of hydrocarbon or water was inferred from the resistivity readings. High deep resistivity readings corresponding to sand units indicated hydrocarbon bearing or freshwater zones while low deep resistivity readings showed water bearing zones.

RESULTS AND DISCUSSION

Five wells were used for the purpose of this study namely Ngammaeast 01, Ngornorth 01, Kinasar 01, Ziye 01 and Murshe 01. These wells were selected on the basis of their surface elevations such that the resulting

Table 1. Average petrophysical parameters for Ngammaeast 01.

Sand unit	Thickness (m)	NGR (%)	Vshale (%)	S_w (%)	S_h (%)	K(mD)	BVW	S_{wirr} (%)	PHIE (%)
Sand 1	62.30	90.67	25.08	54.38	45.62	9986.74	0.12	6.59	22.05
Sand 2	131.63	86.88	15.29	40.44	59.56	18948.68	0.11	6.04	27.41
Sand 3	166.40	5.93	36.65	94.48	5.52	6669.82	0.16	7.06	17.53
Sand 4	32.01	85.70	20.61	97.00	3.00	5903.73	0.21	7.09	20.92

NGR (%) = net to gross ratio; Vshale = volume of shale (%); S_w= water saturation (%); S_h= hydrocarbon saturation (%); S_{wirr}= irreducible water saturation (%); BVW = bulk volume of water; PHIE= effective porosity (%); K = permeability (millidarcy).

Table 2. Average petrophysical parameters for Ngornorth 01.

Sand units	Thickness (m)	NGR (%)	Vshale (%)	S_w (%)	S_h (%)	K(mD)	BVW	S_{wirr} (%)	PHIE (%)
Sand 1	67.37	75.89	24.01	24.58	75.42	72881.72	0.08	4.73	33.10
Sand 2	126.75	90.60	18.05	24.37	75.63	43558.13	0.08	5.34	31.05
Sand 3	217.78	20.47	29.09	51.72	48.28	86860.75	0.17	4.61	32.84
Sand 4	34.17	89.34	23.50	62.89	37.11	90836.50	0.22	4.57	34.74

NGR (%) = net to gross ratio; Vshale = volume of shale (%); S_w= water saturation (%); S_h= hydrocarbon saturation (%); S_{wirr}= irreducible water saturation (%); BVW = bulk volume of water; PHIE= effective porosity (%); K = permeability (millidarcy).

Table 3. Average petrophysical parameters for Kinasar 01.

Sand units	Thickness (m)	NGR (%)	Vshale (%)	S_w (%)	S_h (%)	K(mD)	BVW	S_{wirr} (%)	PHIE (%)
Sand 1	57.71	92.19	20.68	36.13	63.87	36159.31	0.11	5.41	30.53
Sand 2	105.97	82.75	19.77	37.61	62.39	31615.35	0.11	5.48	30.23
Sand 3	225.18	92.30	14.33	27.22	72.78	26829.95	0.08	5.66	30.79
Sand 4	35.32	61.53	16.76	20.67	79.33	29596.23	0.06	5.51	30.89

NGR (%) = net to gross ratio; Vshale = volume of shale (%); S_w= water saturation (%); S_h= hydrocarbon saturation (%); S_{wirr}= irreducible water saturation (%); BVW = bulk volume of water; PHIE= effective porosity (%); K = permeability (millidarcy).

Table 4. Average petrophysical parameters for Ziye 01.

Sand units	Thickness (m)	NGR (%)	Vshale (%)	S_w (%)	S_h (%)	K(mD)	BVW	S_{wirr} (%)	PHIE (%)
Sand 1	51.09	75.01	12.98	4.66	95.34	25666.56	0.02	5.86	29.76
Sand 2	91.62	58.42	19.67	5.65	94.35	23960.88	0.02	5.82	28.37
Sand 3	144.60	94.34	14.30	3.83	96.17	23213.95	0.01	5.75	30.18
Sand 4	50.78	89.29	17.25	34.14	65.86	19625.23	0.01	6.00	28.28

NGR (%) = net to gross ratio; Vshale = volume of shale (%); S_w= water saturation (%); S_h= hydrocarbon saturation (%); S_{wirr}= irreducible water saturation (%); BVW = bulk volume of water; PHIE= effective porosity (%); K = permeability (millidarcy).

topography reflects the overall geomorphology of the Chad basin.

The average value of petrophysical parameters (Tables 1 to 5) such as volume of shale, net-to-gross ratio (NGR), porosity, permeability, water saturation, hydrocarbon saturation, bulk volume of water, etc. (Tables 1 to 5), were obtained for each reservoir unit of interest using appropriate equations as stated in methodology. As a predictive tool, NGR maps were produced for each sand unit across the study wells. The NGR maps help in the determination of the direction and energy of deposition

(that is, from proximal to distal).

Four sand units of interest namely Sands 1, 2, 3 and 4 were delineated and correlated across all the wells of study (Figures 2 and 3) .

Net -to-gross maps

The net-to-gross ratio (NGR) allows for the determination of the energy and direction of deposition of the sands. The higher the NGR value, the better the quality of the

Table 5. Average petrophysical parameters for Murshe 01.

Sand units	Thickness (m)	NGR (%)	Vshale (%)	S_w (%)	S_h (%)	K(mD)	BVW	S_{wirr} (%)	PHIE (%)
Sand 1	35.32	87.19	13.10	18.31	81.69	23881.16	0.05	8.99	25.05
Sand 2	110.38	93.38	12.11	17.66	82.34	25473.98	0.05	5.77	27.08
Sand 3	122.52	70.71	26.98	31.44	68.56	10057.01	0.05	6.49	14.88
Sand 4	67.33	67.26	24.75	57.39	42.61	15900.97	0.10	6.40	17.33

NGR (%) = net to gross ratio; Vshale = volume of shale (%); S_w= water saturation (%); S_h= hydrocarbon saturation (%); S_{wirr}= irreducible water saturation (%); BVW = bulk volume of water; PHIE= effective porosity (%); K = permeability (millidarcy).

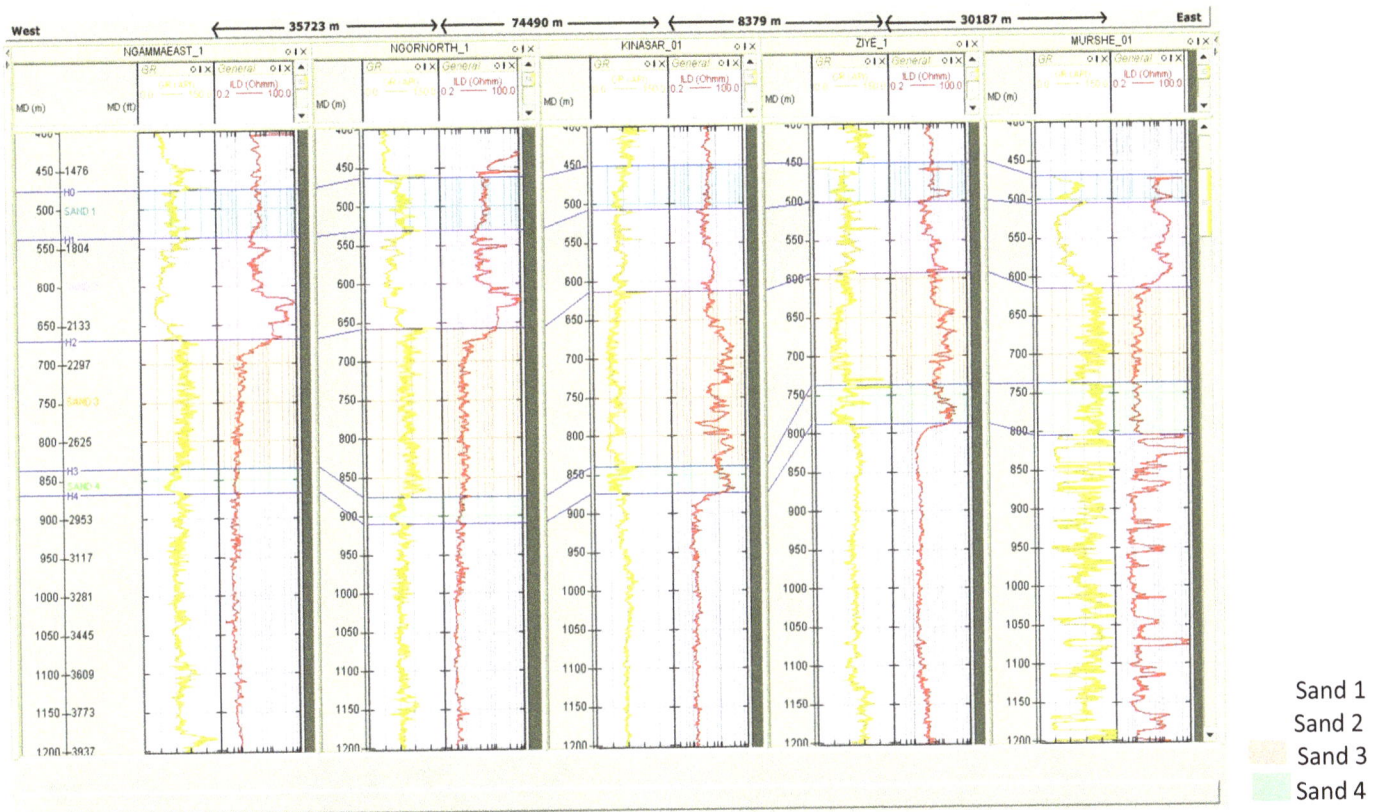

Figure 2. Attempted chronostratigraphic correlation of sands across study wells.

sands. Higher NGR indicates proximity to land that is, proximal source while lower NGR indicates distal source of sediments. Sands are lost on going from proximal to distal as the energy of deposition reduces and the hydro-dynamics of flow tends to zero resulting in the settling out of fine grained materials (Figures 4 to 7).

NGR values obtained for sands investigated in all the wells used for the purpose of this study were contoured using WinSurf® software (Figures 4 to 7). Thus a NGR map was produced for each sand unit. From the maps obtained, the NGR values are fairly consistent across the wells. The least NGR value (20.47%) was recorded in sand 3 in Ngornorth01 well while the highest NGR value (94.34%) was recorded in Sand 3 in Ziye01 well.

No particular trend was observed except in Sand 3 in which the NGR values decreased from about 94.34 % in Ziye 01 to about 5.93 % in Ngammaeast 01 well. On this basis, it can be suggested that deposition of sands with decreasing energy of flow goes from east to west, that is Murshe 01 well to Ngammaeast 01 well (Tables 1 to 5).

Cross-plots

In order to properly characterize the reservoir sands delineated and correlated across the studied wells, a plot of S_w vs. Φ (that is, Buckles plot) was made to determine the grain size range and also to show whether or not the sands are at irreducible water saturation.

To determine the grain size range of each sand unit in

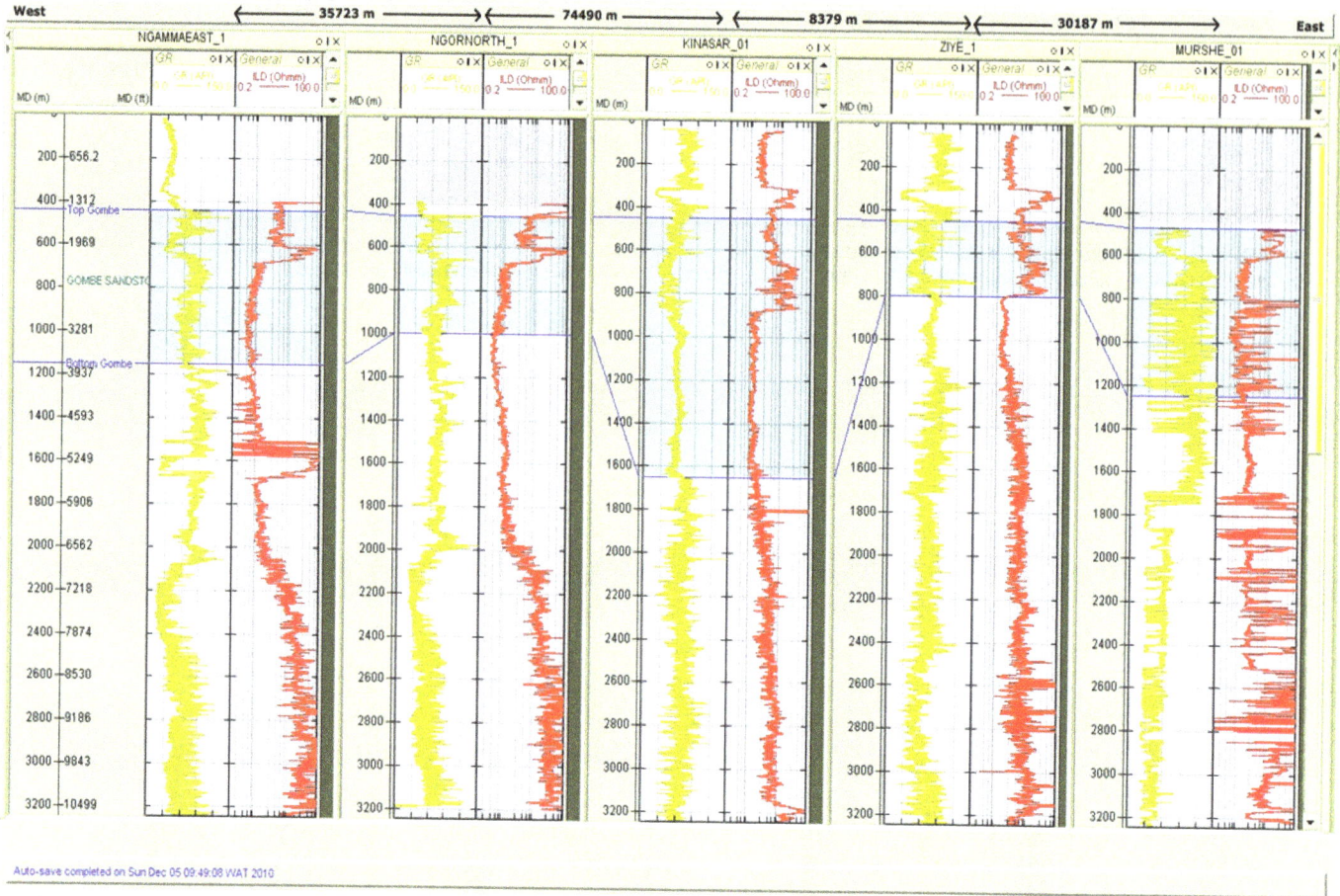

Gombe sandstone

Figure 3. Correlation showing the continuity of Gombe formation across the study well.

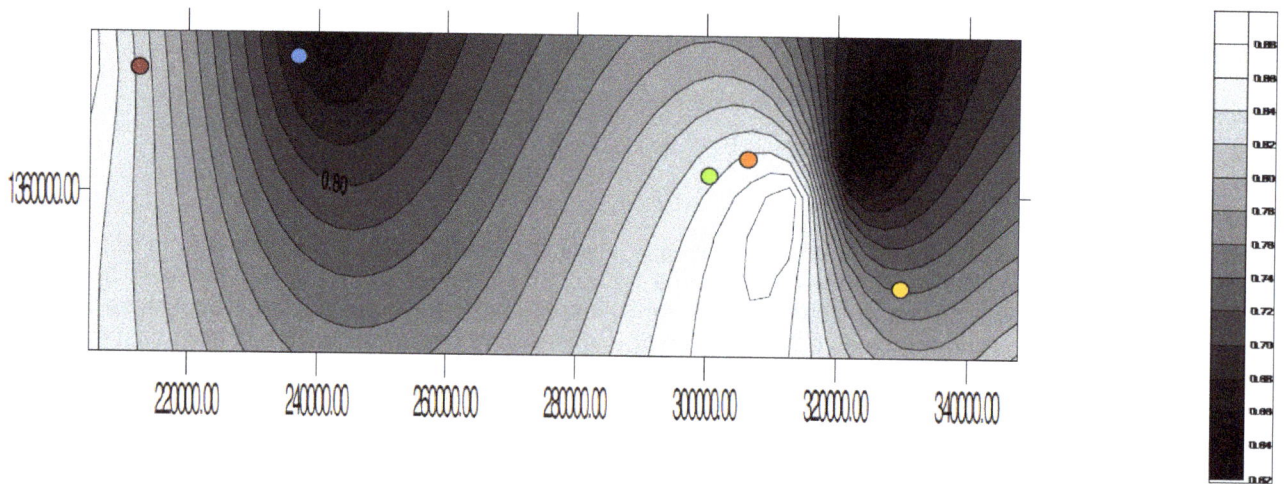

Figure 4. Net-to-gross map of Sand 1.

the studied wells, the average values of porosity are plotted against the average values of water saturation as data points. These points fall within specific fields that represent different ranges of grain size. Figure 8 shows that in Ngammaeast 01 well, the reservoir sands is observed to have very fine grains; sands in Ngornorth 01

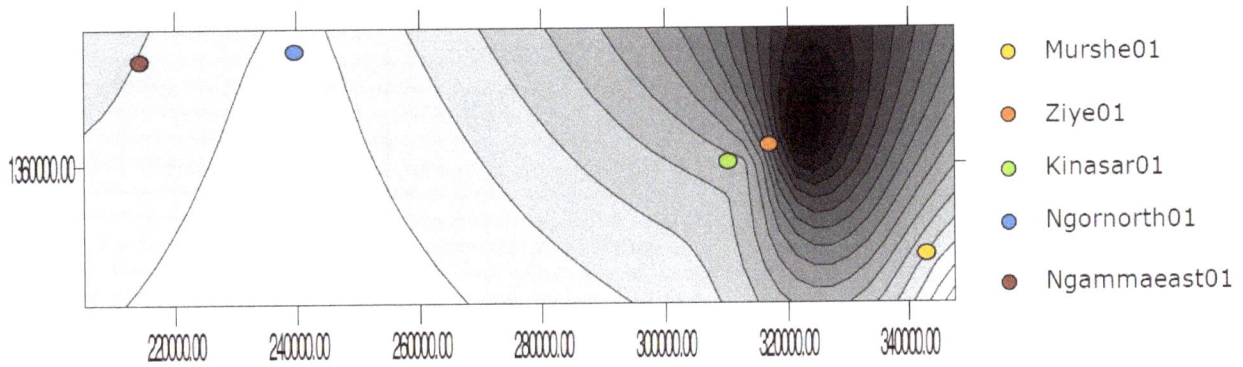

Figure 5. Net-to-Gross map of Sand 2

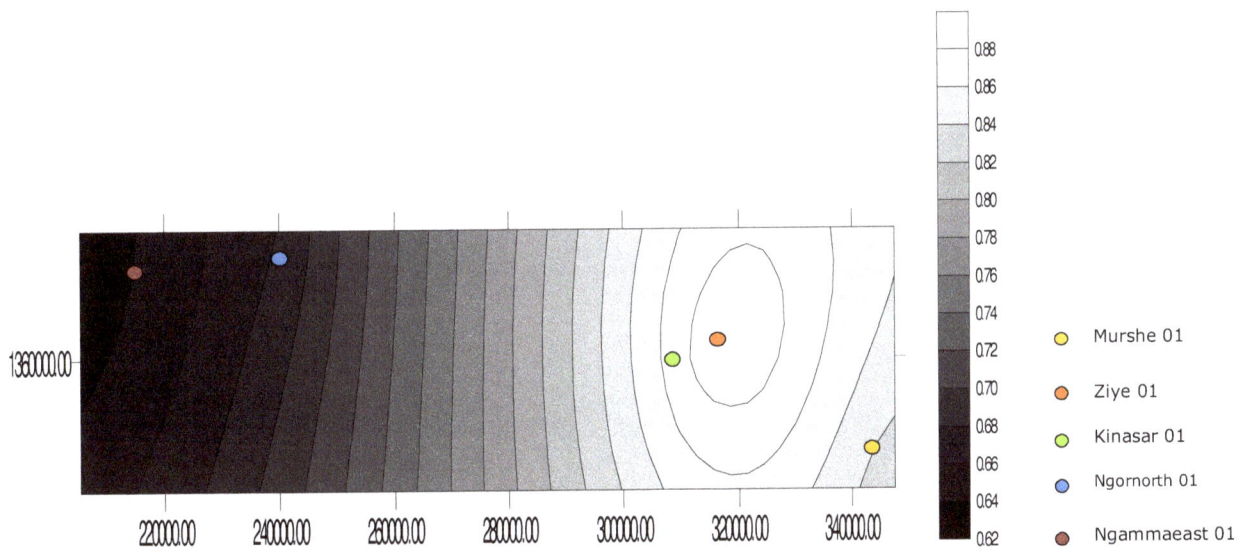

Figure 6. Net-to-Gross map of Sand 3.

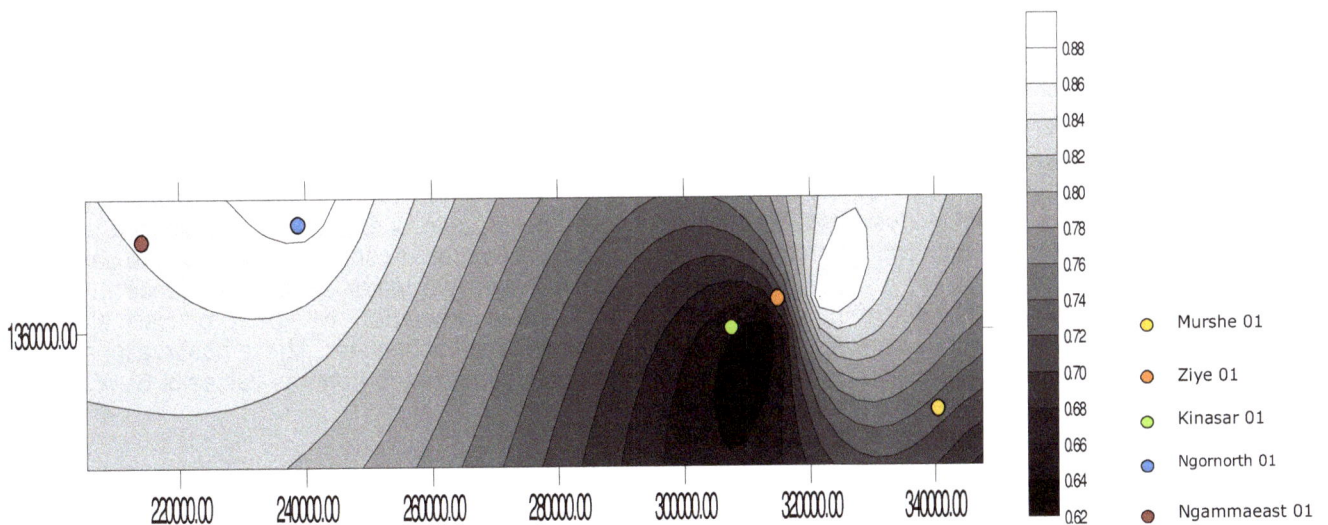

Figure 7. Net-to-gross map of Sand 4.

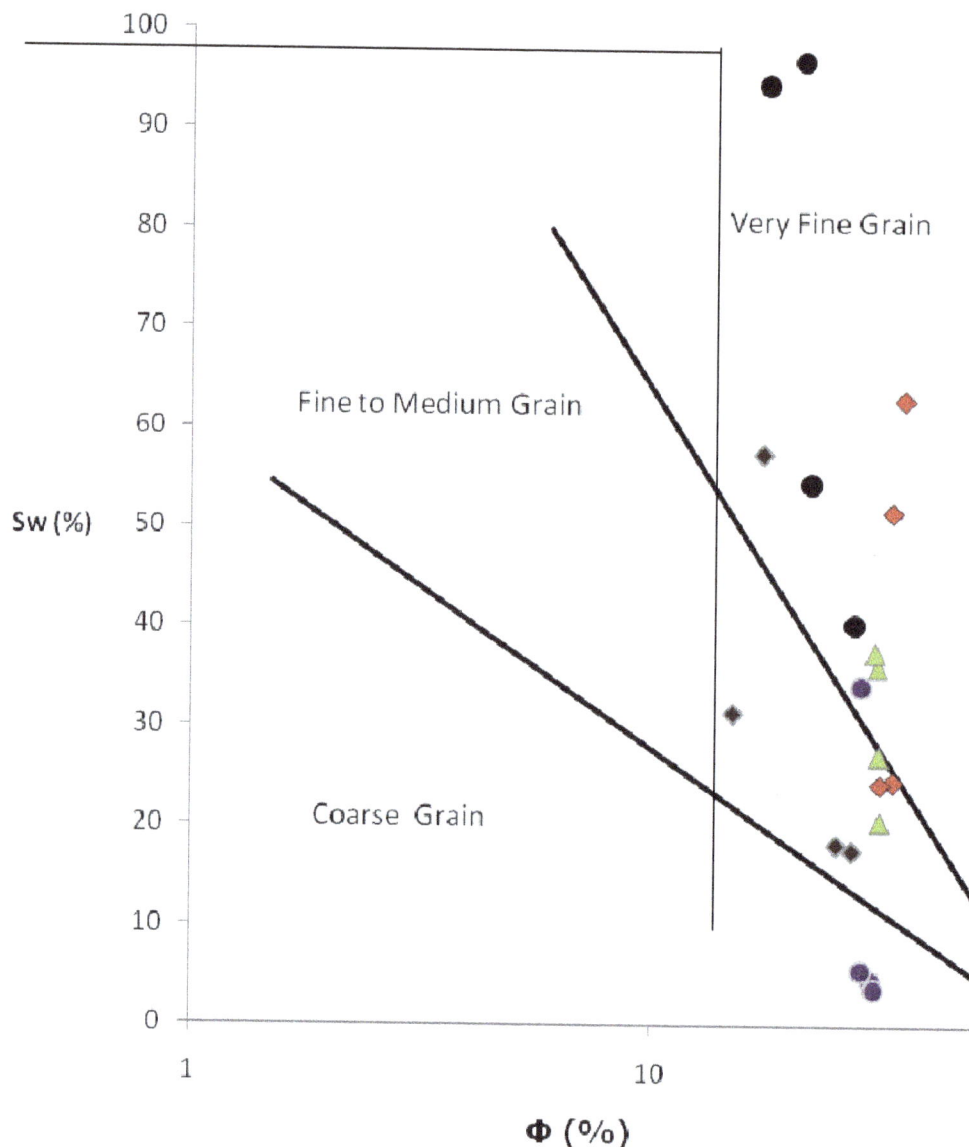

Figure 8. Grain size plot for the investigated sands in the studied wells (After Asquith and Gibson, 1982)

are also very fine grained; for Kinasar 01, well fine grained to very fine grained size reservoir sands are present; most of the reservoir sands in Ziye 1 well are mainly coarse grained; in Murshe 01, the range is from medium to very fine sands.

Using the plot of S_w versus Φ to determine if the sands are at irreducible water saturation, Asquith and Gibson (1982) suggests that if the data points plot along the hyperbolic curves of BVW (Figure 9a), the sands are at irreducible water saturation. However, if the data points are scattered, then the sands are no longer at irreducible water saturation.

From Figures 9b to 9f, it is obvious that with the exception of Ziye 01 well, the sands in each well are not at irreducible water saturation and thus are expected to flow

water along with hydrocarbon. The data points for Ziye 01 as shown on Figure 9e plot approximately along the hyperbolic curve of BVW = 0.02 indicating that the bulk volume of water is consistent and that the sand units are close irreducible water saturation. Hence it is expected that the hydrocarbon will be produced along with a minimal quantity of water. These observations conform to the deductions made from the values of BVW.

Conclusions

Reservoir characterization of the Gombe sandstone (Chad basin) was carried out using five petrophysical well logs. Four prospective reservoirs labeled Sands 1 to 4

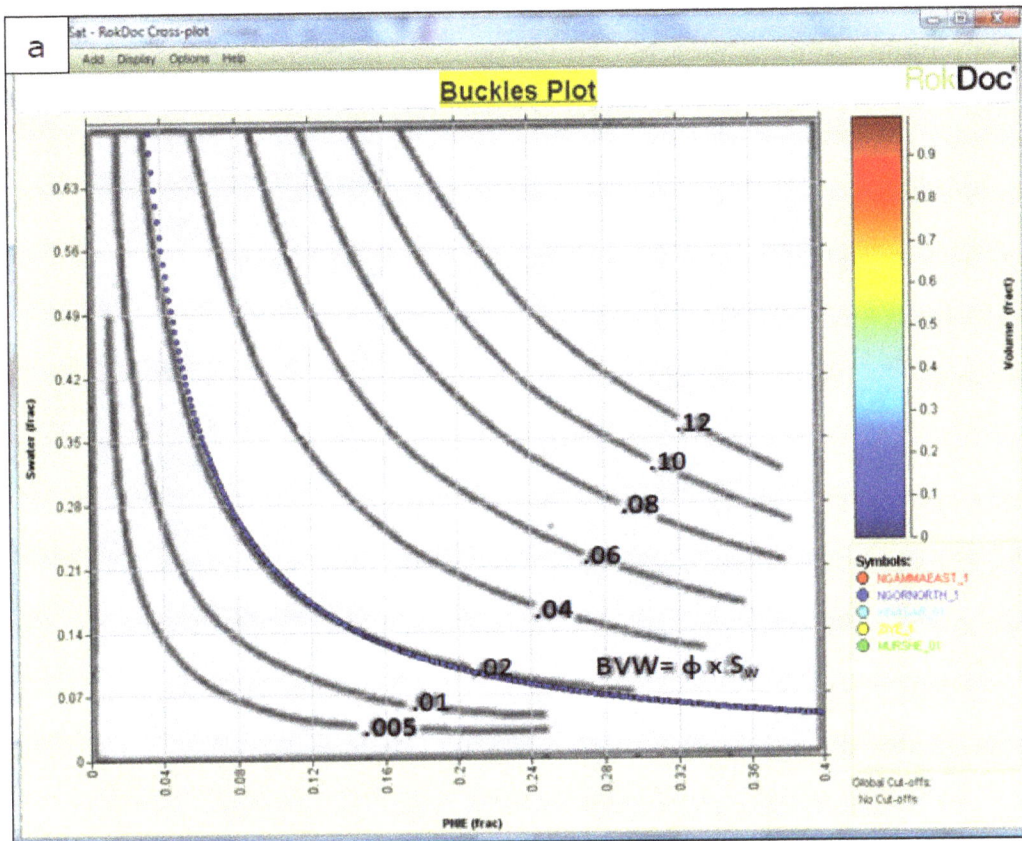

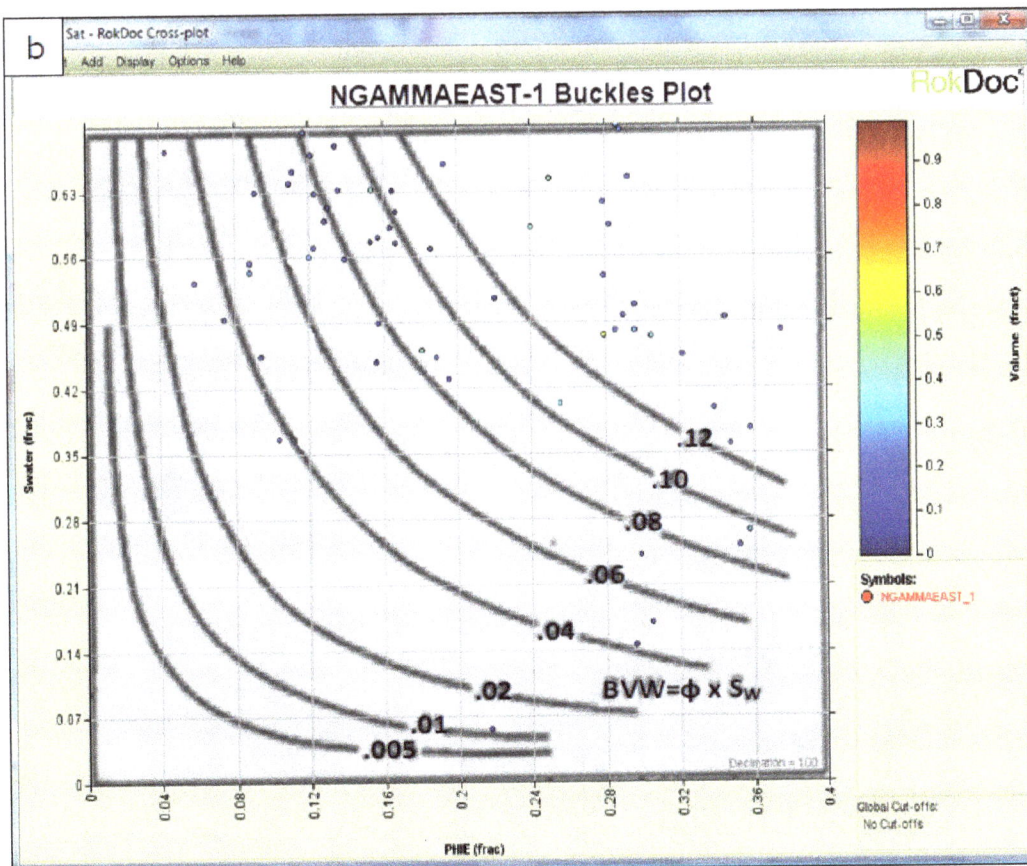

Figures 9a- b. Plot of S_w versus. Φ showing sands at irreducible water saturation.

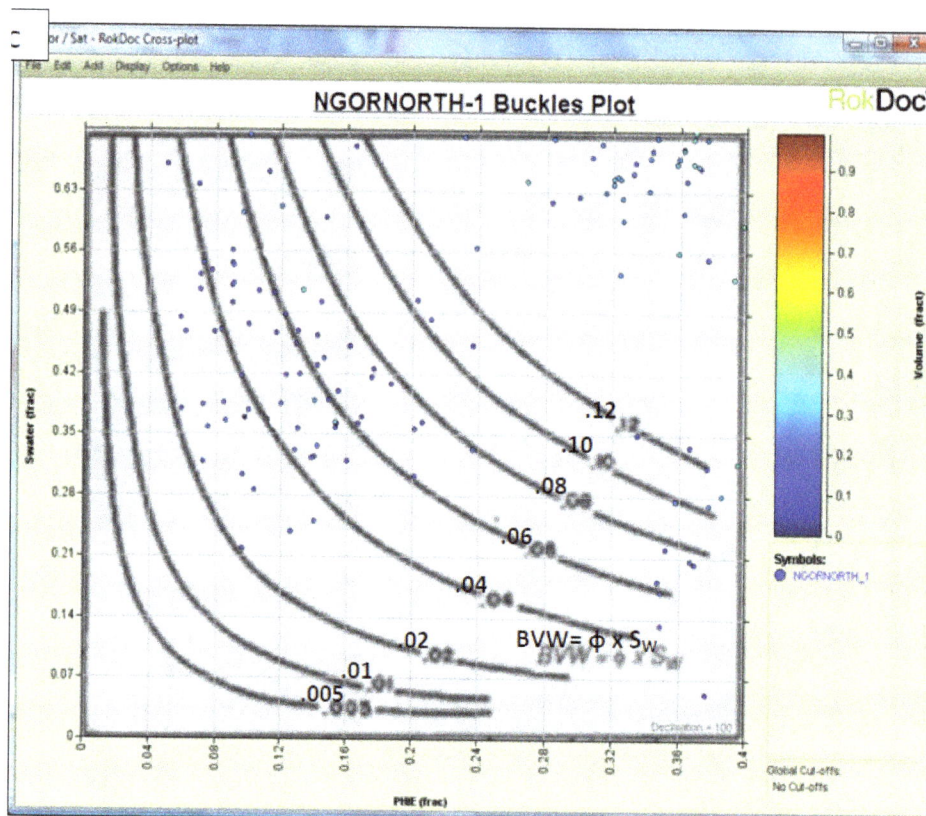

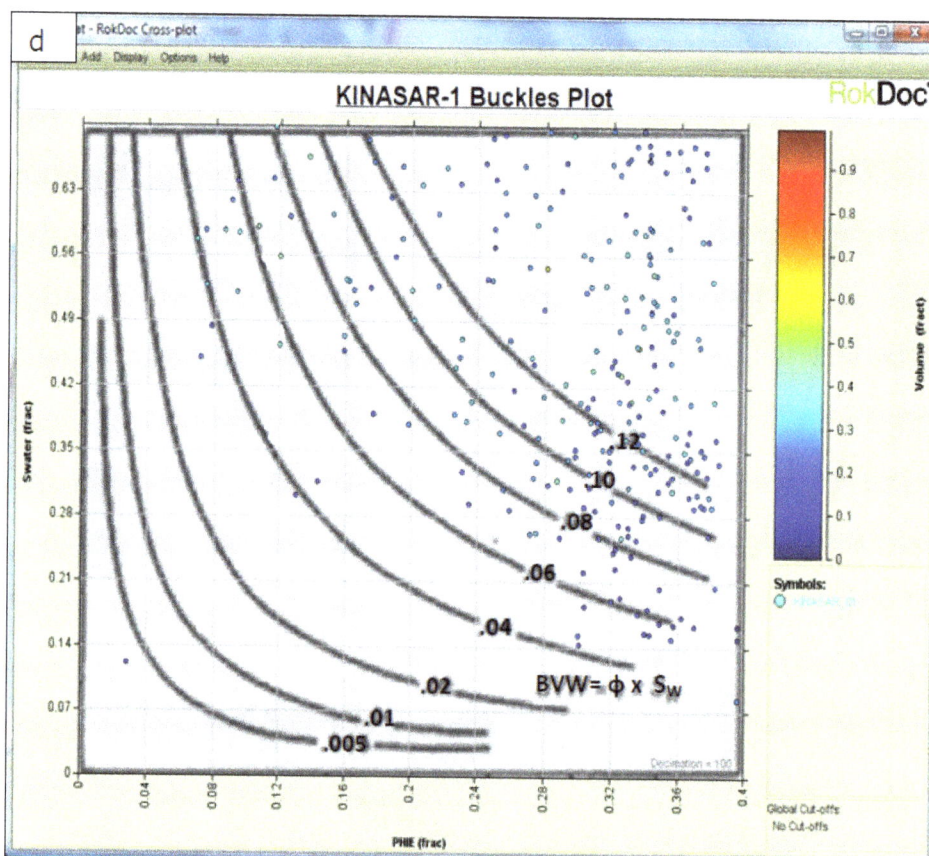

Figures 9c – d. Buckle plots for Ngornorth-1 and Kinasar 01

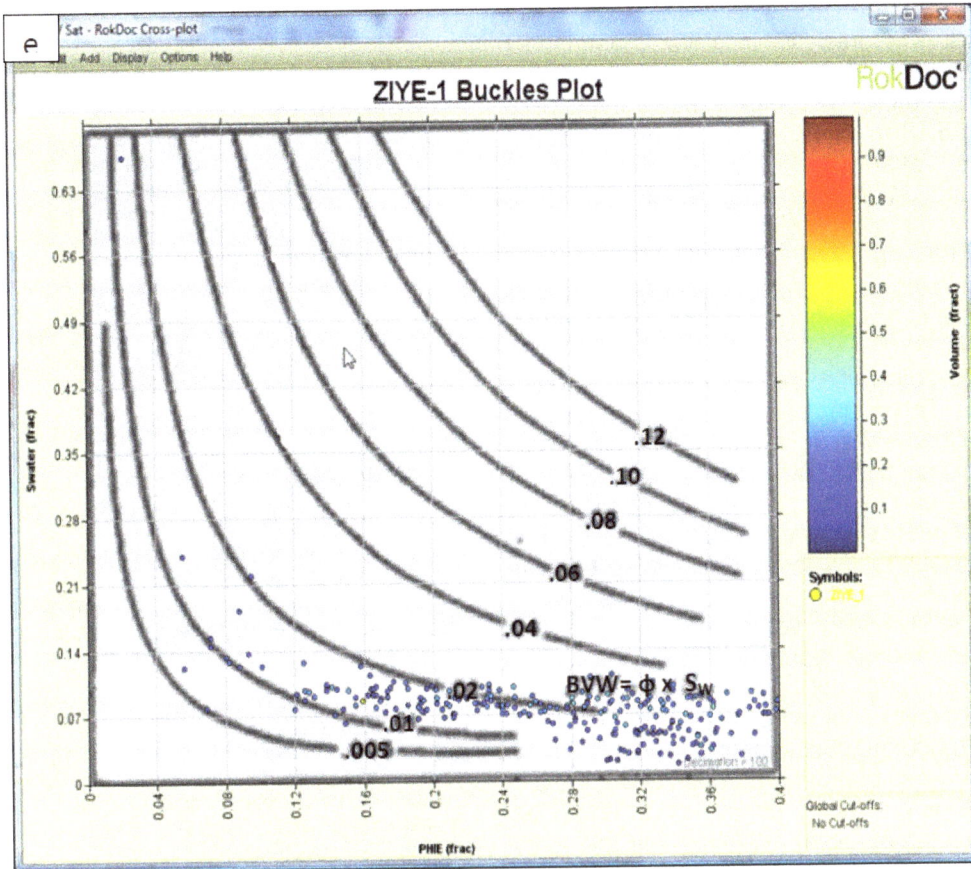

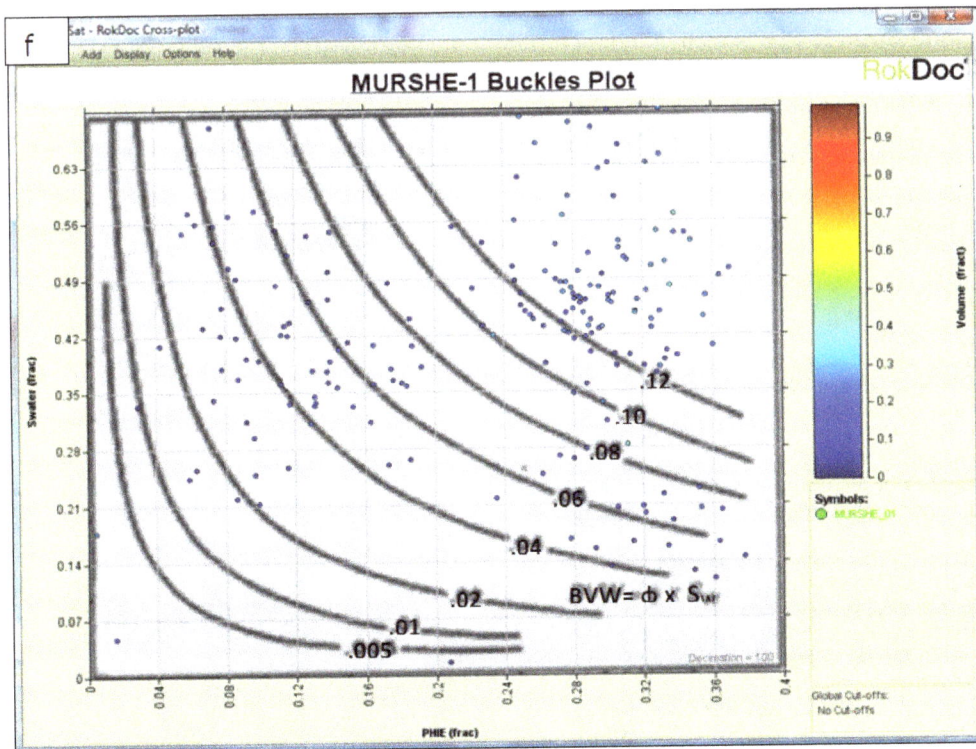

Figures 9e – f. Buckle plots for Ziye 01 and Murshe 01. The sands in all the wells are not at irreducible water saturation, except in Ziye 01 where the data points align along the BVW hyperbolic curve.

were delineated and correlated. The characterization of the four reservoirs of interest within wells Ngammaeast 01, Ngornorth 01, Kinasar 01 Ziye0 1, and Murshe 01 shows that the sand units have thicknesses varying from 32.01 to about 225.18 m; NGR values ranging between 5.93 and 94.34%; average effective porosity ranging between 14.88 and 34.74%; average volume of shale varying between 12.11 and 36.65%; average water saturation between 3.83 and 97.00%; average hydrocarbon saturation values going from 3.00 to about 96.17%; and average permeability varying between 5903 and 90836.5 mD.

The NGR values indicate the presence of quality potential reservoir rocks. The values also helped in suggesting the trend of deposition which is inferred to be going east-west, that is, proximal-distal. Also, from the NGR sand maps, NGR values can be interpolated and predicted for wells drilled at locations between studies wells. The porosity values obtained fall within the stipulated porosity range for sands and sandstone reservoirs (Schlumberger, 1989). These values indicate potential reservoir rocks of optimal porosity. The permeability values fall in the range of permeability values of producing formations which is extremely wide, and this is from 0.1 to over 10,000 mD (Schlumberger, 1989). Thus, it can be conclusively said that the porous sand beds are also permeable, thus enhancing effective fluid flow. The hydrocarbon saturation obtained suggests that the study area might contain hydrocarbon in non-commercial quantities. The petrophysical interpretation of the identified sand units in the study area thus suggests that Gombe formation of the Chad Basin is a prospective hydrocarbon reservoir. The lithology of the study area is essentially fine grained sand, siltstone and local occurrence of coal seams. The Gombe sandstone directly overlies the Fika shale so that the proportion of shale in the upper parts of the studied zone is low and increases as depth increases, that is, towards the bottom of the wells.

REFERENCES

Adegoke OS, Agamanu AE, Benkhelil MJ, Ajayi PO (1986). New Stratigraphic, Sedimentologic and Structural data on Kerri Kerri Formation, Bauchi and Borno States, Nigeria. J. Afr. Earth Sci., 5(3): 249-277

Adepelumi AA, Falebita DE, Olorunfemi AO, Olayoriju SI (2010): Source-rock Investigation of the Turonian-Maastrichtian Fika Shale from Wireline-logs, Chad Basin, Nigeria. Int. J. Petroleum Sci. Technol., 4(1): 19-42.

Ali S, Orazulike DM (2010). Petrophysical parameters of sedimentary formations from Chad basin Nigeria estimated using well log data

Archie GE (1942). The electrical resistivity log as an aid in determining some reservoir characteristics. J. Petroleum Technol., 5: 54-62

Asquith G, Krygowski D (2004): Basic Well Log Analysis: AAPG Methods in Exploration Series. 16p.

Asquith GB, Gibson CR (1982). Basic Well Log Analysis for Geologists. ISSN 0891816526, 2: 216,

Avbovbo AA, Ayoola EO, Osahon GA (1986). Depositional and Structural Styles in Chad Basin of Northeastern Nigeria. AAPG Bull., 70(12): 1787-1798..

Carter JD, Barber W, Wait EA, Jones GP (1963). The Geology of parts of Adamawa, Bauchi and Bornu Provinces in Northeastern Nigeria. Bull. Geol. Surv. Niger., 30: 100

Dessauvagie TFJ (1975), Explanatory Note to the geological map of Nigeria. Niger. J. Min. GeoZ., 9: 1-29.

Du Preez JW, Barber W (1965). The distribution and chemical quality of groundwater in Northern Nigeria. Geol. Surv. Niger. Bull. 36:.38-45.

Ghorab M, Mohmed AMR, Nouh AZ (2008). The Relation between the Shale Origin (Source or Non- Source) and its Type for Abu Roash Formation at Wadi El- Natrun Area, South of Western Desert Egypt. Australian J. Basic Appl. Sci., 2(3): 360-371.

Hilchie DW (1978). Applied Openhole Log Interpretation: Golden, Colorado, D.W. Hilchie Incorporated, 309p.

Larionov VV (1969). Borehole Radiometry: Moscow, U.S.S.R., Nedra.

Morris RL, Biggs WP (1967) "Using log-derived values of water saturation and porosity". Trans. SPWLA Ann. Logging Symp. Paper, 10: 26.

Obaje NG, Attah DO, Opeloye SA, Moumount A (2006). Geochemical evaluation of the hydrocarbon prospects of sedimentary basins in Northern Nigeria. Geochem. J., 40: 227-243

Obaje NG, Wehner H, Scheede G, Abubakar MB, Jauro A (2004). Hydrocarbon prospectivity of Nigeria's inland basins: From the viewpoint of organic geochemistry and organic petrology.

Okosun EA (1995) Review of the Geology of Borno basin. J. Minning and Geo., 31(2): 113-122

Olugbemiro RO, Ligouis B, Abaa SI (1997). The Cretaceous series in the Chad Basin, NE Nigeria: Source Rock Potential and Thermal Maturity. J. Petroleum Geol., 20(1): 51-68.

Petter SW, Ekweozor CM (1982). Petroleum Geology of Benue Trough and Southern Chad basin, Nigeria. AAPG Bull., 66 (8): 1141-1149

Raymer LL, Hunt ER, Gardner JS (1980). An improved sonic transit time-to-porosity transform: Society of Professional Well Log Analysis, 21st Annual Logging Symposium, Transactions .

Schlumberger (1974). Log Interpretation Charts, Schlumberger Educational Services, New York, 83p.

Schlumberger (1989). Log interpretation, principles and application: Schlumberger wireline and testing, Houston Texas, pp. 21-89

Timur A (1968), An investigation of permeability, porosity, and residual water saturation relationships for sandstone reservoirs. Log Analyst, 9: 8 - 17

Wyllie MR, Gregory AR, Gardner GHF (1958). An Experiment investigation of the factors affecting elastic wave velocities in porous media. Geophysics, 23: 459-493.

9

Development of enhanced oil recovery in Daqing

Liu Renqing

Daqing E&P Institute, Daqing, Heilongjiang, China E-mail: renqing.liu1979@gmail.com.

Chemical flooding is playing an important role in stabilizing oil output in Daqing, which is the largest field in China and started its chemical floods since the last 90's. In Daqing Oilfield, the chemical flooding annual output has exceeded 17 million tons; water cut has been dropped down significantly. The success of chemical enhanced oil recovery (EOR) has encouraged the increasing of chemical flooding projects largely in recent years. As a result of investment increase and experience build up, the technology of chemical flooding has been improved, especially polymer injection. The successful experience of Daqing is valuable for the development of chemical flooding EOR in other fields and countries.

Key words: Chemical, polymer, enhanced oil recovery, Daqing.

INTRODUCTION

Early in 1964, Sandiford (1964) first released his research which indicates the mobility of water used in water flooding was greatly reduced by the addition of very small amounts of partially hydrolyzed polyacrylamide (HPAM). Many additional papers sustaining and extending this idea have been published (Du, 2004; Gao and Towler, 2012). In the past forty years (Chauveteau, et al., 1988; Chu, 1994). Researchers around the world have carried out extensive investigation on the mechanism of polymer flooding and laid a solid foundation to the field scale application of polymer flood. Also many field scale polymer flooding projects have been put into production. However, few large-scale successes with these processes have been reported, except in China (Chang et al., 2006; Gao and Feng, 2012). Daqing Oilfield, the largest field of China, is a typical example of chemical flood, especially polymer flooding. Daqing Oilfield started industrial scale polymer flooding since 1996. The oil production rate of Daqing field by polymer flooding has rapidly increased since then (Wang et al., 2002; Feng et al., 2010). The annual oil production from the polymer flood has exceeded 10 million tons (Wang and Liu, 2006). It has become the key technology in stabilizing oil production of Daqing.

In the early years of this century, as the oil price in the international market dropped, the study on enhanced oil recovery in many major oil-producing countries was slowed down. Some pilot tests were suspended since operators stopped investing into the research work.

However, in China the rapid development of economy needs tons of energy to support it. It is important to shift the relying of energy on imported oil. It is essential to increase the oil production of oil fields. Most oil fields in China are developed by water flooding. In Daqing, the recovery efficiency of water flooding is relatively low. Water cut in most mature field is over 80% because of the heterogeneity of reservoirs and high viscosity of oil. The study on enhanced oil recovery had been carried out for more than 10 years before the first commercial injection. Now, notable progress has been made in chemical flooding (Lou and Yang, 1993; Delamaide, et al., 1994). Polymer flooding study focused on the displacement efficiency of polymer PAM, numerical simulation, field project design and prediction technique, which have been used for oil production of industrial scale based on numerous pilot tests. A biopolymer Xanthan gum agent was developed based on Daqing's condition, which can be used under the conditions of high temperature 80°C and high salt concentrations 170,000 ppm. It has been used successfully for profile control of injection wells and is being prepared for pilot flooding test. For the study on surfactant flooding, we have reached successfully the theory and technique of micro-emulsion flooding and micellar flooding is being used for pilot flooding. Recently, alkaline–polymer (AP) and surfactant–alkaline–polymer (SAP) flooding techniques have been developed. The pilot tests of these new chemicals have been conducted successfully (Liao, 2004; Li et al., 2006), that is, in a development test area abandoned, the pilot test's results of evident increase recovery factor and evident decrease in water cut has been achieved (Gao et al., 2010). Meanwhile, non-preflush chemical flooding technique has been developed to manage the chemical injection in the reservoirs with high clay content. The development and production of chemical flooding reagents make it possible to displace reservoir oil with lower cost. Now, we can produce different types of oil field chemical reagents, such as PAM, Xanthan gum, petroleum sulfonate of sodium, etc.

EXPERIENCE BUILD UP

Daqing Oilfield is a large fluvial sandstone reservoir onshore oilfield. In the 20 years' practices of polymer flooding, Daqing Oilfield has developed a complete series of techniques covering reservoir, production, facilities engineering and produced liquid treatment. Polymer flooding has gradually become an important technique to stabilize production capability of Daqing Oilfield. 13 field tests have been conducted successfully since 1989. There are 37 polymer flooding areas which contain nearly 9000 wells by the end of 2005. Polymer flood production has achieved 1×10^9 t/a in the Daqing Oilfield. All the polymer flooding areas get good results of significant water-cut drop, great oil production increase, significant recovery rate increase. Almost every area's water-cut drop more than 20%, even reaches 35% in an area. Compare with water flooding, the recovery rate improve more than10%, with high concentration polymer flooding, the recovery rate is higher.

Based on the research and evaluation, Daqing's reservoir is suitable for polymer flooding. Firstly, Daqing Oil Field is a terrestrial fluvial-delta deposit, and it is mainly fining upward and multi-interval and multi-rhythm with a permeability variation factor between 0.635 and 0.71, which is within the maximum range (0.72) of polymer flooding for enhancing oil recovery. Secondly, Daqing reservoirs have a lower formation water salinity about 7, 000 mg/L which helps polymer solution keep higher viscosity in the reservoir and decrease oil - water mobility ratio greatly.Thirdly, because of the shallow buried oil reservoir and its low temperature, there is no thermo-oxidative degradation with polymer in the reservoir, and therefore economic benefits can be greatly improved due to saving of oxygen exclusion equipment; Fourthly, the result of polymer flooding is closely related with crude oil's viscosity which affects the result if it's too high or too low and at the same geological condition, there is an optimal viscosity range. Crude oil's viscosity in Daqing is about 9 mPa•s which is within the best range of polymer flooding (Niu, 2004).

Based on improving volumetric sweep efficiency and reducing channeling and breakthrough, polymer flooding can yield a significant increase in oil recovery when compared with water flooding. Polymer flooding has been conducted successfully in 13 field tests since 1989 and has been commercially used in the following years. Field tests show that the recovery improved with increasing concentration and injection volume. By applying polymer flooding, the main formation recovery of Daqing Oilfield has reached more than 50%. It is 10% ~ 15% higher compared with other oilfields. In the 20 years' practices of polymer flooding, Daqing Oilfield has developed a complete series of techniques covering reservoir, production, facilities engineering and produced liquid treatment. Polymer flooding has gradually become an important technique to maintain the high and stable production capability of Daqing Oilfield. Daqing oilfield, being the largest polymer flooding field in the world, has 37 polymer flooding areas which contain nearly 9000 wells by the end of 2005. Now, polymer flood production has achieved 1×10^9 t/a in Daqing Oilfield. In this paper, the reservoir condition, pilot test, solution production and injection, production technology and produced liquid treatment technology are described.

Because of the complicated geological conditions in reservoir formations and a widespread distribution of small volume dentrital rocks in the reservoirs, the development of most oil fields in China is being limited by means of the natural water flooding. Instead, the water injection flooding method is employed extensively in these fields.

Laboratory research began in the 1960s, investigating the potential of enhanced oil recovery (EOR) processes in the Daqing Oilfield. For polymer flooding technology, from a single-injector polymer flood with small well spacing began in 1972. During the late 1980s, a pilot project in central Daqing was expanded to a multi-well pattern with larger well spacing.

Favorable results from pilot tests showed that polymer flooding was the one of the most efficient methods to improve areal and vertical sweep efficiency at Daqing, as well as providing mobility control. Consequently, the world's largest polymer flood was implemented at Daqing, beginning in 1996. By 2007, 22.3% of total production from the Daqing Oil Field was contributed by polymer flooding. Polymer flooding should boost the ultimate recovery for the field to over 50% original oil in place (OOIP) with and incremental recovery of 10~12% of OOIP. At the end of 2007, oil production from polymer flooding at the Daqing Oilfield was more than 10 million tons (73 million barrels) per year (sustained for 6 years). Recently, the industrial application has been expanded into the second-class, less-permeable strata.

POLYMER FLOODING

Because of the research efforts devoted in the last two

Table 1. Basic data of polymer flooding pilots.

Pilot	Pattern	Injector	Producer	Distance between injector and producer (m)	Date of injecting	Quantity of injected polymer (ppm × pv)	Dh %OOIP.	Benefit of polymer (ton/ton)
Daqing-x1	Inverted five-spot pattern	1	4	75	1972–1973	163 = 0.2	5.1	153.4
Daqing-HP	Inverted five-spot pattern	4	9	200	1988–1990	272 = 0.3	4.3	81.8
Daqing-PO	Inverted five-spot pattern	4	9	106	1990–1992	506 = 0.3	14	177
Daqing-PT	Inverted five-spot pattern	4	9	106	1990–1992	496 = 0.3	11.6	209
Daqing-G4K	Irregular pattern	3	11	100-160	1986–1989	–	12.7	400

Table 2. Basic data of semi-industrial and industrial projects.

Project	Area (km^2)	Well pattern	Injector	Producer	Distance between injector and producer (m)	Date of injecting
Daqing-G4K	0.86	Irregular pattern	6	12	100–360	1991
Daqing-TP	3.13	Inverted five-spot pattern	21	36	250	1993
Daqing-LP1	1.45	Inverted five-spot pattern	16	25	212	1994
Daqing-LP2	2.09	Inverted five-spot pattern	9	26	300	1994
Daqing-G3E	0.81	Irregular pattern	7	11	100–360	1994
Daqing-G3W	1.03	Irregular pattern	7	18	100–360	1991

decades, experience has been built up, which covers the reservoir engineering, oil recovery mechanisms, solution properties, physical and numerical simulations, and the efficient prediction method of the polymer flooding technology (Gao and Towler, 2011). Based on these laboratory researches, the pilot plant tests have been conducted in several oil fields of various reservoir types, and therefore, have had much experience in commercial fulfillment, like engineering design, surface construction, testing and analysis methods, dynamic control method of drilling and the tracer injection technology (Liu, 1991; Wang et al., 1991; Gao et al., 2010). Since the results obtained from semi-industry tests provide a positive answer for the polymer flooding method, after 1993, this method has been industrialized in our oil fields.

Pilot tests

Table 1 summarizes the test results obtained by the application of the polymer flooding method in the different oil fields of various types. It is found that the polymer solution can apparently improve the trial formatted crude oil from these oil fields always coefficient, consequently, reducing the water cut in these oil reservoirs of heterogeneous geological conditions. The polyacrylamide polymer is used in these oil fields. The reservoir water of law mineralized degree and a not quite high reservoir temperature assure that this polymer can increase the viscosity of the oil-water solution in the reservoir efficiently. Usually, with the injection of 350~380 ppm

pore volume polymer solution, there is an increased profit of 150~200 tons oil production obtained for per ton of polymer injected, and the oil recovery efficiency will be increased in the ranges of 4~14% of OOIP.

Commercial projects

In these projects, we have enlarged the distance between the injection wells and production wells to an average value greater than 200 m. The area involved in each project is larger than the previous one, one project's area is even bigger than 3 km^2, and more than 60 injection and production wells are located in this area. In each project, the water used in the injection or production wells is de-mineralized by the proper treatment technology. The effects of polymer injection are to reduce the water cut and increase sweep efficiency. The success of the pilot test enables us to carry out the semi-industry and industry EOR projects by using this polymer flooding method (Liu, et al., 1995).

Table 2 summarizes the projects executed currently in our oil fields. In these projects, we have enlarged the distance between the injection wells and production wells to an average value greater than 200 m. The area involved in each project is larger than the previous one, one project's area is even bigger than 3 km^2, and more than 60 injection and production wells are located in this area. In each project, the water used in the injection or production wells is de-mineralized by the proper treatment technology. The effects of reducing the water

Table 3. Basic data of combination flooding pilots.

Project	Pattern	Injector	Producer	Distance between injector and producer (m)	Concentration of chemicals			Incremental %OOIP.
					S	A	P	
Daqing -GD4	invertedfive-spotpattern	4	9	50	0.4	1.5	0.1	13.4
Daqing-SSP	invertedfive-spotpattern	4	9	106	0.6	1.25	0.15	20–26[a]
Daqing -2	irregularpattern	4	9	160–190		2	0.1	-

cut and increasing the oil production are observed at all of the production wells in these projects. For example, in the Daqing-TP project, it is found that the oil production is increased after injecting 60 ppm pore volume polymer solution on January 1993, and this increase is maintained until August 1994, at which, the water cut of that reservoir is decreased from 90.7 to 80.1% with a daily production of 1200 tons of oil among which, 408 tons oil is recovered by this polymer flooding method. At present, in the oil fields of our country, the polymer flooding technology has been fully industrialized and its achievement is very promising.

In order to have a sufficient supply of the polymer, a polyacrylamide chemical plant 30,000 tons per year will be opened in the near future (Liu, 1995). It is anticipated that this polymer flooding.

The success of the surfactant–polymer flooding method is proven by the positive results obtained from the past and current pilot tests in our oil fields. However, because of the high consumption rate of surfactant by the micellar solubilization effect, it is found that this surfactant–polymer combination flooding method is not economically feasible in practice. In order to overcome this v high consumption disadvantage, combining with the advantages of the flooding methods of using surfactants, polymer and alkaline solution, we have been developed the alkaline–surfactant–polymer A–S–P. combination flooding method and the alkaline–polymer A–Pcombination flooding method for crude oil with high acid value (Song et al., 1995), and the surfactant–alkaline–polymer combination flooding methodfor crude oil containing natural organic acid (Yang et al., 1995a, b). Because of the characteristics of a higher viscosity value, and also the surface active ability, these kinds of oil-recovery reagents are able to increase the mobility ratio and the displacement efficiency, and therefore, can decrease the interfacial tension efficiently at the oil–water interface. The oil recovery efficiencies of these combination flooding methods are at least one time as high as that of polymer flooding method. Meanwhile, these oil recovery reagents can also combine with the organic acids on the surfaces of crude oil to form the local surface active reactants and when these reactants meet again with the surfactant molecules in the injected reagents, the coordination effect will result, consequently, the effect of decreasing the interfacial tension becomes

pronounced. The alkaline compounds in these reagents can also inhibit the retention loss of the injected chemicals. Thus, under the same displacement efficiency as that of surfactant–polymer combination flooding method, these A–S–P, A–P and S–A–P combination flooding methods will reduce the amount of surfactant consumed by more than 10 times, as well as the capital cost of the surfactants.

In China, we already have the chemical plants to produce these high surface active reagents. The price for the reactant resources of these surfactants is very cheap, and it is very easy to get these materials in our country. These surfactants can activate the surface properties of acid oil and non-acid oil simultaneously, and have the ability to form the ultra-low interfacial tension in a wide concentration range.

Table 3 gives the pilot test results of using these combination flooding methods in our oil fields currently. The pilot test result obtained from a successful injection. Before injecting the chemical slug, the test wells in this oil field belong to the economic-limited and highly developed ones, the water cut of these wells is above 98% and the oil recovery efficiency obtained by the water flooding method is about 54.4% of OOIP., but after injecting the A–S–P combination slug, the water cut of these wells is reduced significantly and the production of crude oil is increased. The increased oil recovery efficiency by this EOR technology is about 13.4% of OOIP.

In order to solve the high clay content problem in some reservoirs of Daqing, we have developed a non-preflush chemical flooding technique, at which the amount of surfactant adsorbed on the reservoir rock surface is reduced by the addition of polyelectrolytes. By using this technology, the pilot test results obtained from the Yumen Oilfield (Yang et al., 1995a, b) are very significant. Hence, according to thecharacteristics of the oil reservoirs, several injection methods have been successfully developed by combining different chemicals, which provide a very promising potential in increasing the oil recovery at Daqing.

CONCLUSION

The heterogeneity of the geological conditions in Daqing reservoirs causes a low oil recovery efficiency using the traditional water flooding technology.

The industrial experience of chemical injections in Daqing shows that the polymer flooding method can be applied successfully in variable reservoirs. Significant progress is obtained from the pilot tests and commercial injections by using the various multiple chemicalsflooding technologies, and these combination methods all have high potential in increasing the oil recovery efficiency.

REFERENCES

Chauveteau G, Combe J, Han D (1988). Preparation of two polymer pilot tests in Daqing oil field, SPE 17632, Presented at the 3rd SPE meeting. Tianjin, China.

Chu KH (1994). The study of polymer enhanced oil recovery technology. Yumen Oil field Technical Report 1–2, 1–14, in Chinese.

Delamaide E, Corlay P, Wang DM (1994). Daqing oil field: The success of two pilots initiates first extension of polymer injection in a giant oil field, SPE 27819.

Feng Y, Ji B, Gao P (2010). "An Improved Grey Relation Analysis Method and Its Application in Dynamic Description for a Polymer Flooding Pilot of Xingshugang Field, Daqing." SPE 128510, presented at the North Africa Technical Conference and Exhibition.

Gao P, Feng Y (2012). "An Improved Grey Relation Analysis Method and Its Application in Dynamic Description for a Polymer Injection". J. Pet. Sci. Technol. 2:1276-1285

Gao P, Feng Y, Zhang X, Luo F (2009). "Investigation of Cyclic Water Injection after Polymer Flood in Xing4-6 Field, Daqing", SPE-126836, presented at the 2009 Kuwait International Petroleum Conference and Exhibition 14-16 December 2009 in Hilton Kuwait Resort, Kuwait.

Gao P, Towler B (2011). "Investigation of Polymer and Surfactant-polymer Injections in South Slattery Minnelusa Reservoir, Wyoming". J. Pet. Exp. Prod. Technol. 30:2243-2551.

Gao P, Towler B (2012). "Integrated Investigation of Enhanced Oil Recovery in South Slattery Minnelusa Reservoir, Part 1: Polymer Injection". J. Pet. Sci. Technol. 30:2208-2217.

Gao P, Towler B, Zhang X (2010). "Integrated Evaluation of Surfactant-Polymer Floods". SPE-129590, presented at the 2010 SPE EOR Conference at Oil and Gas West Asia, 11-13 April 2010 in Muscat, Oman.

Li Yan, Yu Li, Tang S (2006). "A sketch of polymer solution and injection system development of Daqing Oilfield". Ground Engineering of Oil and Gas Field.

Liao G (2004). "Application and experience of industrialized polymer flooding in Daqing Oilfield". Pet. Geol. Oilfield Dev. Daqing 23:1.

Liu H (1995). Polymer augmented waterflooding technique: Status and prospect, Daqing Oil Field. Pet. Explor. Dev. pp. 64–67.

Lou ZH, Yang CZ (1993). Application of biopolymer at high temperature and high salinity conditions, Paper presented at the 1993 Denver ACS.

Niu J (2004). "Practices and understanding of polymer flooding enhanced oil re-covery technique in Daqing oilfield". Pet. Geol. Oilfield Dev. Daqing 23:5.

Song WC, Yang CZ, Han DK (1995). Alkaline–surfactant–polymer combination flooding for improving recovery of the oil with high acid value, SPE 29905, Presented at the SPE meeting, Beijing, China.

Wang DM, Yue-xing H, Delamaide E, Ye ZG, Ha S, Jiang XC (1993). Results of two polymer flooding pilots in the central area of Daqing oil field, SPE 26401, Proceedings of 68thAnnual Technical Conference and Exhibition. Houston, pp. 291–308.

Yang CZ, Han DK (1991). Present status of EOR in the Chinese petroleum industry and its future. J. Pet. Sci. Eng. 6:175–189.

Yang CZ, Han DK, Song WC (1995a). The Alkaline–surfactant–polymer combination flooding and application to oil field for EOR. 8th European Symposium on Improved Oil Recovery. Vienna.

Yang CZ, Han DK, Wong DS (1995b). The application of a new additive in the pilot chemical flooding project in Yumen Oil field. Acta Petrolei Sinica Chin. 16:77–84.

10

Simulation, control and sensitivity analysis of crude oil distillation unit

Akbar Mohammadi Doust, Farhad Shahraki and Jafar Sadeghi*

Department of Chemical Engineering, Faculty of Engineering, University of Sistan and Baluchestan, Zahedan, Iran.

Steady-state and dynamic simulation play important roles in investigation of refinery units. Therefore, simulation can help this investigation and behavior assessment. In this paper, simulation was done by commercial software. In fact, because of solving many state equations simultaneously and using control theory, dynamic simulation has more significant impact than steady-state simulation. Flow, pressure, temperature and level (FPTL) were controlled by Proportional-Integral-Derivative (PID) controllers in the unit. The case study is Kermanshah Refinery. The behavior of the FPTL controllers in dynamic regime were observed after the changing of the crude oil feed flow rate by 3% for 5 h. ASTM D86 boiling points (compositions) of two simulations were compared with experimental data. Finally, system sensitivity to inputs variables was investigated in the MATLAB®/Simulink™ by transferring the dynamic results. Transient responses to changes such as feed temperature, feed flow rates, steam flow rates and the duties of the reboilers of columns in Gasoline unit were plotted. Among of all disturbances, the system is more sensitive to changes in the feed temperature, the duties of the reboilers of columns in gasoline unit and simultaneous combination of above changes.

Key words: Steady-state, dynamic, PID controller, ASTM D86, Sensitivity, MATLAB simulink, transition responses.

INTRODUCTION

Today, distillation of crude oil is an important process in almost all of the refineries. Simulation of the process and analysis of the resulting data in both steady-state and dynamic conditions are fundamental steps in decreasing of the energy costs and controlling the quality of the oil products. The dynamic simulation when adding some Proportional-Integral-Derivative (PID) controllers and setting them to have desired responses, has more significant impacts and challenges than steady-state simulation in crude oil distillation units. A PID controller is a controller that includes three elements (Araki, 2002). PID control systems have exactly the same structure as depicted in Figure 1, where the PID controller is used as

the compensator C(s). The transfer function of a PID controller is:

$$C(s) = K_P \left(1 + \frac{1}{\tau_I s} + \tau_D s \right)$$

(1)

All the three elements are kept in action. Here, K_P, τ_I and τ_D are positive parameters, which are respectively referred to as proportional gain, integral time, and derivative time, and as a whole, as PID parameters. These parameters can be adjusted using some empirical methods. One of them, which is an extension to Ziegler-Nichols method and uses the ultimate gain and frequency for adjustment of the parameters, is Tyreus-Luyben method (Almudena, 2001).

*Corresponding author. E-mail: jsadeghi@hamoon.usb.ac.ir.

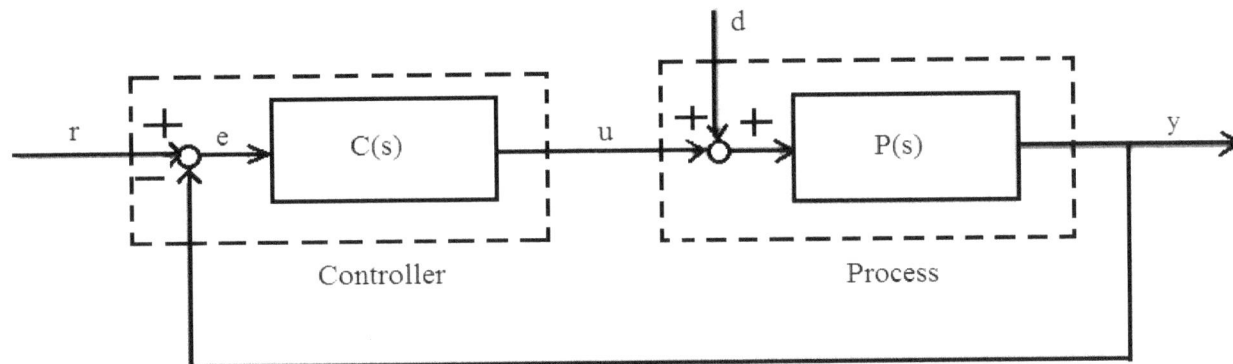

Figure 1. Conventional feedback control system.

Crude oil is a mixture of many thousands of components varying from light hydrocarbons such as methane, ethane, propane, etc., to very high molecular weight components. The compositions of crude oil depend also on the location of exploitation. In the present work, the feed flow rate is 0.046 m³/s (25,000 bbl/day) that is provided by the blending of Crude oils of Ahwaz (60%), Naft-I-Shah (24%) and Maleh-Kuh (16%). Therefore, the feed has very complex compositions. Also the design and optimization of the oil fractionators are very important and complex. In petroleum refining the boiling point ranges are used instead of mass or mole fractions. Four types of boiling point analysis are known: ASTM D86, ASTM D1160, ASTM D158 and TBP (True Boiling Point). Six streams of product were investigated by ASTM D86 from initial boiling point (IBP) to final boiling point (FBP). We studied the system behavior by changing the feed flow rate in the dynamic conditions and MATLAB®/Simulink™. MATLAB software is very flexible for this work, therefore, it was used.

The aims of this work are to investigate the results in steady-state and dynamic simulations, FPTL control while changing the crude oil feed flow rate and comparison of ASTM D86 boiling points (compositions) in two simulations with the correspondent experimental data. At last, sensitivity analysis of crude oil distillation unit in the MATLAB®/Simulink™ was done by transferring dynamic files to it as the basis aim. Directions of transferring files to sensitivity analysis were:

Steady state files ⟶ Dynamic files ⟶ MATLAB®/Simulink™

Physical-mathematical model of the distillation column

In the problems of multiple-stage separation for systems in which different phases and different components play a part, we have to resort to the simultaneous or iterative solution of hundreds of equations. This means that it is necessary to specify a sufficient number of design variables so that the number of unknown quantities (output variables) is exactly the same as the number of equations (independent variables). This number of equation can be found and counted in a mathematical model.

The usual method to mathematically model a distillation process in refining columns is the theoretical stage method. To find the number of the theoretical stages of an existing column, the real number of stages might be multiplied by column efficiency. For each theoretical stage, the mass balance of individual components or pseudo components, energy balance, and vapor-liquid equilibrium equation can be written. The set of these equations creates the mathematical model of a theoretical stage. The mathematical model of a column is composed with models of individual theoretical stages. Finally, thermodynamic model Braun K10 "BK10" was used for the unit, because it is a model suitable for mixtures of heavier hydrocarbons at pressures under 700 kPa and temperatures from 170 to 430°C. The values of K10 can then be obtained by the Braun convergence pressure method using tabulated parameters for 70 hydrocarbons and light gases (Aspen Physical Property System, 2009). At low pressures, the Braun K10 model is strictly applicable to predict the properties of heavy hydrocarbon systems. Using the Braun convergence pressure method by the model at, given the normal boiling point of a component, K value is calculated at system temperature and 10 psia. The K10 value is then corrected for pressure using pressure correction charts. Using the modified Antoine equation one can find the K values for any components that are not covered by the charts at 10 psia and corrected to system conditions using the pressure correction charts (Aspen Physical Property System, 2009).

In existence of a large amount of acid gases or light

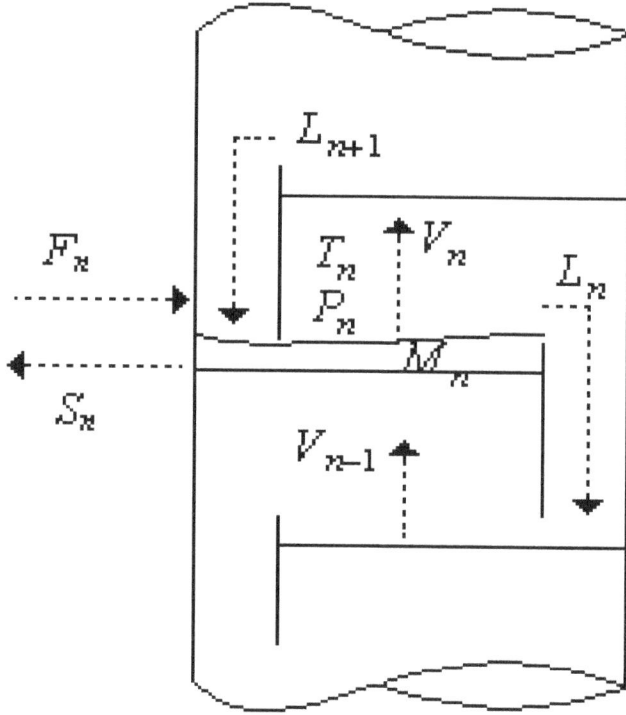

Figure 2. Scheme of a column stage.

hydrocarbons, the accuracy has encountered some problems with this model. All three phase calculations assume that the aqueous phase is pure H_2O and that H_2O solubility in the hydrocarbon phase can be described using the kerosene solubility equation from the API data book (Aspen Physical Property System, 2009).

The above model was solved by commercial software to select BK10 model in the software space. The obtained model was solved by Newton numerical method that is:

$$X_{n+1} = X_n - \frac{f(X_n)}{f'(X_n)}$$

(2)

Mass balance

The following is a representative sketch of any of these stages (Figure 2):

Dynamic general mass balance of stage n:

$$\frac{dM_n}{dt} = L_{n+1} + V_{n-1} + F_n - L_n - V_n - S_n$$

(3)

Liquid holdup on stage n can be calculated as:

$$M_n = \rho_{L,n} \left(A_{T,n} h_{T,n} + A_{D,n} h_{D,n} \right)$$

(4)

In the steady-state space, the left side of Equation (3) is equal zero:

$$0 = L_{n+1} + V_{n-1} + F_n - L_n - V_n - S_n$$

(5)

Dynamic component mass balance of stage n:

$$\frac{d(M_n x_{n,j})}{dt} = L_{n+1} x_{n+1,j} + V_{n-1} y_{n-1,j} + F_n z_{n,j} - L_n x_{n,j} - V_n y_{n,j} - S_n x_{n,j}$$

(6)

In the steady-state space, the left side of equation (6) is equal zero (Lee et al., 1975):

$$0 = L_{n+1} x_{n+1,j} + V_{n-1} y_{n-1,j} + F_n z_{n,j} - L_n x_{n,j} - V_n y_{n,j} - S_n x_{n,j}$$

(7)

Energy balance

Dynamic general energy balance of stage n:

$$\frac{d(M_n h_n)}{dt} = L_{n+1} h_{n+1} + V_{n-1} H_{n-1} + F_n h_f - L_n h_n - V_n H_n - S_n h_n + Q_M - Q_s - Q_{loss}$$

(8)

The changes in the specific enthalpy of the liquid phase are generally very small compared to the total enthalpy of the stage. This means that, normally, the energy balance can be reduced to an algebraic equation which is used as the basis to calculate the flow of vapor from the stage which is made a steady-state space. Finally, the energy balance is as follows (Lee et al., 1975):

$$0 = L_{n+1} h_{n+1} + V_{n-1} H_{n-1} + F_n h_f - L_n h_n - V_n H_n - S_n h_n + Q_M - Q_s - Q_{loss}$$

(9)

Vapor-liquid equilibrium

Vapor-liquid equilibrium of component j for theoretical stage n:

$$y_{n,j} = \frac{\gamma_{n,j} P_{n,j}^{sat}}{P_n \Phi_{n,j}} x_{n,j}$$

(10)

Table 1. The Mass flows of the atmospheric column products.

Product	Mass flow (Kg/s)
Naphtha	19.43
Blending naphtha	0.25
Kerosene	6.55
Atmosphere gas oil	6.38
Atmospheric residue	15.68

Table 2. The Mass flows of the debutanizer column products.

Product	Mass flow (Kg/s)
To fuel	0.38
To LPG unit	0.72
Bottom product	8.2

Table 3. The Mass flows of the splitter column products.

Product	Mass flow (Kg/s)
To flare	0.01
To LSRG Merox	2.1
HSRG to platforming	6.1

This equation is the equilibrium and in real state. If each of vapor or liquid phase is ideal then $\Phi_{n,j}$ or $\gamma_{n,j}$ is unit, respectively. If both phases are ideal then $\Phi_{n,j}$ and $\gamma_{n,j}$ are unit. Therefore, the above equation is converted to Raoult's equation:

$$y_{n,j} P_n = x_{n,j} P_{n,j}^{sat} \qquad (11)$$

Pressure

$$P_n = P_{n+1} + \Delta P \qquad (12)$$

$$\Delta P = \left(\frac{V_0}{K} \right)^2 \qquad (13)$$

Where V_0 the volumetric flow is rate of live stream in m^3/h and K is the proportionality constant in $m^3/bar^{0.5}.h$. The value of K for each geometry is different and has specific value which is chosen by software (Almudena, 2001; Lee et al., 1975).

Steady-state simulation

In this work, distillation unit of Kermanshah Refinery was simulated. The three assays of crude oil were characterized by the TBP (True Boiling Point) data, API gravity and light components.

The unit consists of 5 heat exchangers, 2 coolers, 2 heaters, atmospheric column, debutanizer column, splitter column, valves and pumps. The atmospheric column as the main part of the unit had three side strippers and two pumparounds. Important parameters for the pumparound specification are the drown off and the return stages, mass flow rate and temperature drop. For the side strippers, beside the product flow rate, the specification of the steam flow and parameters, the drown off and the return stages, and the number of stripper stages were entered. The feed flow rate of 0.046 m^3/s (25,000 bbl/day) of crude oil was preheated. Then, it was entered to the 35th stage of the atmospheric column with 38 theoretical stages. Temperature of the feed was 328.11 °C (622.6 °F). Products of the column are naphtha, blending naphtha, kerosene, atmospheric gas oil and atmospheric residue. Table 1 shows their mass flow rates.

The product of kerosene, atmospheric gas oil and atmospheric residue played an important role in preheating of the feed, because they had high temperatures, hence energy optimization was done.

To purify the naphtha, firstly it was cooled to 26.67 °C (80 °C). Then the naphtha stream was entered to a two-phase separator and splitter. Fifty percent of the flow was returned as the reflux stream and the other half was preheated and entered to the debutanizer column. The bottom product preheated the feed and entered to splitter column.

Tables 2 and 3 show the mass flow rates of the products (Tables 2 and 3). Also, Figure 3 illustrates the steady-state simulation scheme of the above steps in continuous forms.

Dynamic simulation

After steady-state simulation to observation the effects of changes the crude oil feed in the products of unit and investigation of results in real processes, we exported the stead-state simulation to dynamic simulation.

Before transferring the steady-state files, dynamic simulation requirements should be entered. In addition, the pressure changers (valves, pumps, etc.) are necessary and sensitive to exporting of steady-state simulation to dynamic simulation by "export dynamic (pressure driven)".

For example dynamic requirements of column are column diameter, tray spacing, tray active area, weir

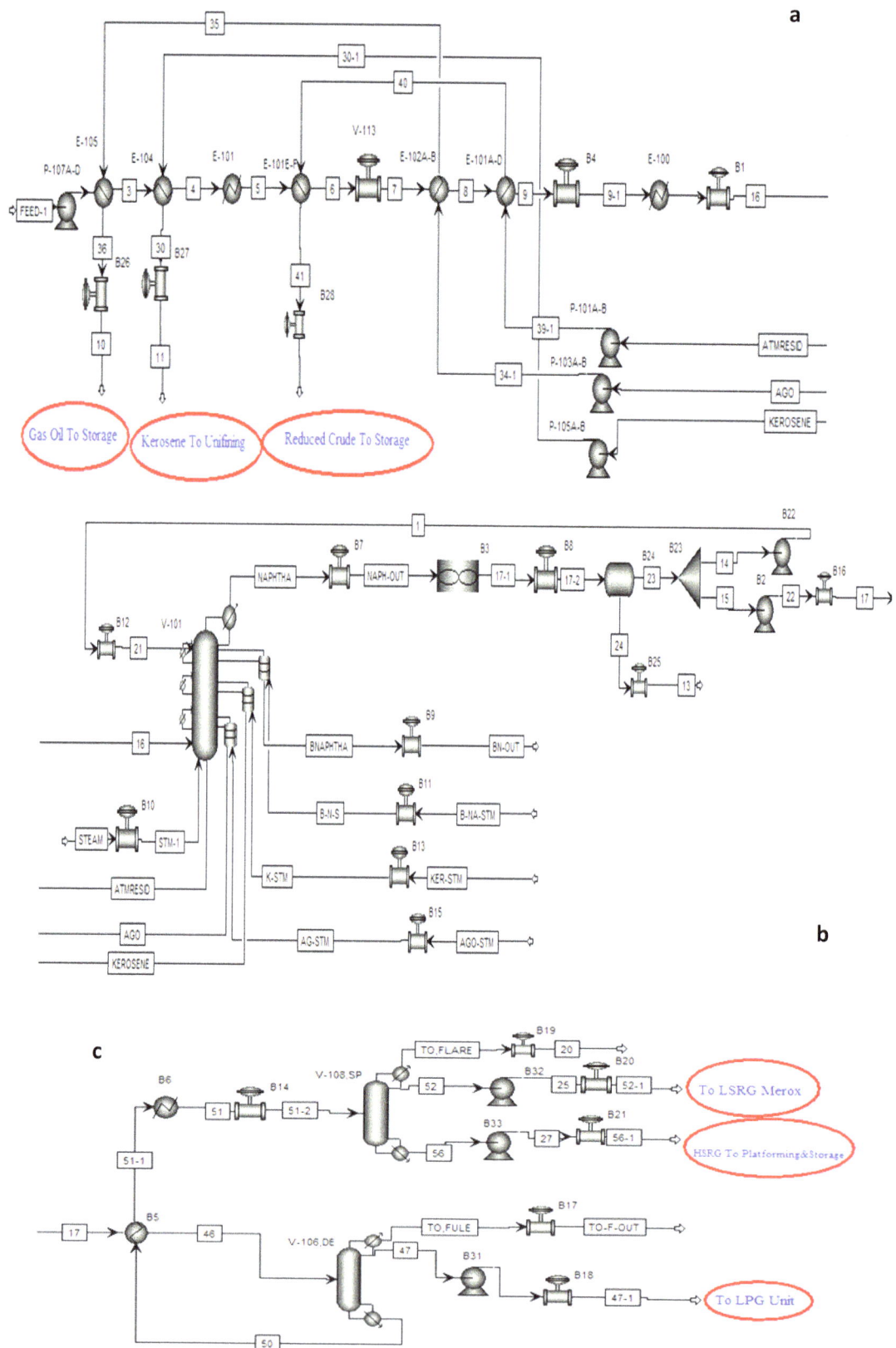

Figure 3. Steady-state simulation scheme of distillation unit; (a) preheating; (b) atmospheric distillation column; (c) Gasoline unit (light and heavy).

length, weir height, reflux drum length and diameter, and sump length and diameter. A "tray sizing" tool can be used to calculate the tray sizes based on flow conditions in the column. Of course, all of dynamic simulation requirements were provided by Research and Development (R&D) Bureau of Kermanshah Refinery.

After entering data and exporting to dynamic simulation in order to control the flow, pressure, temperature and level of streams, especially all products than changing of crude oil feed, controllers should be added in right places in the dynamic space. Dynamic space provides a number of different types of controllers. The PID Incr. model was used for all controllers in the dynamic space. The parameters of each controller (gain, integral time and derivative time) were set to optimal values using the assistance of the "tuning" tool and Tyreus-Luyben method (Luyben, 2006; Juma and Tomáš, 2009). Figure 4 illustrates the dynamic simulation scheme of continuous forms (Figure 4). Streams ID are corresponding to the steady-state simulation scheme.

RESULTS AND DISCUSSION

Distillation temperature ASTM D86

After changing the crude oil feed flow rate, ASTM D86 of six streams (("52-1", light gasoline), ("56-1", heavy gasoline), the feed of debutanizer column (V-106, DE), blending naphtha, kerosene and atmospheric gas oil) in three spaces of experimental, steady-state and dynamic were compared. Experimental data were provided by R&D Bureau of Kermanshah Refinery.

Figures 5 to 10 show a comparison between the experimental ASTM D86 curves with the results of the steady-state and the dynamic simulations. Curves of the feed of debutanizer column (V-106, DE) and atmospheric gas oil stream were in better agreement with the experimental data than the other streams. Of course, maximum difference of other streams was around 12°C. Totally, results of simulations were in good agreement with the experimental data (Kermanshah Refinery, 2009).

2- Sensitivity analysis in the MATLAB simulink

The behaviors of the FPTL controllers in dynamic simulation were observed by increasing the crude oil feed flow rate (+3%). The FPTL were controlled by conventional PID controllers. Set points were set based on Kermanshah Refinery. Twenty-three controllers were applied to control of FPTL of the unit. We tried to set the controller parameters and solved of fluctuations by different control methods to reach a new steady-state. To set the controller parameters, Tyreus-Luyben method

was employed. At last, we investigated of dynamic results by transferring the dynamic files to MATLAB®/Simulink™ Figure 11. The first steady-state then system sensitivity was observed by step changes. Input variables were:

1. Feed temperature (+10°C).
2. Feed flow rates: Ahwaz (+1%), Maleh-Kuh (+1%), Naft-I-Shah (+1%)
3. Steam flow rates: STEAM (interring to atmospheric column, +20%), blending naphtha, steam (+50%), kerosene steam (+30%), atmospheric gas oil (AGO) steam (+30%).
4. The duty of Reboilers: debutanizer column (V-106-DE, +3%), splitter column (V-108- SP, +3%).
5. Mixed of above changes simultaneously.

And outputs were: Stream flow rates: "46" (interring to V-106-DE), blending naphtha, kerosene, atmospheric gas oil (AGO), "39-1" (bottom of atmospheric column), "52-1" (light gasoline, up of V-108-SP column), "56-1" (heavy gasoline, bottom of V-108-SP column), "47-1" (to LPG unit).

Because we wanted to increase the products, increasing of inputs were investigated. After performing above changes, we observed that the major sensitivity was related to feed temperature, the duties of the reboilers of columns in gasoline unit and simultaneous combination of above changes (Figures 12-16). Rest of input changes was not significant to steady-state.

Conclusions

Steady-state and dynamic simulations performed a good investigation into the process and discussing the calculated results. Control of variables in dynamic simulation as a flexible simulator like a pilot, was done very well.

Steady-state and dynamic simulations were in agreement with the experimental data. Any Increment of crude oil feed flow rate, made a complex fluctuations in the FPTL controllers that must be rejected by set of controller parameters and different control methods. Because the feed was a mixture of 3 crude oils and many components, control of system was very complex. The dynamic space demonstrated that temperature controllers were faster and more sensitive than the other controllers. Control of temperature can be replaced by control of the product compositions. In this control structure, small control errors in the FPTL controllers were observed. Therefore, some limitations in dynamic simulation were observed. Because of more flexibility of changing the inputs, disturbances and easier handling of graphs, dynamic files results transferred to

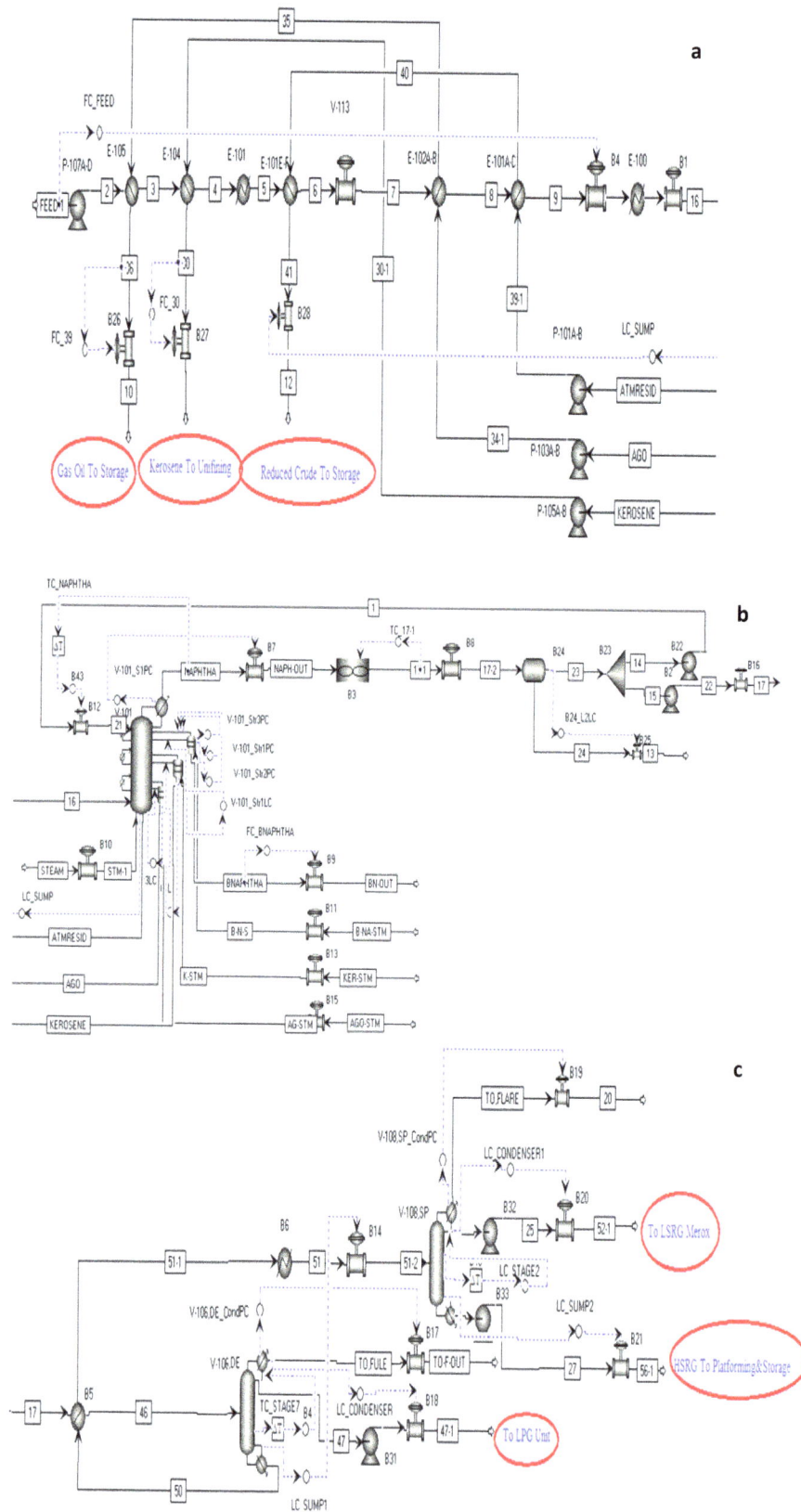

Figure 4. Dynamic simulation scheme of distillation unit; (a) preheating; (b) Atmospheric distillation column; (c) Gasoline unit (light and heavy).

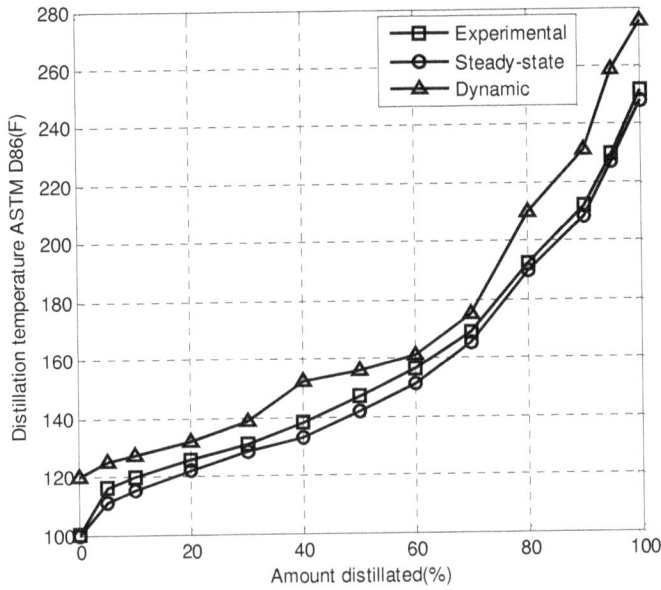

Figure 5. Steady-state, dynamic and experimental ASTM D86 curves of "52-1" stream.

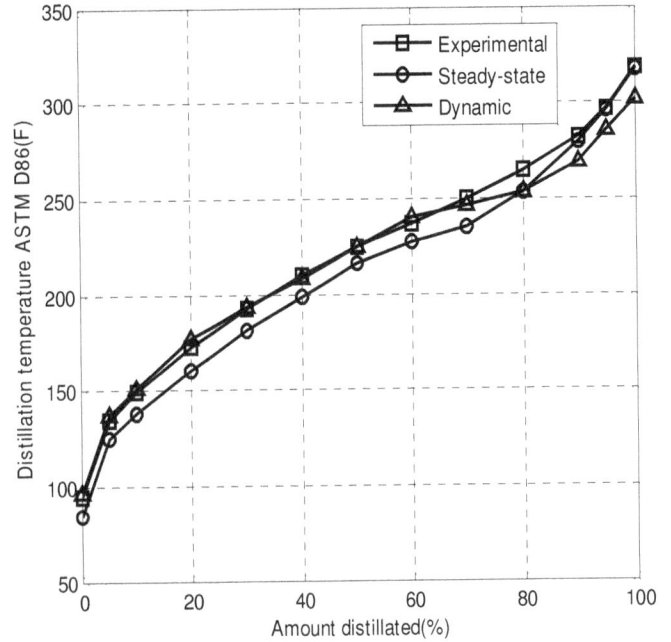

Figure 7. Steady-state, dynamic and experimental ASTM D86 curves of column feed (V-106, DE).

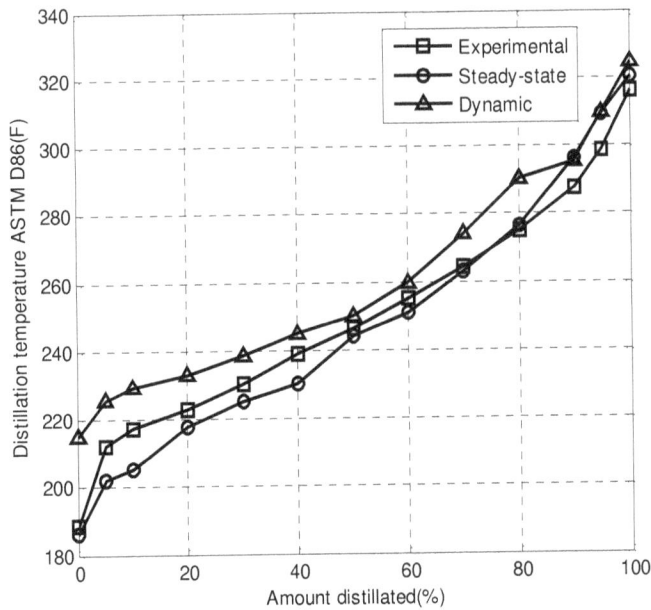

Figure 6. Steady-state, dynamic and experimental ASTM D86 curves of "56-1" stream.

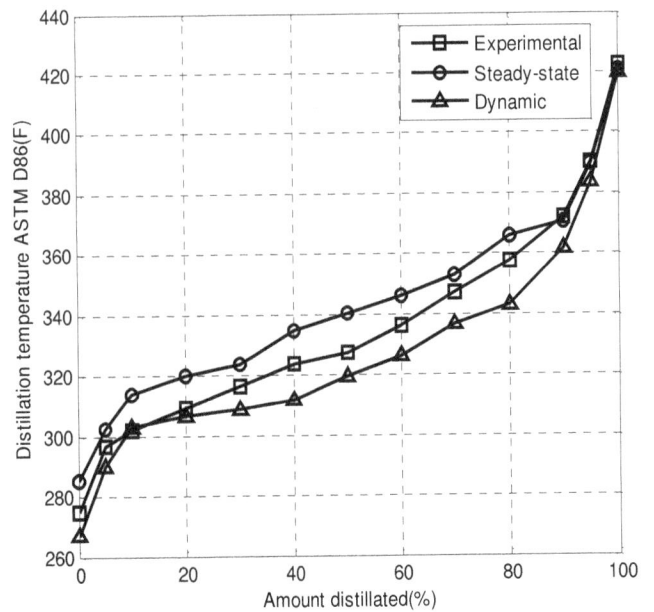

Figure 8. Steady-state, dynamic and experimental ASTM D86 curves of Blending Naphtha (B_NAPHTHA Stream).

MATALB®/Simulink™. Figures 12 to 16 show that more sensitive disturbances were feed temperature, the duties of the reboilers of columns in gasoline unit and simultaneous combination of above changes. Rest of input changes was not significant in transient responses. Therefore, above variables play important roles in the design of distillation units.

ACKNOWLEDGMENT

The financial support provided by the Kermanshah Oil Refining Company is gratefully acknowledged.

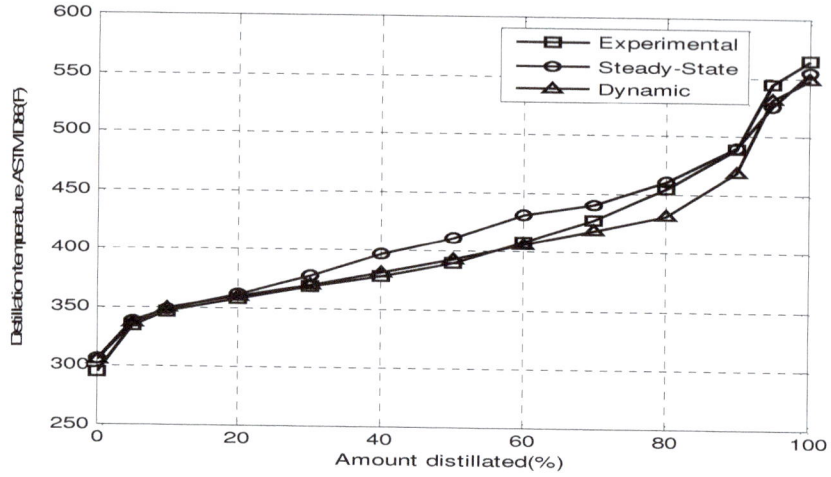

Figure 9. Steady-state, dynamic and experimental ASTM D86 curves of Kerosene.

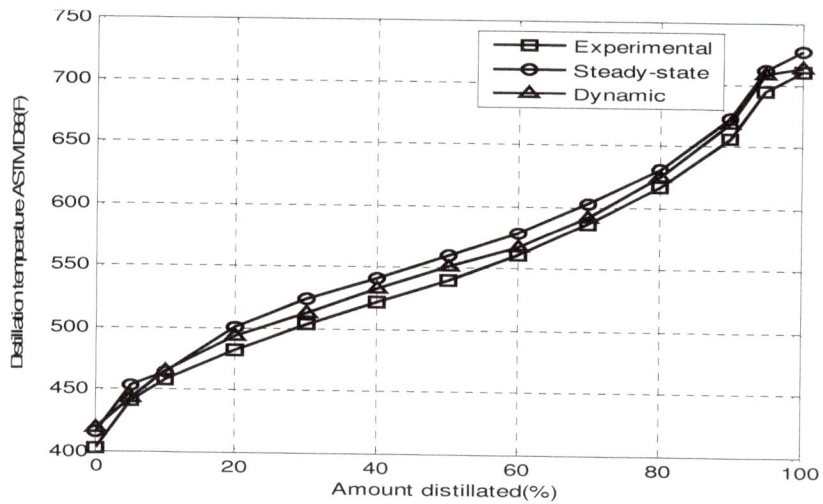

Figure 10. Steady-state, dynamic and experimental ASTM D86 curves of atmospheric gas oil (AGO stream).

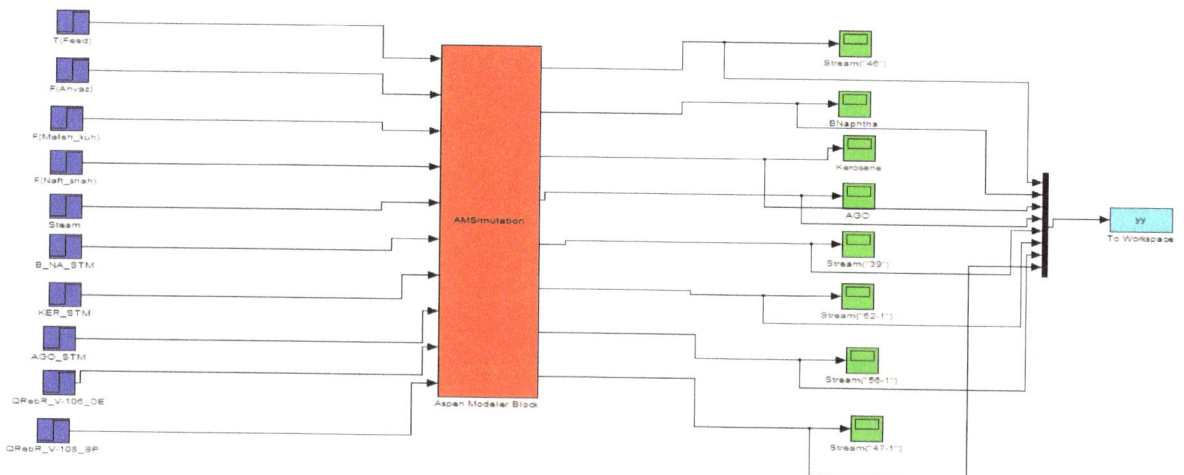

Figure 11. Scheme of Distillation unit in the MATLAB simulink with inputs and outputs.

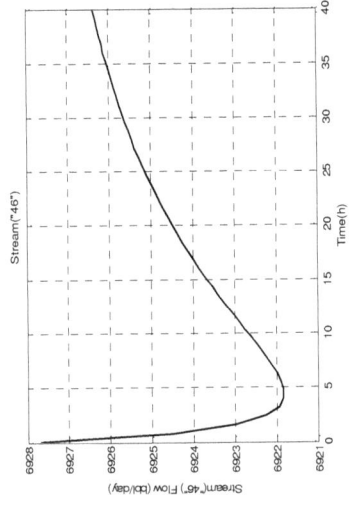

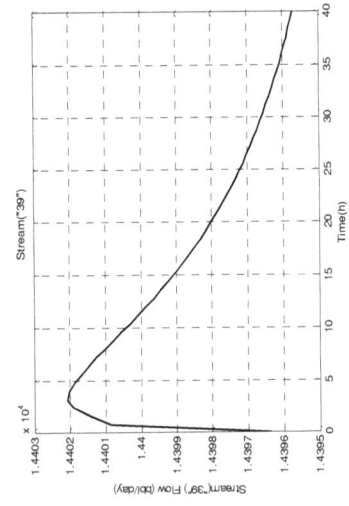

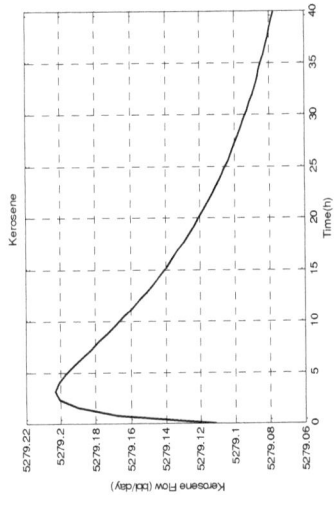

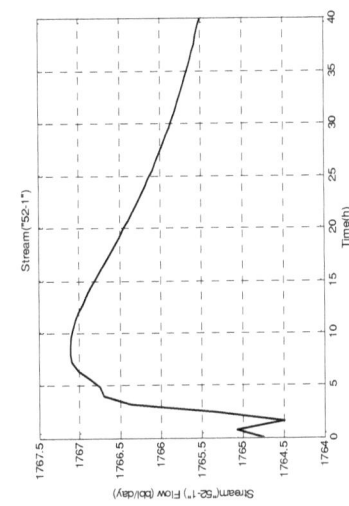

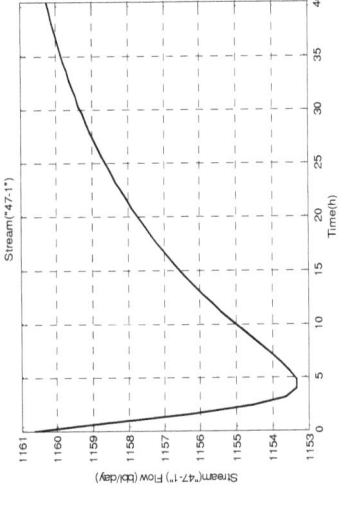

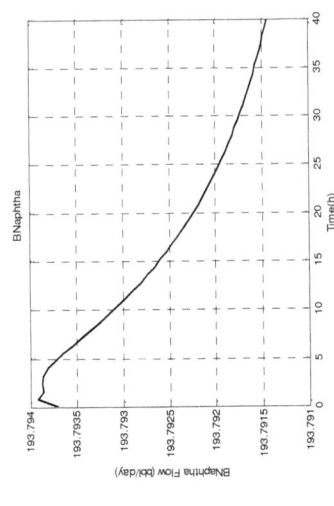

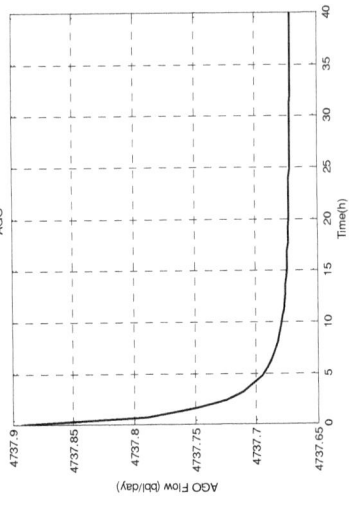

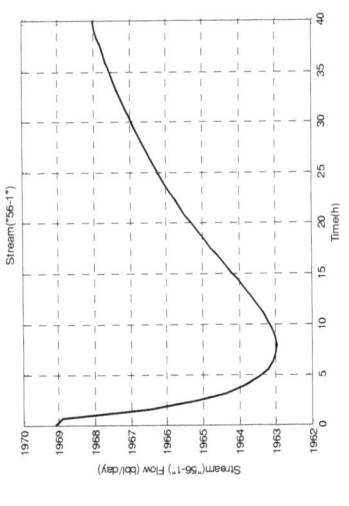

Figure 12. Steady-state curves of stream: 46, B_Naphtha, Kerosene, AGO, (39-1), (52-1), (56-1) and (47-1).

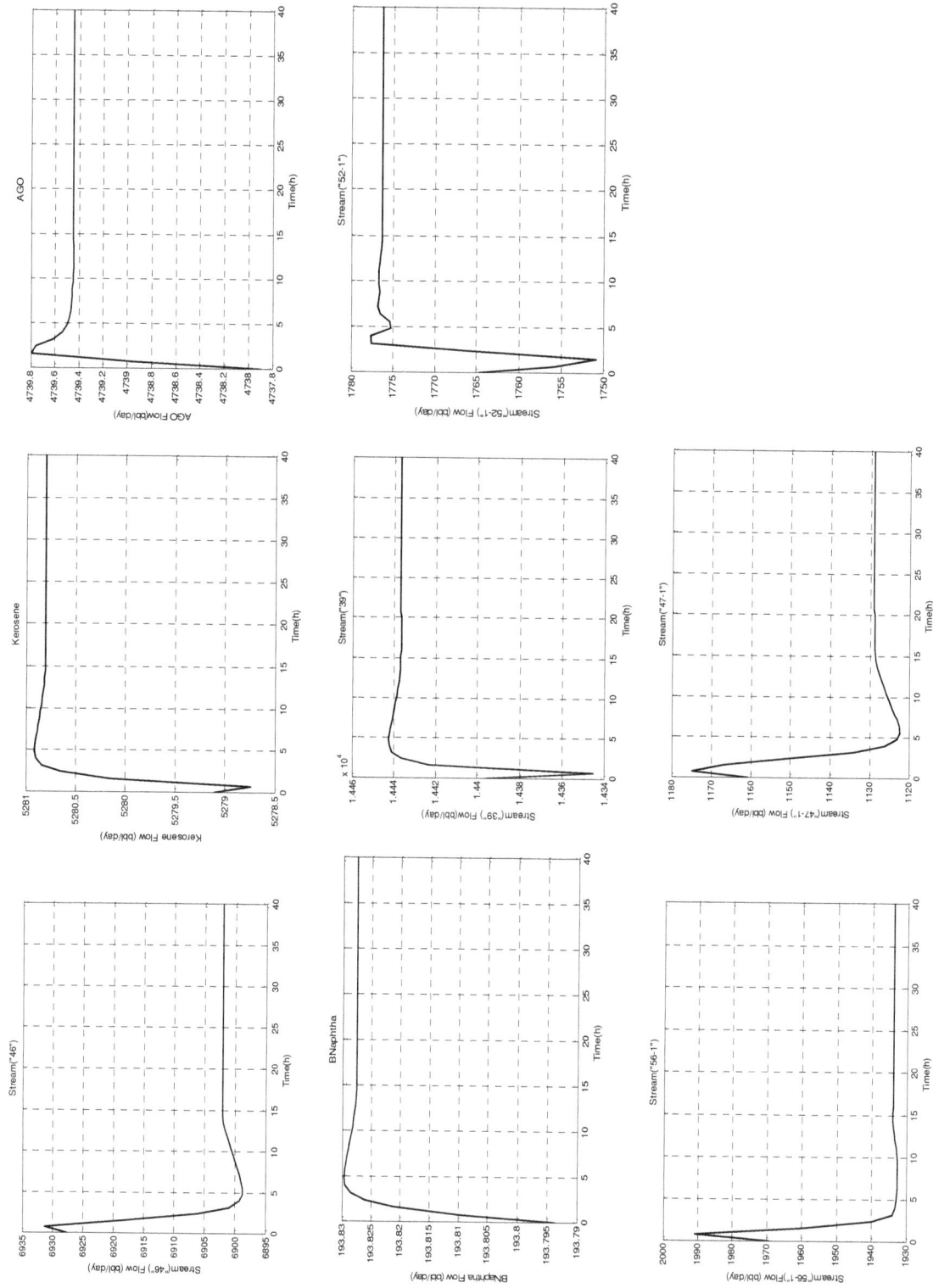

Figure 13. Curves of stream with change of feed temperature (+ 10°C): 46, B_Naphtha, Kerosene, AGO, (39-1), (52-1), (56-1) and (47-1).

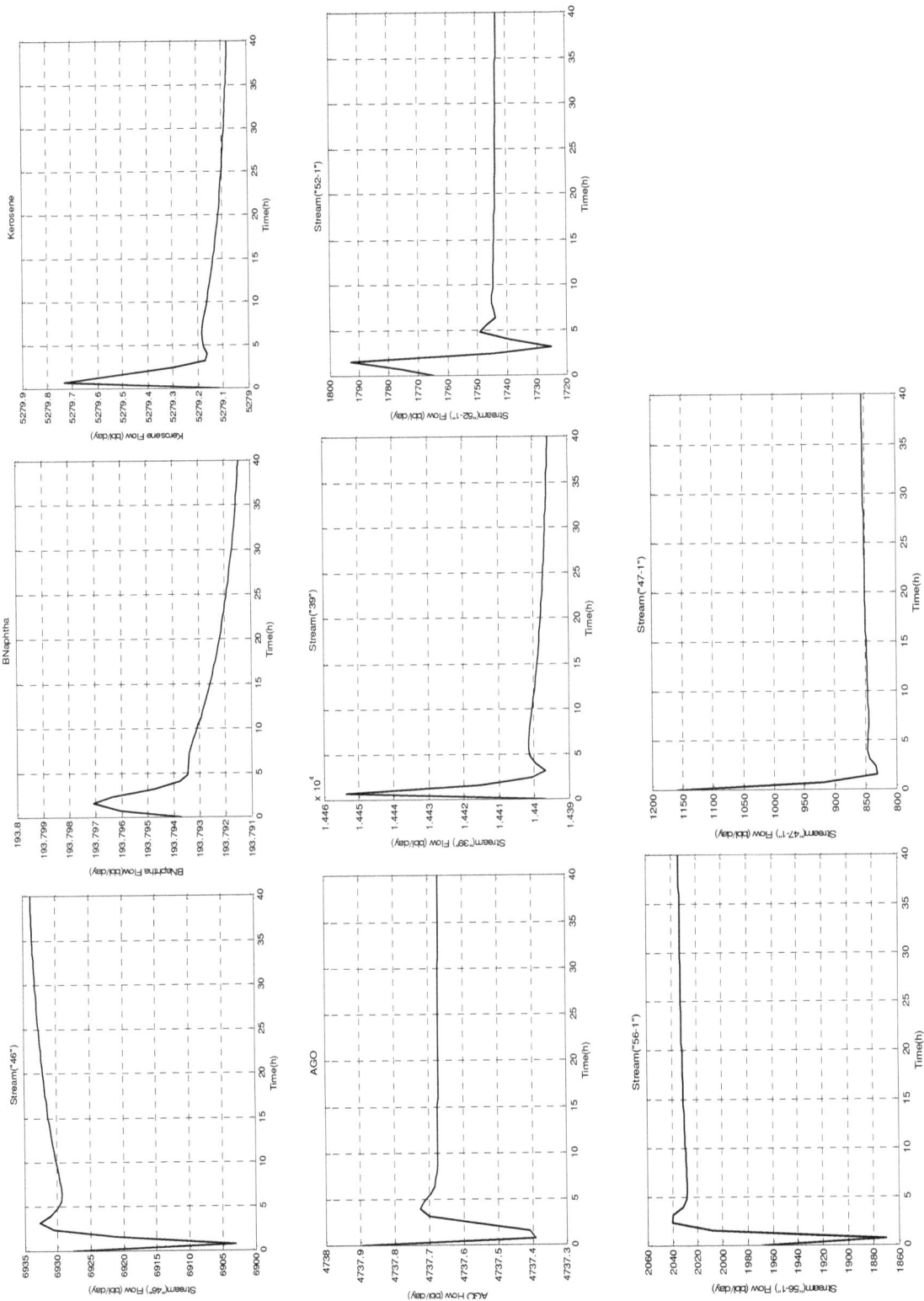

Figure 14. Curves of stream with change of Reboiles duty, V-106-DE (+ 3%): 46, B_Naphtha, Kerosene, AGO, 39, (52-1), (56-1) and (47-1).

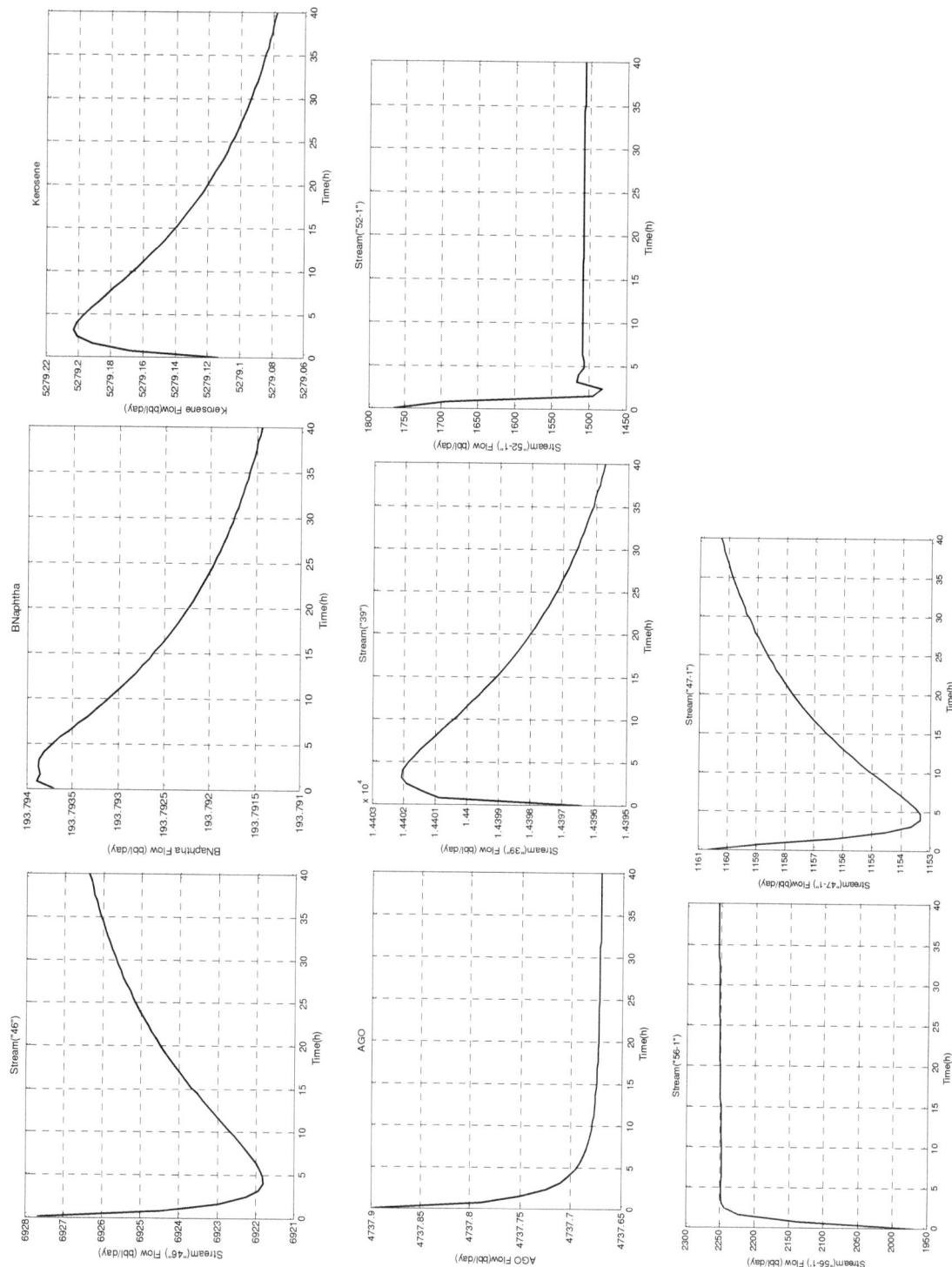

Figure 15. Curves of stream with change of Reboiles'duty, V-108-SP (+ 3%): 46, B_Naphtha, Kerosene, AGO, 39 , (52-1) , (56-1) and (47-1).

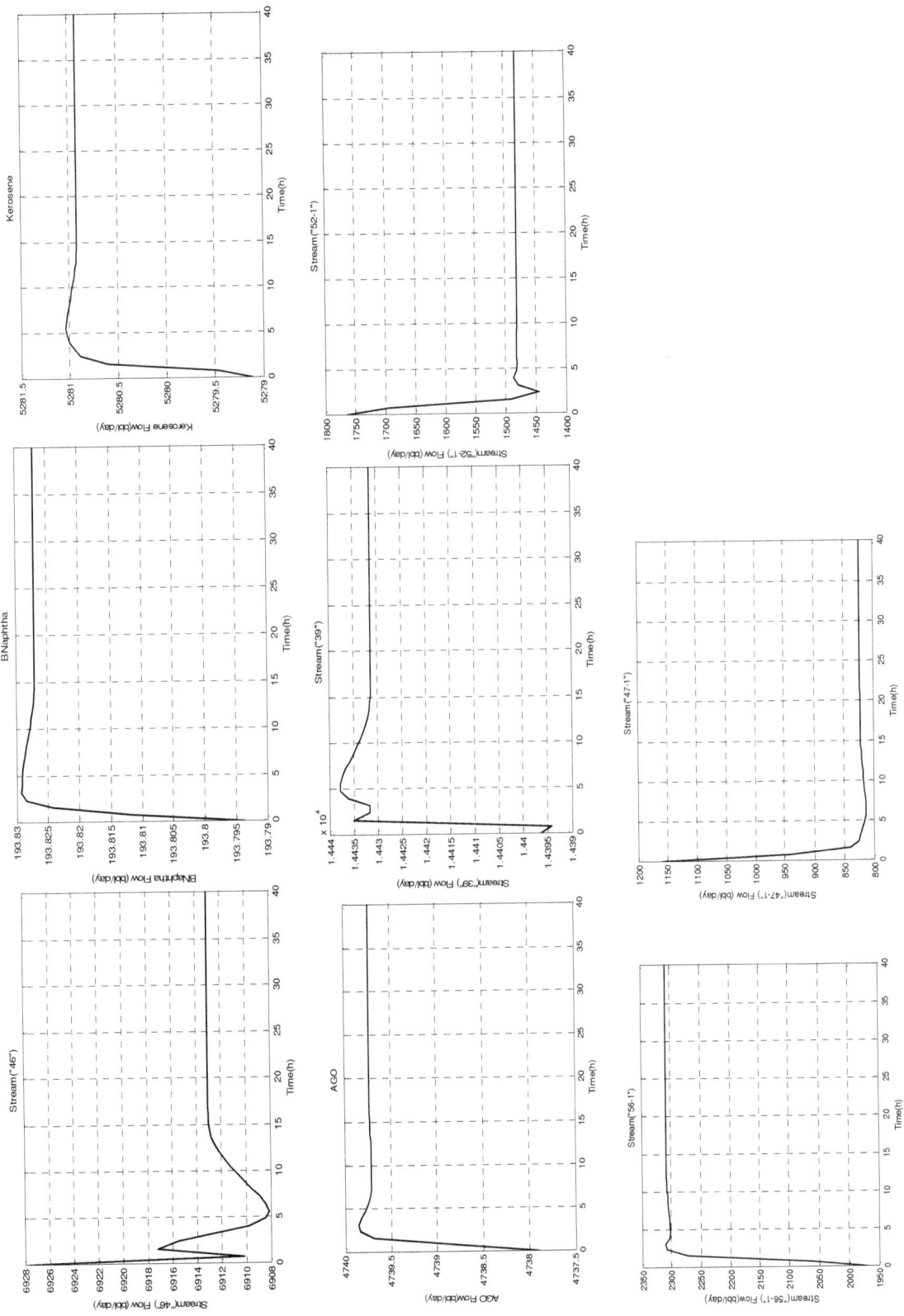

Figure 16. Curves of stream with simultaneous combination of above changes: 46, B_Naphtha, Kerosene, AGO, 39, (52-1), (56-1) and (47-1).

Nomenclature

$A_{D,n}$: surface area of the downcomer [m^2]

$A_{T,n}$: active surface area of the stage n [m^2]

bbl: barrel

C(s): Controller transfer function

D: Load or disturbance

E: Error signal

F_n : molar feed flow onto stage n [$\frac{mol}{s}$]

h_n : molar enthalpy of the liquid on stage n [$\frac{J}{mol}$]

h_{n+1} : molar enthalpy of the liquid from stage $n+1$ [$\frac{J}{mol}$]

H_n : molar enthalpy of the vapor on stage n [$\frac{J}{mol}$]

H_{n-1} : molar enthalpy of the vapor from stage n-1 [$\frac{J}{mol}$]

h_f : molar enthalpy of feed [$\frac{J}{mol}$]

$h_{T,n}$: liquid height on the stage n [$\frac{J}{mol}$]

$h_{D,n}$: liquid height on the downcomer [$\frac{J}{mol}$]

K_P : controller gain

L_{n+1} : the molar liquid that overflows onto stage n from stage $n+1$ [$\frac{mol}{s}$]

L_n : molar liquid flowing from stage n [$\frac{mol}{s}$]

M_n : the liquid mole accumulated on stage n (liquid holdup on stage n) [$\frac{mol}{s}$]

P(s): process transfer function

P_n : pressure on stage n [pa]

Q_M : heat of mixing [$\frac{J}{mol}$]

Q_s : external heat source [$\frac{J}{mol}$]

Q_{loss} : heat losses [$\frac{J}{mol}$]

r: desired value

R&D: Research and Development

S_n : molar side stream from stage n [$\frac{mol}{s}$]

T_n : Temperature on stage n [$^{\circ}C$]

U: Manipulated value

V_{n-1} : the molar vapor flow from stage n-1 [$\frac{mol}{s}$]

V_n : molar vapor flow flowing from stage n [$\frac{mol}{s}$]

$x_{n,j}$: molar fraction of component j in the liquid on stage n

$x_{n+1,j}$: molar fraction of component j in the liquid current from stage $n+1$

Y: Output value

$y_{n-1,j}$: molar fraction of component j in the vapor current from stage n-1

$y_{n,j}$: molar fraction of component j in the vapor current from stage n

$z_{n,j}$: molar fraction of component j in the feed current on stage n

$\rho_{L,n}$: liquid density at stage n [$\frac{mol}{m^3}$]

τ_D : Controller derivative time [s]

τ_I : Controller integral time [s]

REFERENCES

Almudena RF (2001). Dynamic Modelling and Simulation with Ecosimpr of an Ethanol Distillation Column in the Sugar Industry, Madrid, 1: 150-200.

Araki M (2002). Control systems, Robotics and Automation. Kyoto University, Japan, 1: 235-376.

Aspen Physical Property System (2009). Physical property methods and models. Aspen Technol. 1: 356-739.

Juma H, Tomáš P (2009). Steady-State and Dynamic Simulation of Crude Oil Distillation Using Aspen Plus and Aspen Dynamics. Pet. Coal. J. 51(2): 100-109.

Kermanshah Refinery (2009). Operating data of Distillation unit.

Lee BI, Kesler MG (1975). A generalized thermodynamic correlation based on three Parameter corresponding states. AIChE. J. 21(3): 510-527.

Luyben WL (2006). Distillation Design and Control Using Aspen Simulation. John Wiley & Sons. New York, 1: 10-283.

11

Application of piezoelectric energy generated from quartz plus semiprecious metals on wax deposition control

Sulaiman, A. D. I.[1]* Ajienka, A. J.[2] and Sunday, I. S.[2]

[1]Petroleum Engineering Programme, Abubakar Tafawa Balewa University, Bauchi, Nigeria.
[2]Petroleum and Gas Engineering Department, University of Port-Harcourt, Nigeria.

The primary focus of this research is to develop a tool using semi-precious metals and quartz for preventing or eliminating wax deposition in production and transport system. It also focuses on explaining the mechanisms of using piezoelectric energy generated from mixtures of semi-precious metals and quartz to prevent wax deposition. Five samples of waxy crude were sourced from Niger Delta and characterized for the purpose of this experiment. Quartz was sourced from Bauchi State of Nigeria while two semi-precious metals, zinc and lead sourced from Jos Metallurgical Development Centre, Plateau State of Nigeria were used. The metals were grounded to smaller sizes and mixed with the quartz. Molten aluminium was used to fabricate the tool. The tool was then fixed into the flow line. Each of the crude samples was flown through the flow system under reduced temperature and pressure (below its wax appearance temperature) and the quantity of wax deposited was recorded before the tool and after the tool. The results show that immediately after the condenser, higher percentage of wax deposit was recorded but as the fluid pass through the fabricated tool the quantity of deposits reduced drastically. This is an indication of effectiveness of the fabricated tool in preventing wax deposits along the flow lines. It was also noticed that the effectiveness is more pronounced with zinc plus quartz crystal with over 32% success in achieving deposit reduction on the most potential problematic crude sample while lead achieved only 17.9% success on the same sample.

Key words: Piezoelectric material, piezoelectric energy, semi-precious metals, wax appearance temperature, wax deposition.

INTRODUCTION

Wax deposition in petroleum reservoirs and production equipment has been a continuous problem for hydrocarbon industry. It can occur and present a lot of problems in oil production, transportation and storage. These cause the reduced crude oil pumping rate, severe startup problems after shutdown and wax deposition. As oil development sites move to deeper and therefore colder water, the appearance of wax deposition on pipe walls and other production facilities becomes inevitable. Therefore, finding economic, technical and environmen-tally friendly solutions for the prevention, management, and remediation of wax deposition problem has become a necessity. Among the two existing handling methods of wax crudes, Removal (curative) and Preventive methods (Ajienka, 1990), preventive may be the best. Within the preventive methods are mechanical, chemical, electrical, thermal and combination of others. Of recent, there emerged the use of magnetic technology for the control of wax deposition (Nguyen et al., 2001).

The choice of any method of treatment depends on the severity of the problem, operational environment, and the location in flow system where the deposition takes place. However, none of the methods are without limitations. For example the chemicals are expensive, environmentally hazardous. They are also very sensitive, and can work effectively only on specific crudes. Thermal

*Corresponding author. E-mail: ibrsule@yahoo.com

treatment such as hot oiling can cause formation damage and contraction/expansion of facilities.

Although efforts were made in the past to use semi precious metals and quartz to generate piezoelectric energy to prevent wax deposition (Enercat Tool), the application of locally sourced materials (from Nigeria) for this purpose and the technical processes has not been reported in the literature. Therefore the major challenge in this work was to explain the technical processes involve in using piezoelectric energy generated from mixtures of semi-precious metals and quarts to prevent wax deposition in production and transport system.

REVIEW OF MECHANISM OF PIEZOELECTRIC ENERGY ON WAX DEPOSITION CONTROL

A survey of literature shows that tremendous amount of research work has been done ranging from experimental studies to analytical and simulation studies in order to predict, manage, prevent and treat wax deposition in oil wells, flow lines and in gathering tanks. Also substantial literature on wax composition, factors affecting its deposition and the effect of its deposition on the economy of production, processing, storage, and transport operations have been published.

Factors affecting wax deposition

The mechanism and extent of wax deposition in a flowing system have been studied by many researchers (Jessen et al, 1958; Tronov, 1969; Patton et al., 1970; Armenskii et al., 1971; Bott et al., 1974; Eaton et al., 1976). The basic factors that contribute to the extent of wax deposition in flowing system include: flow rate, temperature differential, cooling rate, and surface properties.

Piezoelectric effect

Piezoelectricity is the ability of some materials (notably crystals and certain ceramics, including bone) to generate an electric potential in response to applied mechanical stress. This may take the form of a separation of electric charge across the crystal stress. If the material is not short circuited applied charge induces a voltage across the material. The piezoelectric effect is reversible in that materials exhibiting the direct piezoelectric effect (the production of an electric potential when stress is applied) also exhibit the reverse piezoelectric effect (the production of stress and/or strain when an electric field is applied). The effect finds useful applications such as the production and detection of sound, generation of high voltages, electronic frequency generation, microbalance, and ultra fine focusing of optical assemblies. It is also the basis of a number of

scientific instrumental techniques with atomic resolution, the scanning, probe microscopes and everyday uses such as acting as the ignition source for cigarette lighters and push-start propane barbecues. Now it has found application in hydrocarbon industry.

MATERIALS AND METHODS

Piezoelectric materials

Piezoelectric material produces an electric charge when subjected to a force or pressure. The piezoelectric material such as quartz or polycrystalline barium titanate, contain molecules with a symmetrical charge distribution. Therefore under pressure, the crystals deform and there is a relative displacement of the positive and negative charges with the crystals. In this research work, quartz crystal is used, being one of the most stable piezoelectric materials. Figure 1 shows the displacement of electrical charge due to the deflection of the lattice in a naturally piezoelectric quartz crystal. The larger circles represent silicon atoms, while the smaller ones represent oxygen.

Mechanism of piezoelectric effects

In a piezoelectric crystal, the positive and negative electrical charges are separated, but symmetrically distributed, so that the crystal overall is electrically neutral. Each of these sides forms an electric dipole and dipoles near each other tend to be aligned in regions called Weiss domains. The domains are usually randomly oriented, but can be aligned during poling (not the same as magnetic poling), a process by which a strong electric field is applied across the material, usually at elevated temperatures.

When a mechanical stress is applied, this symmetry is disturbed, and the charge asymmetry generates a voltage across the material. Figure 2 shows the crystal structure and charges distribution in piezoelectric material.

Piezoelectric materials also show the opposite effect, called converse piezoelectric effect, where the application of an electrical field creates mechanical deformation in the crystal. The lack of conclusive literature data on the effects of piezoelectric energy on the phase behaviour of paraffins can be explained by the complex nature of crude, in which a lot of components, having different chemical and physical properties are found.

Experiments

Crude oils

Five samples of waxy-crudes (Dead Oil) were collected from different oil fields in Niger Delta, Nigeria. Addax Petroleum Nigeria provided three samples of crudes (samples OSS-1, OSS-2, and OSS-3) while Shell Petroleum Development provided the remaining two samples (A&B) for the purpose of this research. The specification of the samples remained undisclosed by the donor companies.

Measurement of STO properties

The densities of samples were determined using hydrometer method. The API was determined using ASTM D287. The specific gravities were measured at 60°F using ASTM D1298. The corresponding pour points were measured following the standard

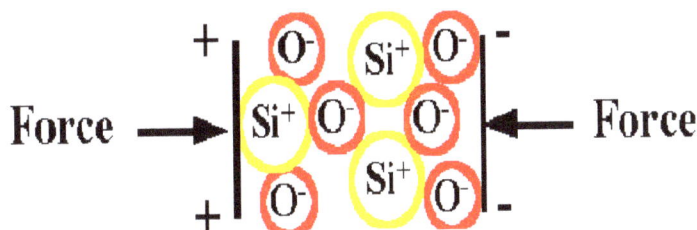

Figure 1. Displacement of electrical charge due to the deflection of the lattice in piezoelectric quartz crystals.

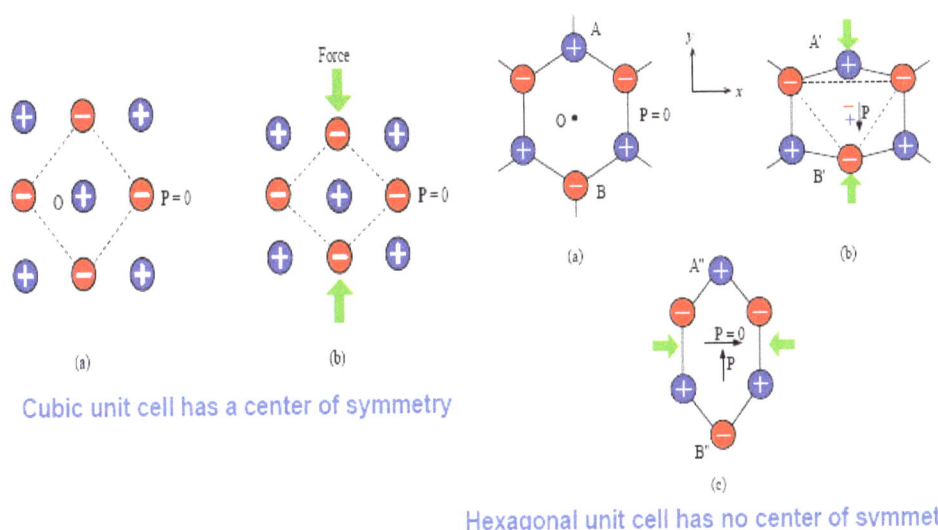

Cubic unit cell has a center of symmetry

Hexagonal unit cell has no center of symmet

Figure 2. Crystal structure and charges distribution of piezoelectric material.

ASTM D97 procedures. The kinematic viscosities were determined using ASTM D-4624. The results obtained in the above analysis are in Table 1.

Fabrication of the tool (Aluminum Jacket)

Aluminum jacket was fabricated using molten aluminum. The inner pipe has its internal diameter very close to the external diameter of the Pyrex glass through which the crude was passed. The external aluminum jacket is shorter but has wider internal diameter. The space between the external and internal jacket and the tight cover, created space for putting the mixture semi-precious metals and grounded quartz particles (Figure 3).

Wax deposition experiment

Quartz was grounded to crystal form using crushing machine and added to semi-precious metals (zinc, lead and cobalt) which were also chopped into pieces. Each of this mixture (zinc +quartz and zinc + lead) was put in the fabricated tool and covered with aluminum lid. The tool was then fixed in flow line. An electric ring heater in the oil tank was used to heat each of the oil samples to about 75°C to dissolve all the wax and other impurities in the crude. The oil was then flown through the coolant where it was allowed to

cool below its wax appearance temperature (WAT). The flow was re-circulated using boaster pump for 18 h, after which the system was stopped.

The temperatures were recorded with six thermocouples and a recorder (meter). Four thermocouples recorded the upstream and downstream temperatures of the oil and the water; one thermocouple was placed inside the oil reservoir for temperature control, and the other was placed in cold water reservoir for recording the temperature. A rotometer was used to measure the flow rate of waxy crude along the flow line.

Prior to each use, the equipment was cleaned by pumping technical grade toluene through it for one hour. The system was then drained and air dried. The Pyrex glass tubing was tilted to allow easy draining of oil by gravity. However, since there is a limit to which a glass can be tilted, an opening was provided in the flow line closed to the oil tank and a valve fixed. A foot pump model "Bee" was used to push out any oil that could not come out as a result of imperfect tilting of the Pyrex glass. The coolant pump was turned on and the flow rate was adjusted with a flow control valve to give the desired oil temperature drop along tubing. As the ring heater was turned on, the system was allowed to reach thermal equilibrium. This took about 30 min. An electronic temperature controller, model Jetec, JTC-903 was used to maintain the temperature of the oil in reservoir at 70°C (180°F). The oil temperature at the condenser was set at 70°C and the outlet temperature at 15°C (50°F). Then the flow was measured and adjusted to the proper rate. The flow was

Table 1. Properties of samples OSS-1, OSS-2, OSS-3, A and B.

S/N	Test	Test Method	Unit	OSS-1	OSS-2	OSS-3	A	B
1	Pour Point	ASTM D 97	°C	-6	-8	-2	3	-12
2	Cloud point	ASTM D 2500	°C	13	7.5	14	-	-
	Viscosity @							
	40		Cst	5.91	10.4	12.4		
3	60	ASTM D445		3.22	6.03	8.24	3	-12
	80			1.09	0.93	2.36		
4	Wax Content	UCP 46	Mass%	8.19	5.27	8.61	<5	<5
5	SG	ASTM D287	°F	0.9174	0.9063	0.9477	0.8817	0.8717
6	Density	HYDROMTER	Kg/l	917	906	948	-	-
7	Heat of combustion(gross)	ASTM D240-02	MJ/kg	33.18	40.24	38.13	-	-
8	Heat of combustion(net)	ASTM D240-02	MJ/kg	31.29	38.27	36.22	-	-

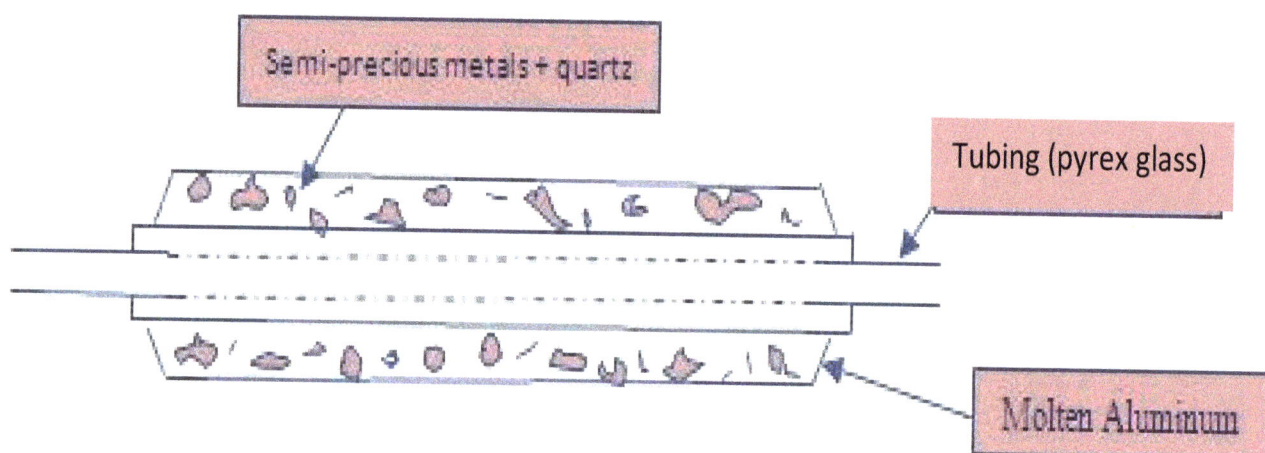

Figure 3. Detailed diagram of the fabricated tool.

continued for 18 h. As the flow continued, the temperatures of the inlet and outlet were taken to ascertain the effectiveness of the improvised condenser. Then, the flow was stopped. In the first instance, it was zinc + crystal quartz in the fabricated tool. The cold water was maintained at 0° ±0.5°C in an insulated 7 liters reservoir. The flow rate of the coolant (water) was measured using SOCAM flow meter model ISO4064. PEDROLLA 0.5 Hp pump model PKM60 was used for the continuous circulation of the crude oil and the coolant in the flow system.

At the end of the experiment the glass tubing was drained at ambient temperature, and then cut into sections, each about 14 cm long. The deposited wax was melted out of the tubing by heating the preweighed sections at 180°F (82°C) for 10 min. The amount of mass lost by the tubing was compared with the weight of the wax obtained. The difference was considered to be due to the volatization of lighter components. The same process was repeated for samples OSS-2 and OSS-3. The readings were taken and the obtained results compared. At the end of each experiment, the oil reservoir and the flow system were flushed using toluene and air dried before the next sample is re-run.

After the three samples were ran, the fabricated tool was removed, loosed, and zinc and crystal quartz were poured out and the tool thoroughly cleaned. Lead plus crystal quartz was then placed in the tool and replaced back into the flow line. The experiment was run again for all the three samples (OSS-1, OSS-2 and OSS-3). The same procedure was followed as did with zinc plus crystal quartz.

RESULTS AND DISCUSSIONS

Stock tank oil properties

Table 1 summarizes the measured densities, pour point, cloud points, and viscosities of samples OSS-1, OSS-2, OSS-3, A and B. As can be seen right from the table, sample OSS-3 has the highest molecular weight, asphaltene and wax content and the largest C_{36+} fraction. This gives the clear justification why sample OSS-3 has the highest cloud point (WAT) value as indicated in Table 1. Sample OSS-2 on the other hand, has the lowest asphaltene content and the smallest C_{36+} fraction, thereby, exhibiting lowest clout point (WAT) value. From this analysis it can be understood that samples A and B are not likely to cause much problem to the production system. Also it is obvious that to access the performance of the fabricated tool, the outlet temperature at the condenser is not supposed to be much more than the WAT for each sample.

Compositional analysis

Table 2 summarizes the composition of crude oil samples

Table 2. Results of compositional analysis of waxy crudes.

Laboratory No. Sample code	LPEC/1/08 Oss-1	LPEC/2/08 OSS-2	LPEC/3/08 OSS-3	LPEC/4/08 A	LPEC/5/08 B
Nitrogen, mole %				0.32	0.14
Light ends, mole %					
C_2- Hydrocarbons	60.37	58.73	71.21	70.11	63.4
C3- Hydrocarbons	5.43	5.28	9.32	9.12	7.64
IC4 (Isobutene)	2.99	2.84	1.98	2.18	4.22
NC4 (Normal Butane	1.92	1.67	1.36	1.85	3.98
IC5 (isopentane)	1.13	1.01	0.74	0.97	2.24
NC5(normal pentane)	0.86	0.83	0.38	0.42	1.83
Sum C2 - C5, mole %	72.82	70.43	84.99	84.65	83.31
Light Naphthene (bp<175 ^{o}F) mole %	18.15	16.14	19.86	20.23	16.21
Medium naphtha(175<bp<250°F mole,%	5.9	5.17	11.43	11.92	5.61
Heavy naphtha 9250<bp<375F mole %	2.1	1.76	9.22	8.96	3.09
molecular weight (g/mol) C6-C10	194	187	223	212	209
Kerosene(375<bp<65°F) mole %	1.3	1.06	7.24	7.28	1.83
Sum C10-C12 mole %	28.15	24.13	47.75	48.39	26.74
LGO (530<bp<650) mole %	0.96	0.87	5.87	5.84	1.27
Molecular weight (g/mol) C28 +	360	298	429	425	330
PGO(650<pb<1049F) mole %	0.68	0.53	3.68	4.13	0.98
Residual Oil (bp>1049F) mole %	0.43	0.36	2.97	2.04	0.64
Molecular weight of C36+ (g/mol)	537	515	693	615	519
Av. Molecular weight of whole oil (g/mol)	164	150	238	185	148

OSS-1, OSS-2, OSS-3, A and B using ASTM D5297 method. From this compositional analysis, information about n-paraffin, iso-paraffins and naphthenes concentration in the dead oils is established. It is also an evident that oils collected from the same field but different reservoir/well has relatively different composition. The differences may be attributed to geological factors and or variation in temperature profiles and production histories for different wells. The implication of the observed variations in oil composition could be interpreted in relation to measured STO property.

Wax deposition experiment

Visual observation of inside of the Pyrex tube (test section) at the end of tests showed that, as the oil flowed up the section, pecks of wax were formed on the pipe wall. Also up the section, a complete continuous film of deposit was observed (solid points) which increased in thickness from there to the top. It was difficult to determine the exact location where the pecks of wax began.

The quantity (in grams) of deposits per time at the section before the fabricated tool was far more than the quantity of deposits inside the tool and even after the crude has passed the tool. The quantity was estimated by visual observation and also was measured using G&G

electronic scale model JJ1000. This is also a confirmation that the tool is working effectively. Figures 1 and 2 show the comparism between the percentage wax deposits before and after the samples have passed through the fabricated tool during the test.

Conclusions

1. A tool that controls wax deposition was developed, tested and prooved to be effective in the laboratory.
2. The mechanism of semiprecious metals (Zinc, Colbat and Lead) plus quartz was studied and found to effectively reduce/eliminate wax deposit in the production system through the generation of piezoelectric reaction. Materials (Zinc, Lead and Quartz) were used in their natural form.
3. From this studies, Zinc plus quartz has proven to be more effective than Lead plus quartz crystals. Zinc was able to achieve 32.35% success on sample OSS-3 which is the most potentially problematic crude oil in the samples used for this research work; while lead could achieve only 17.86% on the same sample.
4. Among the five samples collected for the purpose of this studies, two were dropped because early investigations on the sample indicated that they have no potential capability of depositing wax easily. The two samples (A&B) have very low pour point and wax content

(Table 1). Sample OSS-3 was found to have the highest cloud point and wax content (14°C and 8.61 respectively) thereby would cause more flow problems in the production, flowlines and pipelines. This is followed by OSS-1 and the lowest, interms of cloud point and wax content is OSS-2. However, it was generally observed that the major factor in the deposition of wax from crude oil is the temperature falling below its wax appearance temperature (WAT). This is in agreement with the findings of other reasearchers (Agarwal, 1989; Ajienka, 1990 and Armenskii, et al,1971).

REFERENCES

Agarwal KM (1989): "Influence of Waxes on the Flow Properties of Bombay High Crudes," Fuels Vol. 68, P937

Ajienka JA (1990): "The effect of Temperature on Rheology of Waxy Crude Oils and its Implications in Production Operations". PhD Dissertation, University of Port Harcourt, Nigeria.

Ajienka JA, Ikoku CU (1990): "Waxy Crude Oil in Nigeria. Practices, Problems and Prospects". Energy Sources Vol.12 pp463-478.

Armenskii EA, Novoselov VF, Tugunov PI (1971): "Paraffin Deposition in Short Pipelines."Izv. vyssh. Ucheb. Zaved. Neft Gaz. Vol.14, No.7, P71.

Bott TR, Gudmundsson JS (1977): "Deposition of Paraffin Wax from Flowing system". Inst. Petroleum Tech. pp IP-77-007.

Bott, TR, Walker RA (1974): Conference in Heat Transfer and Design and Operation of Heat Exchangers Johannesberg.

Eaton PE, Weeter GY (1976): "Paraffin Deposition in Flow Lines." Paper No. 76-CSME/CSChE-22 presented at the 16th Natl. Heat Transfer Conference, St Louis, Aug. 1976.

Jessen FW, Howell JN (1958): "Effect of Flow Rate on Paraffin Accumulation in Plastic, Steal and Coated Pipes." Trans., AIME . 213, P 80.

Patton CC, Casad BM (1970): "Paraffin Deposition From Refined Wax-Solvent system". SPE Journal. pp17

Tronov VP (1969): "Effect of Flow Rate and Other Factors on the Formation of Paraffin Deposits," Tr. Tatar. Neft, Nauch.- issled. Inst. Vol.13, pp 207

12

Experimental study of new improved oil recovery from heavy and semi-heavy oil reservoirs by implementing immiscible heated surfactant alternating gas injection

Mehdi Mohammad Salehi*, Eghbal Sahraei and Seyyed Alireza Tabatabaei Nejad

Chemical Engineering Department, Sahand University of Technology, Tabriz, Iran.

After thermal processes, steam injection is the most common enhanced oil recovery (EOR) methods in heavy oil reservoirs. It is common in heavy and semi heavy oil reservoirs to inject gas in conjugate with surfactant solutions to enhance the effect of gas on reservoir fluid which is called surfactant alternating gas (SAG) injection. The interest in immiscible and miscible SAG process has been grown recently. In this work, an experimental study of immiscible heated SAG injection in a sand pack was performed. This new method is a combination of SAG and thermal process and can be used in heavy and semi-heavy reservoirs. The experiments were performed with sand pack under certain temperature, pressure, constant rate and 1.2 pore volume (PV) injected. Result shows that method effectively improves oil recovery in comparison with SAG injection and other method [gas flooding, water flooding and water alternating gas (WAG)]. Conducting these experiments indicate that using heated surfactant and heated nitrogen instead of unheated surfactant solution and nitrogen and other method can lead to interfacial tension reduction, oil swelling and viscosity reduction. Thereafter, immiscible heated SAG injection can be used as an effective and feasible EOR method in heavy and semi-heavy oil reservoirs.

Key words: Surfactant alternating gas (SAG), sand pack, heated SAG, foam.

INTRODUCTION

Petroleum industry was introduced three stages for improving the recovery of hydrocarbon reservoirs, that is:

1. Improving recovery in primary production such as drilling infill wells.
2. Secondary methods of enhanced production such as immiscible injection methods including water or gas injection or water alternating gas injection.
3. Enhanced oil recovery (EOR) methods such as thermal, chemical and miscible injection methods.

During recent years, great effort has been directed towards investigating different methods of enhanced oil recovery from the hydrocarbon reservoirs. One of these methods is surfactant alternating gas (SAG) injection. In this process which is formed by combination of the two older and traditional methods of alternative injection of surfactant solution and gas. Certain volumes of surfactant solution and gas are alternatively injected into the reservoir (Viet and Quoc, 2008). The main aim of surfactant alternating gas injection is increasing the recovered oil volume in the reservoirs. It is notable that this method has the essential potential for improving microscopic displacement efficiency, mobility, and swapped area. Also contact of surfactant solution and gas produce foam. Foam inside porous medium is defined as a dispersion of gas in liquid such that the liquid phase is continuous and at least some part of the gas is made discontinuous by thin liquid films called lamellae. The foam occurs as gas disperses within a surfactant solution and the motilities of gas and the aqueous phase are reduced (Falls et al., 1988).

Composing foam is a phenomenon which can improve sweep efficiency during gas injection, and several field applications of it have been reported (Hoefner and Evans, 1994; Patzek, 1996; Renkema and Rossen, 2007).

Foams for gas diversion can be placed in the reservoir by continuous co-injection of surfactant solution and gas or injecting alternating slugs of surfactant solution and gas. Different foam-injection strategies have been used in field trials due to stratigraphic differences, foam behavior and operational concerns (Xu and Rossen, 2003).

Several alternatives have been proposed to increase sweep efficiency of CO_2 injection in the field or in experimental works, such as injecting water alternating gas (WAG) (Christensen et al., 1998), direct CO_2 thickeners (Heller et al. 1983), and injecting surfactant solution alternating gas (Tsau and Heller, 1992). The benefits of using SAG to improve the efficiency of CO_2 displacement have been reported by several investigators (Skauge et al., 2002; Yaghoobi et al., 1998).

Laboratory and field studies indicate that foam potentially presents an efficient method of reducing CO_2 mobility (Tsau et al., 1998; Bernard and Holm, 1964). A possible advantage of SAG over WAG for mobility improvement is that it can contain higher gas saturation (over 85 to 95% gas). This means that a relatively small amount of water was used to decrease CO_2 mobility. Foam has other properties that are favorable to oil recovery, particularly by CO_2 flooding. The apparent foam viscosity is greater than the viscosity of its components which increases oil recovery due to improved mobility ratio. It also increases trapped gas saturation and decreases the oil saturation. In addition, high trapped gas saturation usually reduces gas mobility. All of these unique properties of foam indicate that it should be useful in CO_2 flooding. Foam properties may also cause unfavorable increases in injectivity and chemical costs (Syahputra et al., 2000).

The SAG phase operations were conducted without major problems. SAG injection has proved to be an efficient injection procedure. SAG is operationally similar to WAG and requires little additional effort. Injection should be performed below fracturing pressure (Blaker et al., 2002).

As indicated, earlier use of thermal recovery method is one of the most important methods for enhanced oil recovery. This has a more effective role in increase of hydrocarbon fluids recovery in heavy and semi-heavy oil reservoirs. Therefore combining SAG injection and thermal methods can result in an increase in final oil production from these reservoirs.

In this experimental study in addition to immiscible SAG injection method which is considered as an effective EOR method, immiscible heated surfactant alternating heated nitrogen gas (immiscible heated SAG) injection was employed too. In this new process aim of injecting heated fluids is affecting heavy and semi-heavy fluids.

MATERIALS AND METHODS

All the equipment and fluids which were used in immiscible SAG injection, immiscible Heated SAG injection and other EOR method was described below.

Fluids

(i) Bangestan crude oil was used in all experiments. The crude oil is intermediate (24°API). Purified gas (nitrogen) was used in experiments.
(ii) Aqueous phase was made due to sodium dodecyl sulfate dissolution in water.

Apparatuses

Fluid injection system

During the experiments a pump with high performance liquid chromatography was used to displace fluids in the sand pack.

The operating fluid of the pump is twice-distilled water and it was injecting into the pipes and fittings with constant flow rate of infusion into the bottom of fluid accumulator (brine water, surfactant solution, crude oil or nitrogen). Therefore, the accumulator fluid was injected into the sand pack with constant flow rate.

Accumulators: They were used to provide high pressure for injection. The distilled water is transported from the pump to the bottom of the accumulator to move the piston upward and compact the contained fluid.

Core holder: Core holder was made of anticorrosion stainless steel (grade 316) of 5 cm diameter and 15 cm height.

Heater: For heating the injected surfactant solution a heater was placed on route. Over this heater, a vessel containing a high boiling point material was placed.

Pressure differential gauge: It was used to measure the pressure drop along the sand pack.

Back Pressure Regulator (BPR) and effluent collector: A backpressure regulator (BPR) was used to provide a constant backpressure during core flood experiments. One of the BPRs which were installed at the outlet of the apparatus was operated at 156×10^5 Pa.

Separator and produced fluids measurement system: Separator was constructed from a steel pipe with an opening at the top for fluid entrance and two openings for fluid departure. An outlet at the top for gas and another at the bottom for draining the liquids were designed. The effluent was collected to measure oil recovery using a fractional collector. Figure 1 depicts a graphical schematic of injection system of this experiment.

Experiments were carried out on a conventional sand pack and in the following procedure.

Experimental study of new improved oil recovery from heavy and semi-heavy oil reservoirs...

95

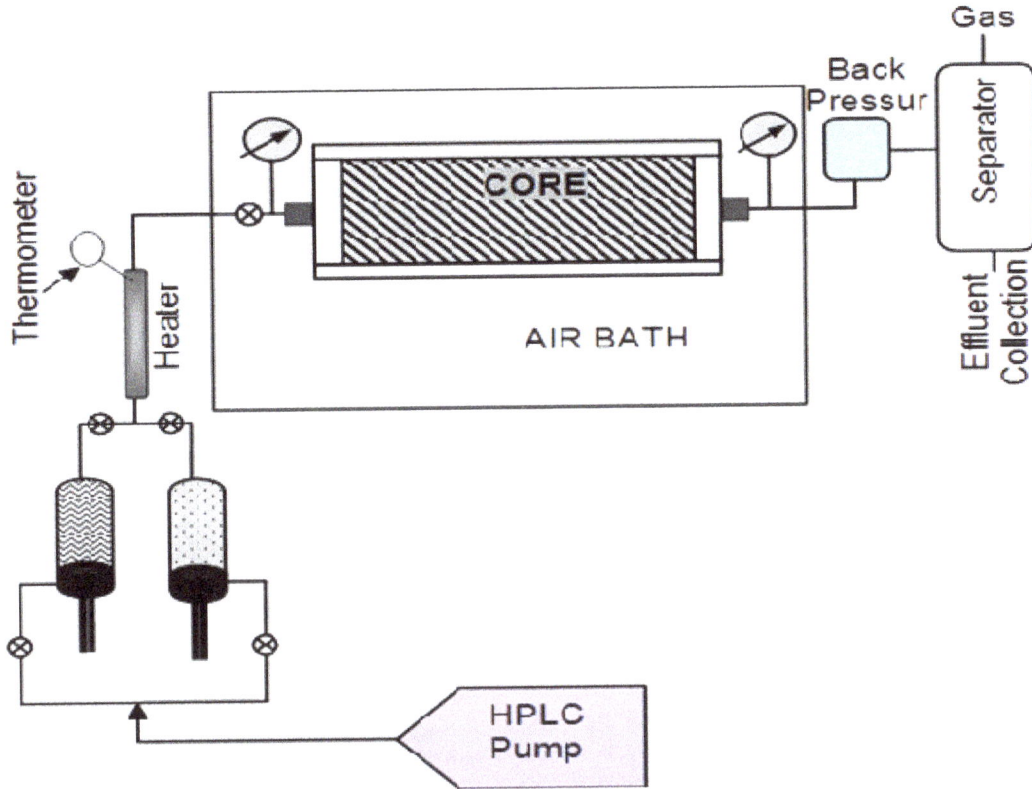

Figure 1. Injection System Schematics.

Sand pack preparation

Silica grains with size distribution of 80 to 250 μm were used for preparing sand pack to obtain a homogeneous model with appropriate permeability. The silica's seeds strew into the core holder after washing. The core holder which was contained the sand pack was put into the shaker to squash the fluids. Screen and glass fiber were installed at the inlet and outlet of core holder to prevent removal of silica.

Porosity measurement

In this work the weight method was employed to determine porosity. In this method the sand pack (moreover metallic sheath) was measured in dry state initially, then it was saturated with distilled water and the mass was measured again. The difference between two measured mass was equivalent to the mass of water which was saturating the sand pack. So the pore volume of the sand pack can be calculated regards to water density. With distinguishes of bulk volume, the porosity can be determined using Equation 1:

$$\phi = \frac{V_{fluid}}{V_{total}}$$

(1)

Permeability measurement

The sand pack permeability was measured with brine solution after porosity measurement. Permeability measurement was based on Darcy's law, which can be rearranged as the following Equation 2:

$$\frac{q\mu}{A} = k\frac{\Delta P}{L}$$

(2)

Where q is the flow rate; μ represents the viscosity of fluid; A is the cross-sectional area of the sand pack; k is the permeability; ΔP represents the pressure drop along the sand pack; and L is the length of the sand pack. Normally pressure drops at different flow rates were measured. Then qμ/A was plotted versus ΔP/L. A straight line which was crossed through the origin can be fitted to the data. The slope of the line represents the permeability of the sand pack. If the data deviate significantly or systematically from the linear trend, there may have been an experimental artifact in the data.

Sand pack saturation procedure

Since the tests are carried out under irreducible water saturation, first the sand pack must be saturated with water and then with oil. Therefore for saturating sand pack with water, the lower core holder valve was kept open so water can be entered from the bottom and saturates the sand pack to 100%. Then oil was injected into core holder through its top valve. In this stage, initial level of saturation of the oil in sand pack was 83%, and irreducible water saturation was 17%. After preparation of core holder and sand pack it were placed horizontally inside the air bath chamber for injection tests. Finally nitrogen with surfactant solution specific flow rate as indicated in the article was injected into the sand pack. Oil recovery factor in each stage was measured. Details of the conventional sand pack are indicated in Table 1.

Table 1. Properties of conventional sand pack.

Property (unit)	Quartz sand
Core diameter (cm)	5
Core height (cm)	15
Bulk volume (cc)	294.37
Pore volume (cc)	85.36
Porosity (%)	29
Permeability (md)	350

Core flooding experiments

Experiments were carried out on a conventional sand pack and in the following order.

Scenario 1: Immiscible surfactant alternating nitrogen gas injection

After sand pack preparations, the oil saturated sand pack at presence of irreducible water for immiscible SAG injection was placed horizontally in the air bath system which was set at 70 °C. Volume of each slug was equal to 15% of the pore volume. Ratio of injected surfactant solution and gas volume was set as one. First the gas and then surfactant solution were injected. Alternative injection of surfactant solution and gas was continued to 1.2 PV. During injection operation, flow rate was 0.2 cc/min. After each injection stage amount of oil recovery was measured.

Scenario 2: Immiscible heated surfactant alternating heated nitrogen gas injection

After heating the surfactant solution and nitrogen to 120 °C, they were injected in the core in the same condition with last scenario. Figure 2 depicts the oil recovery for SAG and heated SAG.

Scenario 3: Comparison of SAG and heated SAG methods with water flooding, gas flooding and WAG

In this section the oil recovery of SAG and heated SAG were comprised with oil recovery of gas flooding, water flooding and WAG. All last three injections had flow rate of 0.2 cc/min. All experiment was done at 70 °C and 144.74×10^5 Pa.

Water flooding: In this experiment, first sand pack was saturated with oil at irreducible water saturation, and then 1.2 PV of water was injected at secondary recovery stage with rate of 0.2 cc/min.

Gas flooding: In gas flooding process after saturating sand pack with oil at irreducible water saturation, 1.2 PV gas continuously was injected in sand pack at rate of 0.2 cc/min. Injection pressure was less than Minimum Miscible Pressure (MMP) of nitrogen, therefore gas flooding was immiscible process.

Water alternating gas: In this experiment water and gas were injected alternatively with rate of 0.2 cc/min in volume ratio of 1:1. Figure 3 shows oil recovery versus PV injected for water flooding, gas flooding and WAG.

RESULTS AND DISCUSSION

As mentioned earlier, aim of this laboratory experiment was comparison between heated SAG with other EOR methods.

Comparison of heated SAG with SAG, water flooding, gas flooding and WAG

Figure 4 compares oil recovery factors of heated SAG injection with SAG, WAG, water flooding and gas flooding process. This figure illustrates that recovery factors of heated SAG, SAG, WAG, water flooding and gas flooding are about 87, 74, 62, 56 and 50%.

Corresponding to experiments, gas flooding has lower recovery factor than other methods. It is due to immiscibility of injected nitrogen as the MMP of nitrogen is 344 to 551×10^5 Pa in contrast with injection pressure of 144.74 × 10^5 Pa (in spite of normal microscopic efficiency, immiscible gas flooding has low macroscopic efficiency).

Water flooding after gas flooding has least recovery factor. This is mainly because of injection of no gas in this process. Although, water flooding has low microscopic efficiency, high macroscopic efficiency of this method provides higher recovery factor than gas flooding method.

The recovery of WAG injection is more than water and gas flooding. In this method, injected water controls mobility and gas stability of front. Since, gas works better than water in microscopic displacement and water works better in macroscopic displacement, combining water and gas alternatively can increase microscopic and microscopic displacement. Also water alternating gas injection decreases fingering and its irritability control.

The recovery of SAG injection inclusively is more than WAG, water and gas flooding. This is because of composing foam in contact with nitrogen and surfactant. The composed foam increases viscosity of nitrogen and the contact time of oil and nitrogen. This increases the microscopic efficiency and oil recovery factor, consequently.

This comparison demonstrates heated surfactant alternating nitrogen injection has the highest recovery. This is because of composing foam in contact with nitrogen and surfactant. The composed foam increases viscosity of nitrogen and the contact time of oil and nitrogen. This increases the microscopic efficiency and oil recovery factor, consequently. Also using heated surfactant solution and nitrogen instead of cold surfactant solution and nitrogen can lead to interfacial tension and viscosity reduction, and oil swelling. Therefore heated SAG method has highest recovery factor in comparison with SAG, water flooding, gas flooding and WAG.

Conclusion

1. Experimental studies showed that recovered oil in the

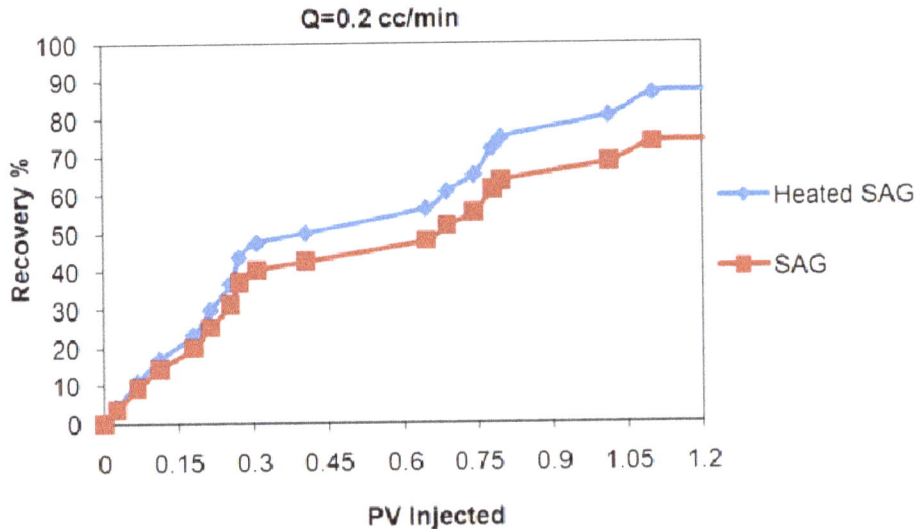

Figure 2. Oil recovery factor in immiscible heated SAG injection and immiscible SAG.

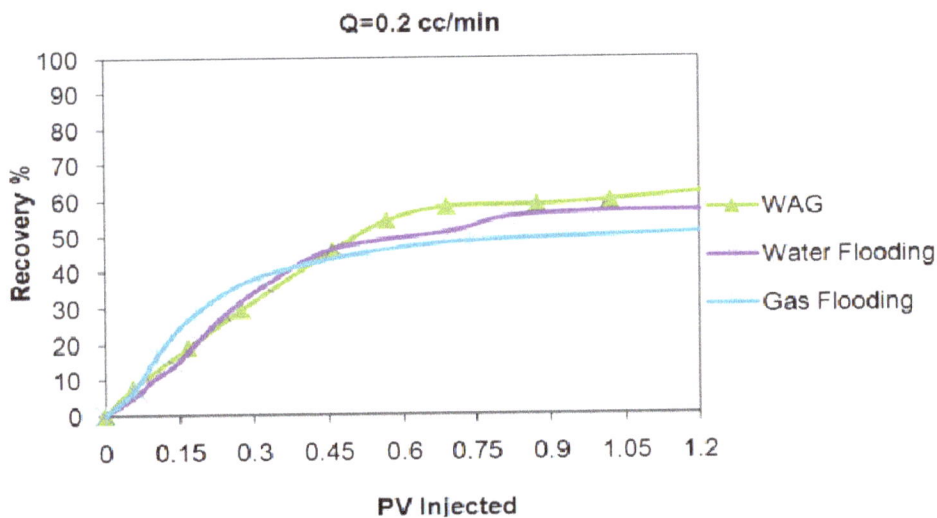

Figure 3. Oil recovery factor in WAG, gas and water flooding.

case of alternating injection of heated surfactant and heated nitrogen is more in comparison with SAG, water flooding, gas injection and water alternating gas. The conducted experiments results in the recovery factors of 87% for alternating injection of heated surfactant and nitrogen (in 1:1 volume ratio), 74% for SAG, 62% for WAG process, 56% for Water flooding and 50% for gas flooding.

2. When the gas is injected into the reservoir, the existing free gases in the porous media lower the relative permeability of aqueous phase in three phase regions under the relative permeability of aqueous phase where there is only oil and aqueous phase. As a result aqueous phase transfers from two phase region into the three phase region and removes a greater volume of oil. So the use of alternative heated surfactant solution and nitrogen injection method increases the volume of the swapped oil by aqueous phase after gas injection.

3. The heated nitrogen gas, because of its mobility, enters areas which are not accessible during ordinary injection.

4. Alternative heated surfactant and nitrogen injection in the conventional reservoirs bearing heavy and semi-heavy oils can be used as an appropriate method for enhanced oil recovery.

ACKNOWLEDGEMENT

The authors would like to thank Dr. Ali Daneshfar of the

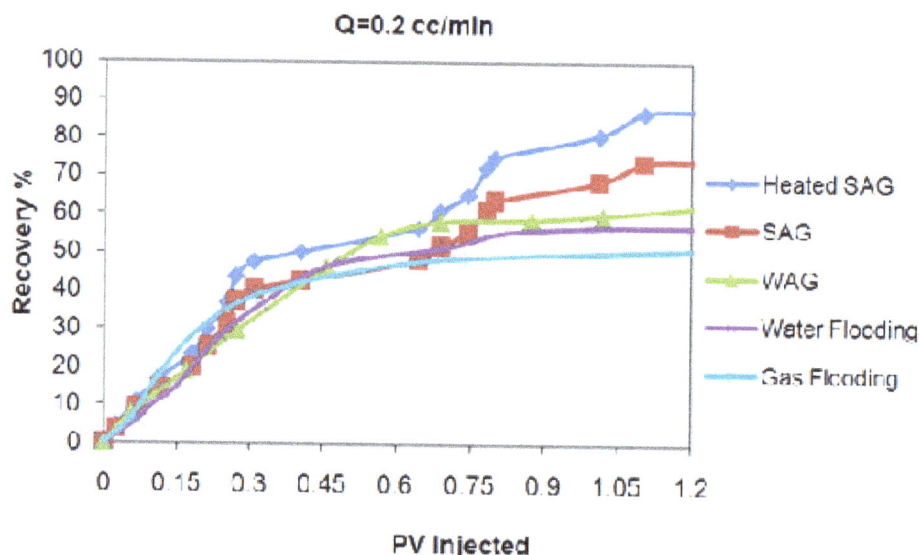

Figure 4. Comparison of oil recovery Heated SAG with SAG, WAG, Water and Gas Flooding.

Department of Chemistry, Faculty of Science, Ilam University, Iran on his useful discussions on surfactant.

REFERENCES

Bernard GG, Holm LW (1964). Effect of foam on permeability of porous media to gas. SPEJ., pp. 267-274.

Blaker T, Aarra MG, Skauge A, Rasmussen L, Celius HK, Martinsen HA, Vassenden F (2002). Foam for gas mobility control in the Snorre Field: The FAWAG project. Paper SPE 78824 was revised for publication from paper SPE 56478, first presented at the 1999 SPE Annual Technical Conference and Exhibition, Houston, 3–6 October.

Christensen JR, Stenby EH, Skauge A (1998). Review of WAG field experience. Paper SPE 39883 prepared for presentation at the SPE International Petroleum Conference and Exhibition of Mexico held in Villahermosa, Mexico.

Falls AH, Hirasaki GJ, Patzek TW, Gaugliz DA, Miller DD, Ratulowski T (1988). Development of a mechanistic foam simulator: The Population Balance and Generation by Snap-Of. SPERE., p. 884.

Heller JP, Dandge DK, Card RJ, Donaruma LG (1983). Direct thickeners for mobility control in CO_2 floods. Paper SPE 11789 presented at the 1983 International Symposium on Oilfield and Geothermal Chemistry, Denver.

Hoefner ML, Evans EM (1994). CO_2 foam: Results from four developmental field trials. Paper SPE/DOE 27787 presented at the 1994 SPE/DOE Symposium on Improved Oil Recovery, Tulsa.

Patzek TW (1996). Field applications of foam for mobility improvement and profile control. SPE Reservoir Engr., pp. 79-85.

Renkema WJ, Rossen WR (2007). Success of SAG foam processes in heterogeneous reservoirs. Paper SPE 110408 prepared for presentation at the 2007 SPE Annual Technical Conference and Exhibition held in Anaheim, California, U.S.A.

Skauge A, Aarra MG, Surguchev L (2002). Foam-Assisted WAG: experience from the Snorre Field. Paper SPE 75157 prepared for presentation at the SPE/DOE Improved Oil Recovery Symposium, Tulsa.

Syahputra AE, Tsau JS, Grigg RB (2000). Laboratory evaluation of using lignosulfonate and surfactant mixture in CO_2 flooding. Paper SPE 59368 prepared for at the 2000 SPE/DOE Improved Oil Recovery Symposium held in Tulsa, Oklahoma.

Tsau JS, Heller JP (1992). Evaluation of surfactants for CO_2-foam mobility control. Paper SPE 24013 presented at the 1992 SPE Permian Basin Oil and Gas Recovery Conference, Midland.

Tsau JS, Yaghoobi H, Grigg RB (1998). Smart foam to improve oil recovery in heterogeneous porous media. Paper SPE 39677 presented at the 1998 SPE/DOE Improved Oil Recovery Symposium, Tulsa.

Viet QL, Quoc PN (2008). A novel foam concept with CO_2 dissolved surfactants. Paper SPE 113370 prepared for presentation at the SPE/DOE Improved Oil Recovery Symposium held in Tulsa, Oklahoma, U.S.A.

Xu Q, Rossen WR (2003). Experimental study of gas injection in surfactant-alternating-gas foam process. Paper SPE 84183 prepared for presentation at the SPE Annual Technical Conference and Exhibition held in Denver, Colorado, U.S.A.

Yaghoobi H, Tsau JS, Grigg RB (1998). Effect of foam on CO_2 breakthrough: is this favorable to oil recovery? Paper SPE 39789 presented at the 1998 SPE Permian Basin Oil and Gas Recovery Conference, Midland.

13

Flue gas enhanced oil recovery (EOR) as a high efficient development technology for offshore heavy oil in China

Dong Liu[1] and Wenlin Li[2]*

[1]State Key Laboratory of Offshore Oil Exploitation (CNOOC Research Institute), China.
[2]E&P Institute, Sinopec, Beijing, China 100101

China Offshore oilfield is rich in heavy oil, M oilfield in Bohai Bay is the most thick heavy oil field in offshore by far with the formation oil viscosity as high as 450 ~ 950 mPa.s. Using conventional oil field development (depletion and water flooding), the heavy oil field exposed its limits, such as low productivity of single well, low production rate and low predicted recovery. As is well known, the development effect of thermal recovery is not only related with the thermal recovery methods (steam stimulation, steam flooding, etc.), but also related with the thermal medium (steam, hot water, flue gas etc.). A new thermal media called multi-thermal fluid, which contains steam, hot water and flue gas was researched for horizontal well stimulation, which is to inject N_2 and CO_2 at the same time when injecting steam. This multi-thermal fluid could employ various mechanisms of each fluid, including reducing oil viscosity by heating and dissolving gas, increasing pressure by injecting gas, expanding heating range, reducing heat loss and gas assisting gravity drive etc. With the advantages of single thermal injection, this method made up the disadvantages of single thermal fluid injection and can improve steam stimulation effect in some special formation conditions. In this article, the author first analyzed the characterization of multi-thermal fluid, and then researched its stimulation theory. Based on this theory, the single element of geological data, such as dip angle, permeability, rock compression coefficient, formation thickness, oil viscosity, and injection data, such as injection volume, gas-water-ratio and CO_2 concentration was analyzed using numerical simulation software. Based on this result, selecting data that make more contribution in incremental oil, production rate, and enhanced oil recovery (EOR) to make sensitivity study to summarize formation that multi-thermal fluid could be applied. Based on the theoretical research, a multi-thermal fluid stimulation pilot test was carried out at M oilfield applying compact thermal recovery equipment. The first cycle of the pilot test showed some optimistic results. Multi-thermal fluid stimulation becomes a new efficient development model for offshore heavy oil, opening up a domestic precedent of thermal development of offshore heavy oil (except shallow sea).

Key words: Heavy oil, Bohai Bay, flue gas, offshore.

INTRODUCTION

Heating heavy oil, enhancing fluidity of formation oil by injecting high temperature and high pressure steam into formation is an effective method to exploit heavy oil. It is also the primary method for heavy oil extraction in China. Steam injection has two stages. The first is steam stimulation, and the second is steam flooding. Currently,

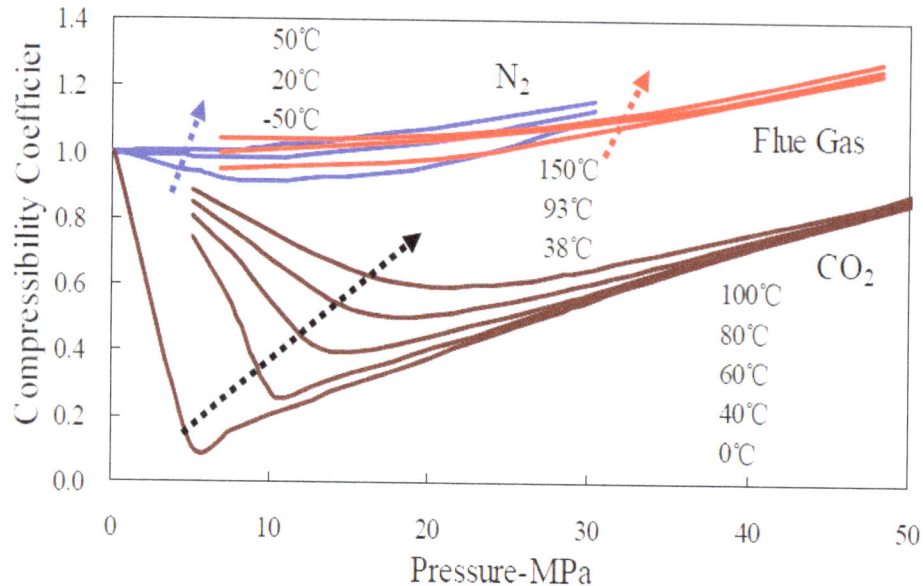

Figure 1. Compressibility Coefficient under Different Pressure and Temperature.

the domestic method is steam stimulation.

As the heat medium of conventional steam stimulation, steam has the characteristics of producing high heat. The main mechanism of steam is to reduce oil viscosity, to eliminate the plugging, to reduce the interfacial tension and to have thermal expansion effect of fluid and rock (Zhao et al., 2010; Li et al., 2012). As relative high heat loss along the way, high formation pressure, and injection pressure in bottom hole close to or above the critical pressure during steam injection process, the steam quality in bottom hole is low. Although the use of a high efficient heat insulation tube can reduce heat loss and improve the steam quality. It can also increase the cost of the steam injection; and increasing steam injection rate is limited by reservoir fracture pressure (Liu et al., 2000). In addition, during the steam injection process, steam overlap and steam channeling is easily aroused due to gravity differences to reduce sweep efficiency (Fu, 2001; Zhao et al., 2011).

To improve development by simply increasing the amount of cyclic steam injection is limited by the economical oil-gas ratio. So currently, the more feasible approach is that with same amount of steam injection, to inject non-condensate gas to change the distribution of formation fluid to improve the heating range and development efficiency (Lookeren, 1983; Liu, 2001; Gao and Towler, 2011). Domestic and international lab and field test studies have shown that injection of flue gas mixed with steam is an effective method to develop heavy oil reservoir (Chung et al, 1988; Peng, 2009).

To improve the development efficiency of offshore heavy oil and enhanced recovery, this article studied to inject N_2 and CO_2 at the same time when injecting steam,

through assistance of multi- thermal media to exploit the offshore heavy oil field.

CHARACTERISTICS OF FLUE GAS

Flue gas is the general designation for burnt gas from industry, and it is mainly produced by diesel engine on well site (Peng, 2009; Gao and Towler, 2011). The chemical composition of flue gas generally contains 80 ~ 85% of N_2, 10 ~ 15% of CO_2 and impurity of the rest. For the steam stimulation process, the sweep area of flue gas which is used as non-condensate gas is much larger than that of steam injection only in the same injection volume.

The relationship of compressibility between pressure and temperature of N_2, CO_2 and flue gas is shown in Figure 1. The curve shows that the compressibility coefficient increases with pressure. At the same temperature, the compressibility coefficient of N_2 is the highest and higher compressibility coefficient is good for oil displacement. The compressibility of flue gas is close to N_2, as both of them are less affected by temperature.

N_2 has weak solubility in both fresh and salt water. CO_2 and natural gas dissolve much more easily in water than N_2. This characteristic is very useful for keeping reservoir pressure by injection N_2 into the reservoir. Temperature has influence on solubility at a certain extent, but less effective when it becomes stable (Figure 2). Pressure and salinity are the main influence factors on solubility of N_2 in water. Solubility decreases with salinity and increases with pressure (Gao and Towler, 2011).

The behavior of flue gas depends on the proportion of N_2 and CO_2 mixed in the flue gas. Previous research

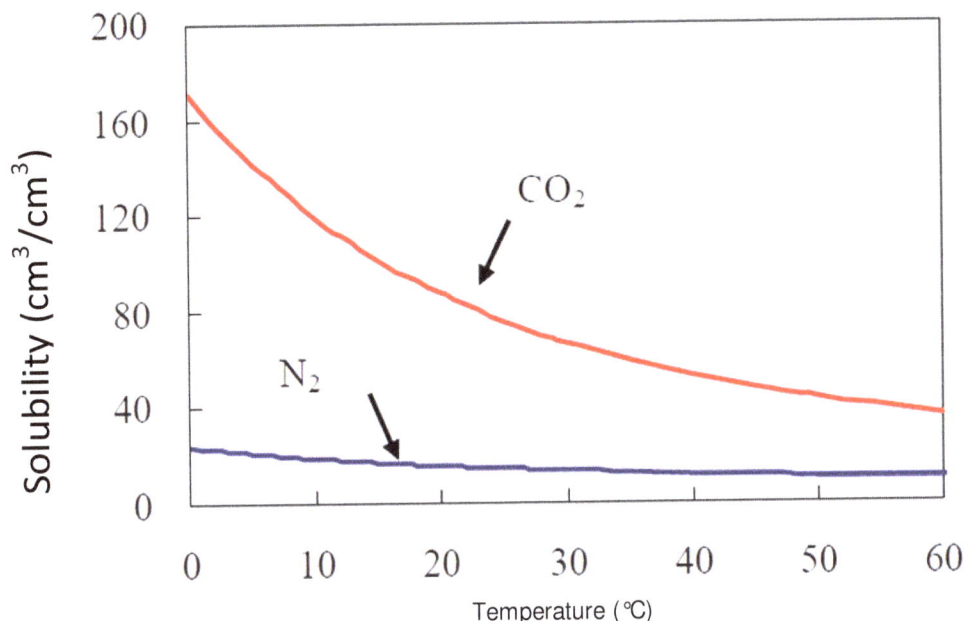

Figure 2. Solubility of N_2 and CO_2 In Water under Standard State.

Table 1. Steam Heat Chamber Reserch Programe.

Scenarios	Group1	Group 2	Group 3	Group 4
Assistant Steam stimulation	Steam	Steam+N_2	Steam+CO_2	Steam+Flue gas
Assistant Water stimulation	Water	Water +N_2	Water +CO_2	Water +Flue gas

Flue gas contains 85% of N_2 and 15% of CO_2.

shows that the mechanism of multi-thermal fluid is a combined result of two competitive mechanisms-a free-gas mechanism provided by N_2 and a solubilization mechanism provided by CO_2.

STIMULATION MECHANISM OF FLUE GAS

The essential mechanism of multi-thermal fluid stimulation technique is to exploit crude oil by the synergistic effect of gas and steam. This multi-thermal fluid could employ various mechanisms of each fluid, including reducing oil viscosity by heating and dissolving gas, increasing pressure by injecting gas, expanding heating range, reducing heat loss and gas assisting gravity drive (Feng et al., 2010).

Reducing oil viscosity by dissolving gas

Because of the N_2 and CO_2 in the multi-thermal fluid, gas

dissolving in crude oil under a higher pressure could reduce oil viscosity and increase expansion coefficient of oil. According to the laboratory experiment of oil sample in M oil field, gas dissolving in the multi-thermal fluid can reduce oil viscosity through dissolving in heavy oil, in which N_2 leads to viscosity reduction of approximate 10% and CO_2 viscosity reduction of 70% (Figures 3 and 4).

Expanding heating rage

A radial model for single-well steam cyclical of horizontal well is built up by CMG STARS to simulate injection steam alone, steam mixture with N_2, steam mixture with CO_2 and steam mixture with flue gas separately. In order to compare with effects of steam stimulation and muti-thermal fluid stimulation, four more scenarios are designed, which is shown in Table 1. Each scenario is designed to inject thermal medium for 30 days, then shut in well for 5 days, and to produce for 1 year.

The temperature of thermal medium injected is 250°C

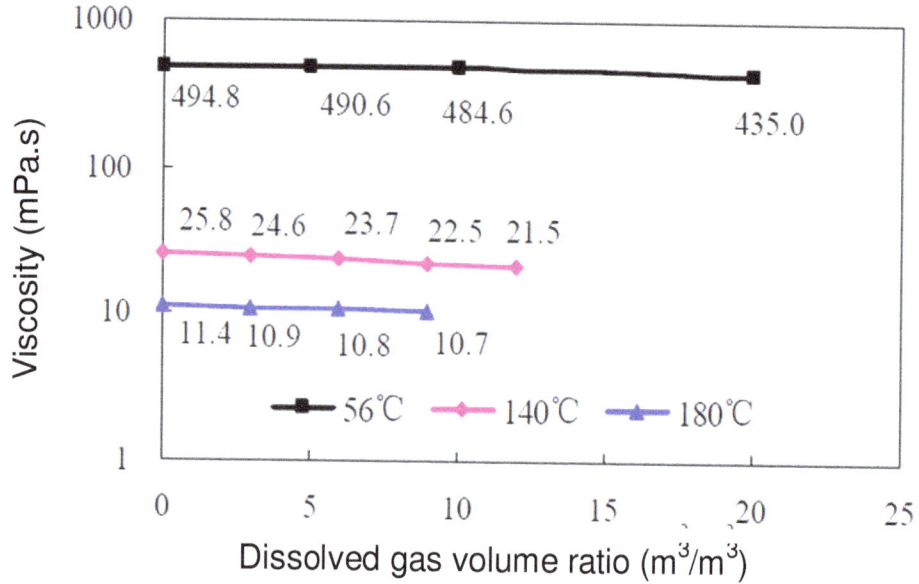

Figure 3. Relationships between Dissolved Gas-Oil Ratio and Viscosity of N_2 - Heavy Oil Mixture.

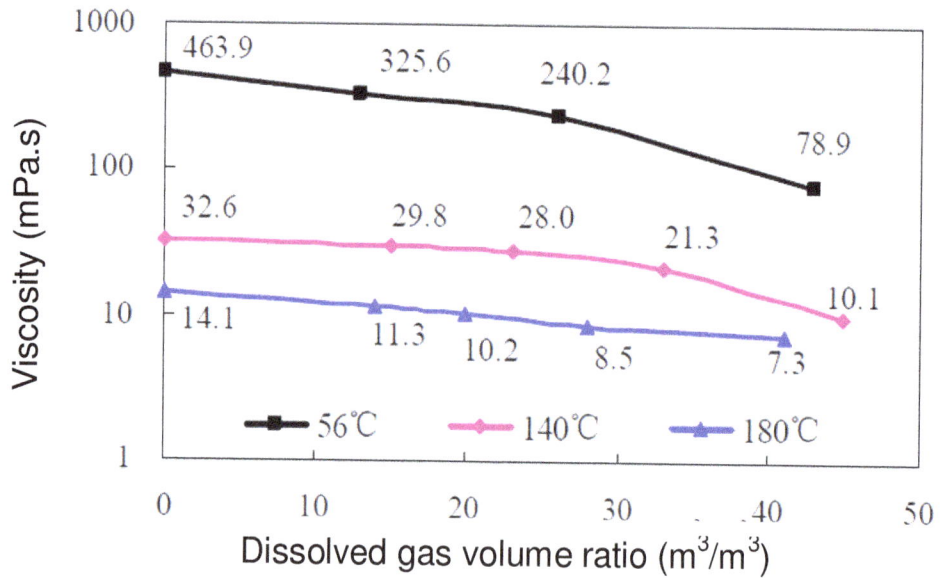

Figure 4. Relationships between Dissolved Gas-Oil Ratio and Viscosity of CO2 -Heavy Oil Mixture.

and injection volume of steam and water separately is 6,000 m³ each cycle. The volume of non-condensate gas injected in each scenario is 300,000 m³ in the standard state every cycle. After thermal media injection for 30 days, the heating range and the average formation pressure are shown in Figures 5 and 6.

The heating chamber volume of multi-thermal fluid injection is two times of that of the steam injection. N_2, as an inert gas, has a low thermal conductivity coefficient (Table 2) and a lower density than steam. So the N_2 would spread upward and crate a heat preservation zone at the top of formation. On one hand, it would obviously reduce heat loss of steam injected to rock in the top of layer and improve heat efficiency; on the other hand, it would reduce overlapping of steam, which might extend steam chamber laterally and enhance swept volume. N_2 injection in the casing tubing annulus at low temperature would have affect heat insulation, which could reduce the

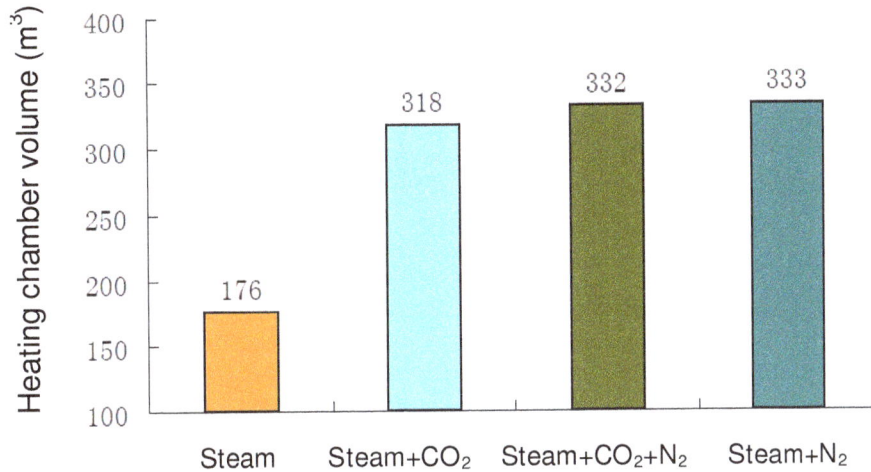

Figure 5. Heat chamber volume.

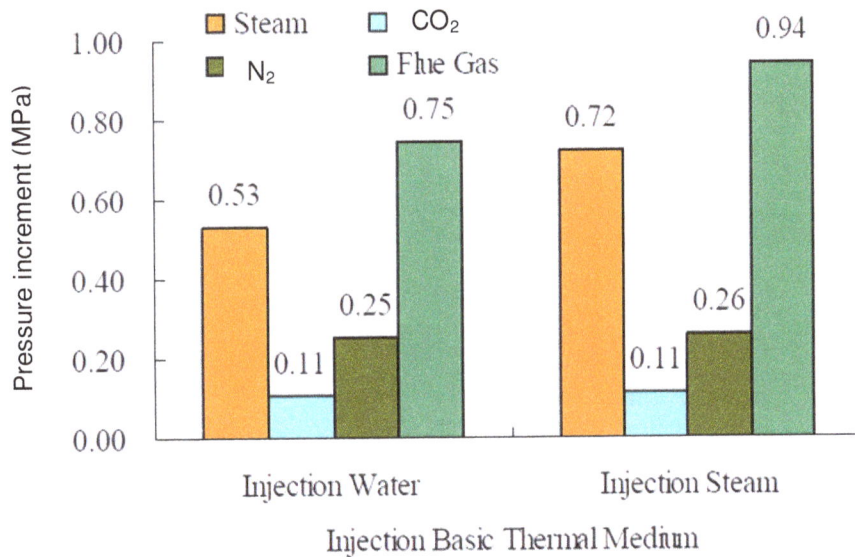

Figure 6. Formation Pressure Maintenance.

heat loss of steam injected in the casing tubing annulus.

Pressure maintenance

Multi-thermal fluid have a significant influence on pressure maintenance (Figure 6). The pressure increment of injection steam with quality of 0.4 is much higher than that of injection water with quality of zero. The order of pressure contributions is N_2> CO_2> steam>water. In the formation of high-pressure gas chamber, the average pressure of gas chamber is up to 0.2 ~ 2.0 MPa. Steam stimulation is depletion development, maintaining reservoir pressure is needed in the later stimulation. The pressure maintenance role of multi-thermal fluid can help to maintain reservoir pressure. The pressure maintenance effects of non-condensate gas mixed with steam injection is better than the type mixed with hot water.

Reduce interfacial tension

The interfacial tension between the fluids or fluid and rock in reservoir directly affect the fluids distribution in the rock, capillary force and fluid flow. The interfacial tension

Table 2. Thermal Conductivity of Varied Thermal Media.

Thermal Media	Thermal conductivity W/(m.K) Heavy Oil0.5~0.8
Rock (No Oil)	2.0~3.5
Rock (Contains Oil)	1.5~2.5
Water	0.4~0.5
Steam	0.02~0.025
CO_2	0.01~0.25
N_2	0.01~0.05

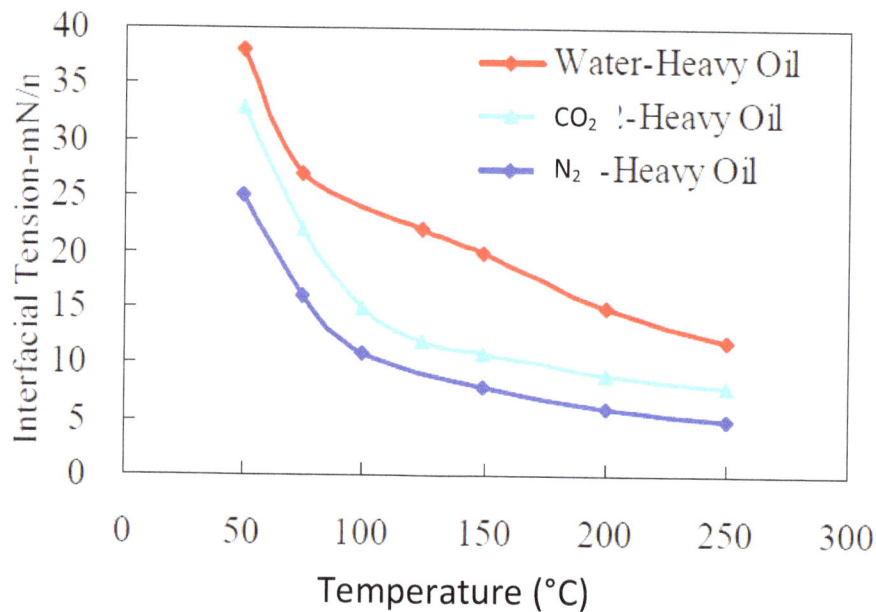

Figure 7. Interfacial tension of different temperature.

between oil and gas is nearly 30% of the interfacial tension between oil and water (Figure 7), which thereby improved the displacement efficiency.

Gravity displacement

Under the oil-water two-phase flow conditions, for oil, the driving force of the gravity is as follows:

$$Gow = (\rho w - \rho o) \, g.hst \qquad (1)$$

Under oil - Gas (including steam) two-phase flow conditions, the driving force of the gravity to the crude oil as follows:

$$Gog = (\rho o - \rho g) \, g.hst \qquad (2)$$

In high-pressrue displacement laboratory experiments, the displacement pressure is 0.2 ~ 1.0 MPa, Therefore, the role of gravity displacement is only 2.74% of the role of pressure displacement. Unless the reservoir pressure is depletion, gravity displacement effect is not insignificant.

SENSITIVITY ANALYSES

To analyze the main geological parameters and injection parameters which influence multi-thermal fluids stimulation effects and the oil increment effects under different geological conditions and injection conditions for multi-thermal fluid, a sensitivity analysis is carried out by thermal simulation software of STARS module of CMG. Base model has a number of 41 × 41 × 14 grids, and grid step is 10 × 10 × 1, and other parameters Table 3.

Numerical simulation model

Through the establishment of the single-well model,

Table 3. Rock and fluid thermal parameters in numerical model.

Parameter name	Value	Parameter name	Value
Rock compressibility 1/KPa	5×10^{-6}	Rock Thermal Conductivity kJ/(m.day.C)	163
Rock Volume Heat Capacity kJ/(m³.C)	2575	Upper and Lower Rock Volume Heat Capacity kJ/(m³.C)	2200
Upper and Lower Rock Thermal Conductivity kJ/(m.day.C)	105	Crude Oil Relative Density (f)	0.956
Reservoir Temperature D	50	Oil Compressibility (1/MPa)	5.3×10^{-4}
Oil Thermal Expansion Coefficient m³/m³·D	1.6×10^{-4}	Specific Heat for Oil (kJ/kgD)	2.12

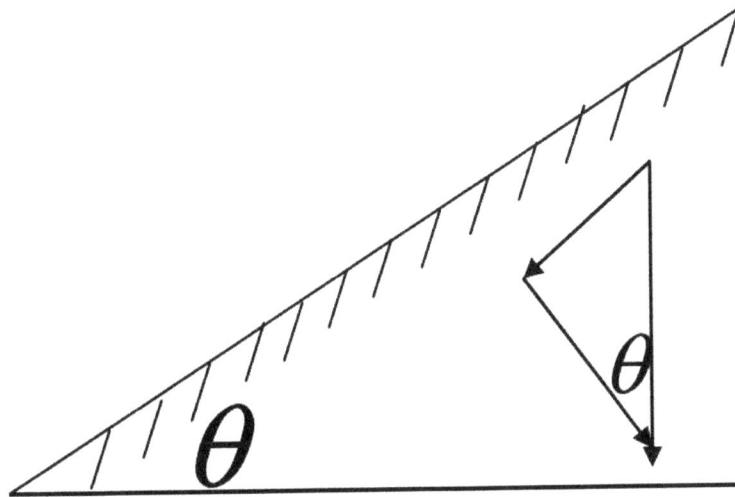

Figure 8. Formation dip schematic diagram.

sensitive factors of the multi-thermal fluids stimulation and steam stimulation are researched. Sensitive factors contain different geological conditions, such as dip, reservoir thickness, and permeability and its rhythmic, vertical permeability and horizontal permeability ratio, compression coefficient, and formation crude oil viscosity and so on.

Formation dip

Based on single-well model, five scenarios for formation dip angle (Figure 8) cases such as 1°, 5°, 10°, 15° and 20° are studied.

With steep reservoir structure, the multi-thermal fluid stimulation can get higher recovery than steam stimulation (Figure 9). Large dip is a favorable factor for multi-thermal fluid. As the dip angle increases, the recovery of multi-thermal fluid increases, at the same time, the incremental oil increase on basis of steam stimulation. In the case of structure with a dip angle, the

gravity displacement driving force of oil - gas two-phase flow of crude oil is as follows:

$$G_{og} = (\rho_o - \rho_w) gh_{st} \sin \theta \qquad (3)$$

The greater the formation dip angle, the stronger the gravitational differentiation is. When inject gas from down-dip parts of the structure, the gas entered into the structure of high position to form secondary gas top by gravity differentiation, which will drive the top of the remaining oil to the well in lower part to obtain a higher oil recovery.

Formation permeability

Based on single-well model, eight scenarios for different formation permeability cases such as $1,000 \times 10^{-3}$, $2,000 \times 10^{-3}$, $3,000 \times 10^{-3}$, $5,000 \times 10^{-3}$, $10,000 \times 10^{-3}$, $15,000 \times 10^{-3}$, $20,000 \times 10^{-3}$ and $30,000 \times 10^{-3}$ µm² are researched.

Higher permeability has higher percentage recovery

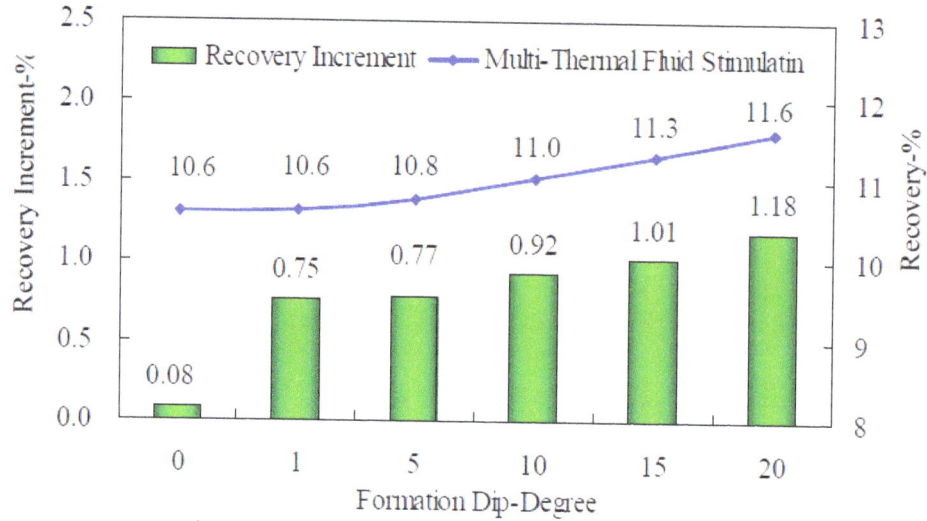

Figure 9. Recovery of different formation dip.

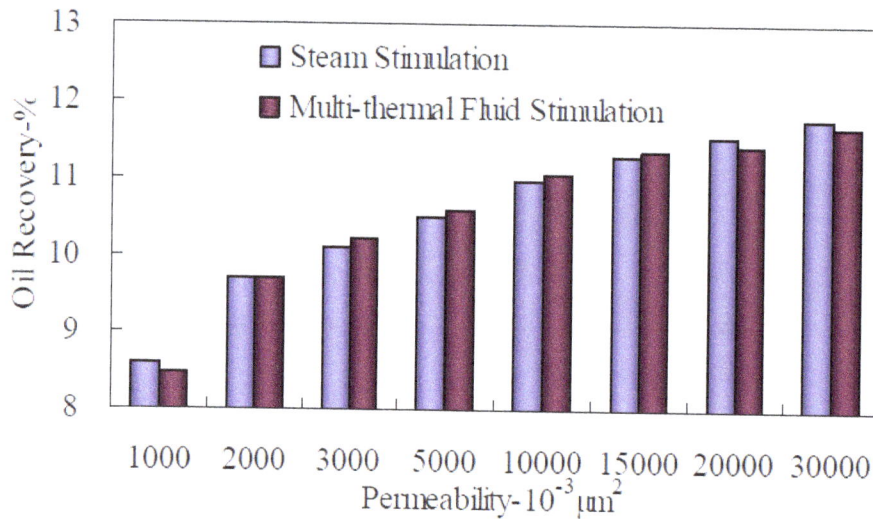

Figure 10. Recovery of different formation permeability.

(Figure 10) with both steam stimulation and multi- thermal fluid stimulation, the recovery percent increased with the formation permeability.

When permeability is under $2,000 \times 10^{-3} \ \mu m^2$, the effect of steam stimulation is better than that of multi- thermal fluid stimulation. The lower permeability restrains the overlap of non-condensate gas. When $2,000 \times 10^{-3} \ \mu m^2 < K < 15,000 \times 10^{-3} \ \mu m^2$, multi-thermal fluid stimulation gets better recovery than that of steam stimulation. The higher permeability makes non-condensate gas easily to expand to the top of the formation, which play the role of thermal insulation and restrain stream overlap. When $K > 15,000 \times 10^{-3} \ \mu m^2$, the effect of steam stimulation is better than that of multi-thermal fluid stimulation. Excessive permeability

makes gas channeling when non-condensate gas is injected. Therefore the most suitable permeability range for multi-thermal fluid stimulation is from 2,000 to 15,000 $\times 10^{-3} \ \mu m^2$.

Permeability rhythm

The permeability in homogeneous model is $5,000 \times 10^{-3} \ \mu m^2$ and the permeability in positive rhythm model is $1,000 \times 10^{-3}$, $2,000 \times 10^{-3}$......$8,000 \times 10^{-3}$ and $9,000 \times 10^{-3} \ \mu m^2$ from bottom to top successively. For reverse rhythm model, the permeability is $9,000 \times 10^{-3}$, $8,000 \times 10^{-3}$ ...... and $1,000 \times 10^{-3} \ \mu m^2$ from bottom to top

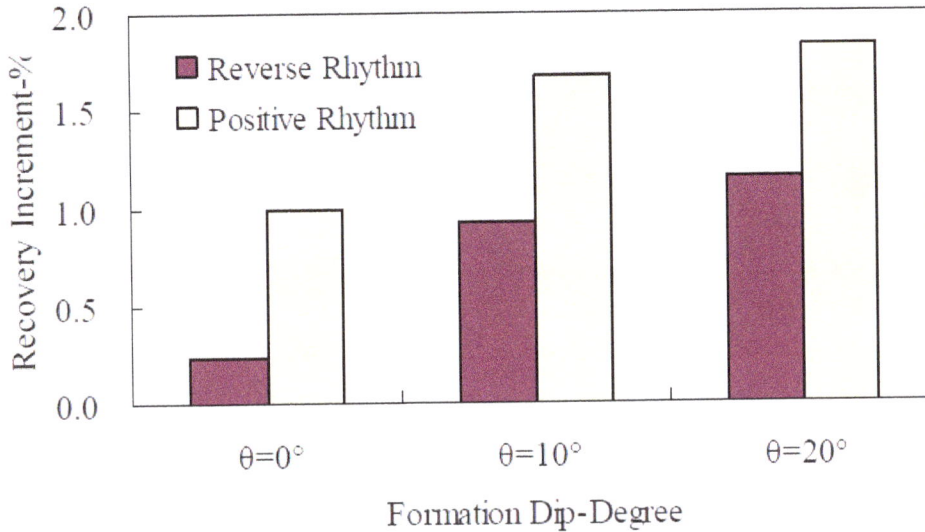

Figure 11. Recovery of different permeability rhythm.

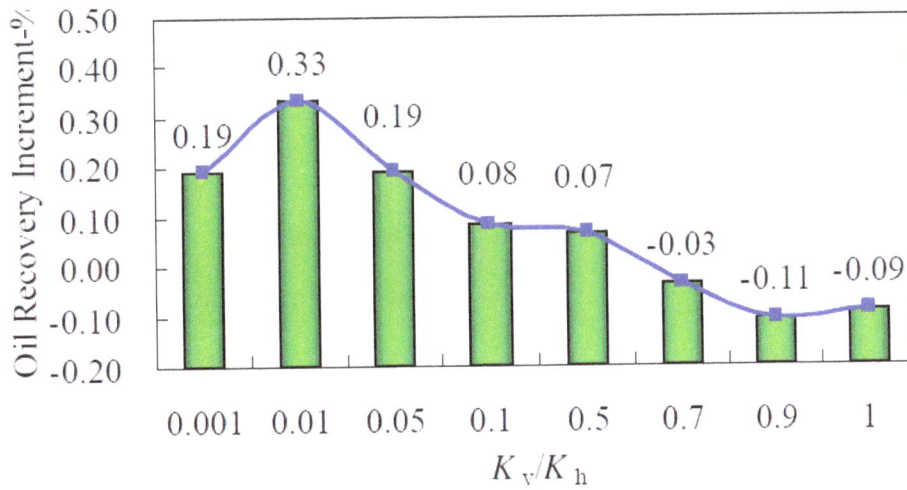

Figure 12. Impact of Kv/Kh.

successively.

Multi-thermal fluid stimulation in positive rhythm formation gets higher recovery than that in negative rhythm formation (Figure 11). When formation inclination angle increases, the oil increment on the base of steam stimulation improves in both positive and negative rhythm formation. The reason is that steam can cause stream overlap in both positive and negative rhythm formations. When the formation dip increases, the effect of stream overlap increases. Non-condensate gas mixed in multi-thermal fluid diffuses upward quickly and forms a barrier layer at the top of the formation, but the steam injected

mainly diffuses downward.

Kv/Kh

Based on the single-well model, seven scenarios for ratio of vertical and horizontal permeability cases such as 0.001, 0.01, 0.1, 0.5, 0.7, 0.9 and 1.0 are researched, and the simulation results are shown in Figure 12.

When Kv/Kh is less than 0.01, the recovery improvement varies with the increment of the Kv/Kh. This indicates that multi-thermal fluid stimulation increases

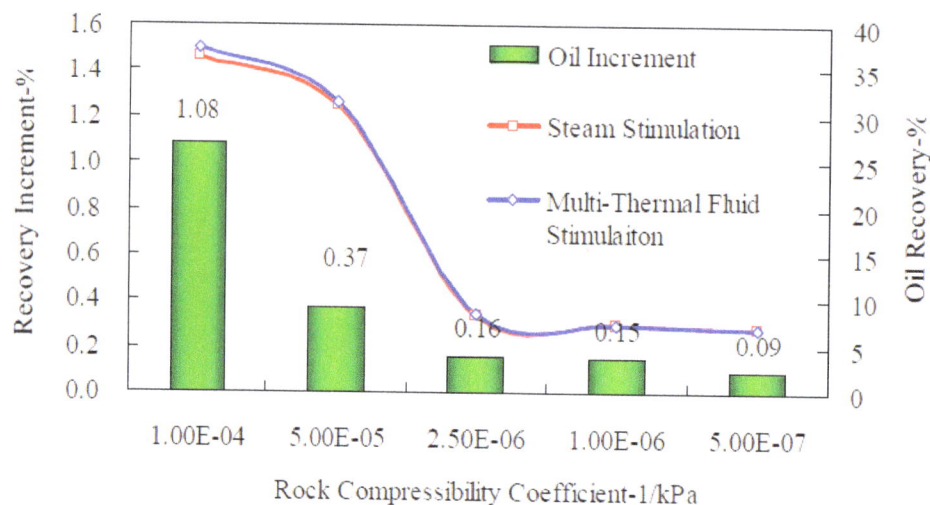

Figure 13. Impact of rock compressibility coefficient.

swept area and displacement efficiency as the vertical permeability increases. The case is opposite when Kv/Kh >0.01. When vertical permeability is too high, the injected gas is badly overlapped and both the swept area and oil production decrease.

High permeability formation requires higher vertical permeability to get a higher recovery this depends on the gas gravity differentiation mechanism of flooding. It can achieve lower displacement efficiency and easily cause early gas breakthrough. When Kv/Kh <0.01, the effect of cyclic steam stimulation is better than that of cyclic multi-component hot fluid stimulation.

Rock compressibility coefficient

Rock compressibility coefficient is used to reflect the formation elastic energy. The higher rock compressibility coefficient, the larger the elastic energy is. Based on single-well model, five rock compressibility values including 1×10^{-4}, 5×10^{-5}, 2.5×10^{-6}, 1.0×10^{-6}, and 5×10^{-7} 1/kPa are researched.

With both steam stimulation and multi-thermal fluid stimulation, the recovery reduced with the formation energy decreases, and the oil increment efficiency of multi-thermal fluid stimulation is reduced (Figure 13). The reason is that higher rock compressibility formation can store more elastic energy during the heat injection process. It can also add more energy in the production process.

Formation thickness

Based on single-well model, six formation thicknesses

such as 4, 6, 8, 10, 20 and 30 m are researched.

Cumulative oil production of both the steam stimulation and multi-thermal fluid stimulation increases with the thickness increasement (Figure 14). When the formation thickness is less than 10 m, the steam stimulation can has a better development effect, but the opposite is true when the formation thickness is more than 10 m.

Formation crude oil viscosity

Seven different crude oil with viscosity of 300, 500, 2,000, 4,000, 6,000, 8,000, and 10,000 mPa.s are used to research the exploitation effect of multi-thermal fluids at diverse range. The recovery percentage is lower with increased oil viscosity (Figure 15).

When the formation oil viscosity lower than 2,000 mPa.s, multi-thermal fluid stimulation recovery changes a little; when oil viscosity is in the range of 2,000 ~ 8,000 mPa.s, multi-thermal fluid stimulation recovery reduced with the increment of formation oil viscosity. When the formation oil viscosity is higher than 8,000 mPa.s, the stimulation effect is little sensitive to formation oil viscosity.

INJECTION PARAMETERS INFLUENCE

Cycle steam injection volume

Based on single-well model, the improve steam stimulation by multi- thermal fluids stimulation under five steam injection volume of every cycle such as 3000, 6000, 9000, 12000 and 15000 m^3 are researched with the same gas water ratio of 50. With the steam injection

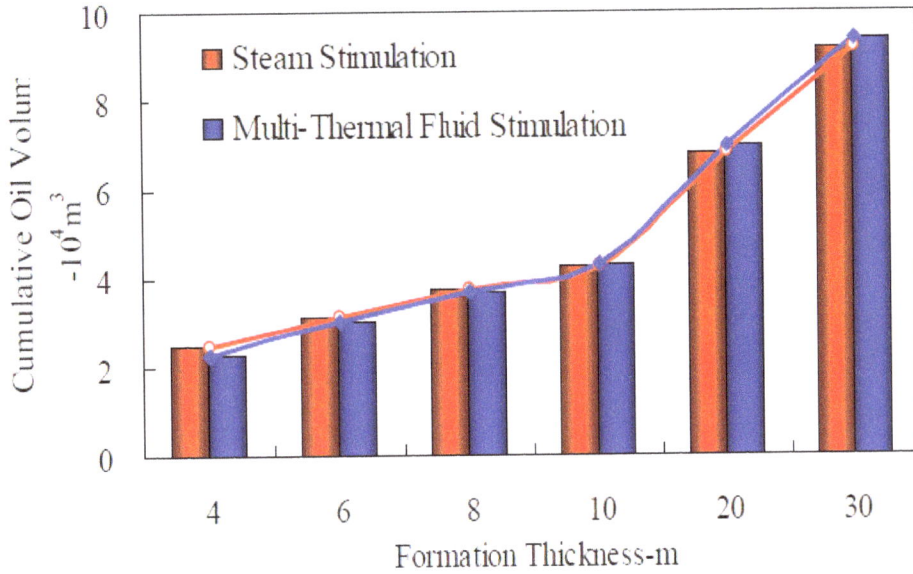

Figure 14. Impact of formation thickness.

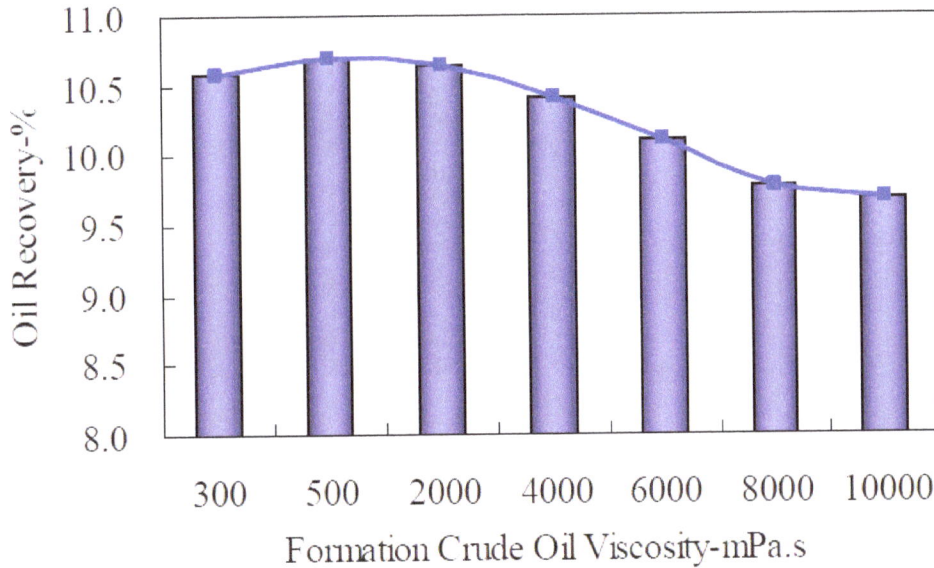

Figure 15. Impact of formation crude oil viscosity.

volume increases, the recovery percent is increased; when the steam injection volume is more than 9,000 m³, steam stimulation get higher recovery than that of multi-thermal fluid (Figure 16).

Gas water ratio

Based on single-well model, the improve steam stimulation by multi-thermal fluids stimulation under five gas water ratio value such as 10, 50, 100, 200 and 300 with the same steam injection of 200m³/day are researched. As the gas water ratio increases, multi-thermal fluid stimulation recovery increased gradually, and the water back recovery also increased (Figure 17). When the gas water ratio is greater than 200, the rate of oil recovery increment become decrease, and the water back recovery increase further.

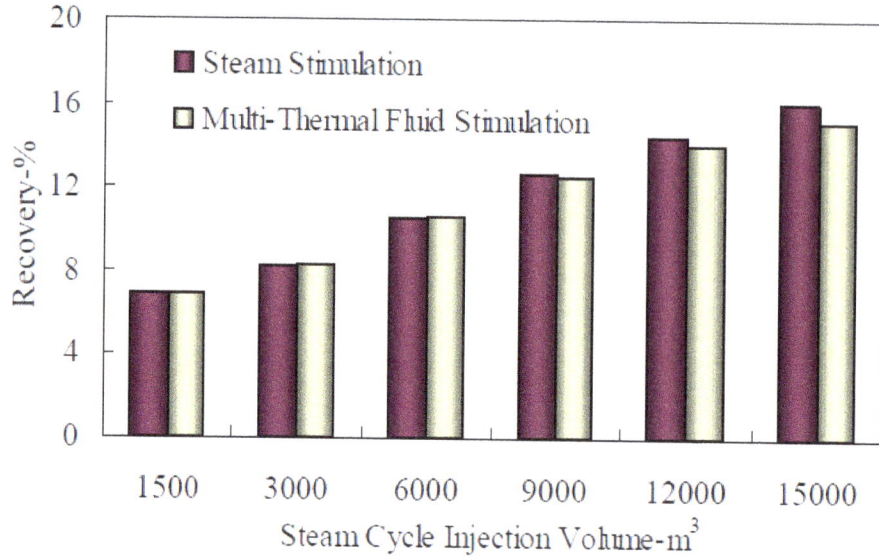

Figure 16. Impact of cycle steam injection volume.

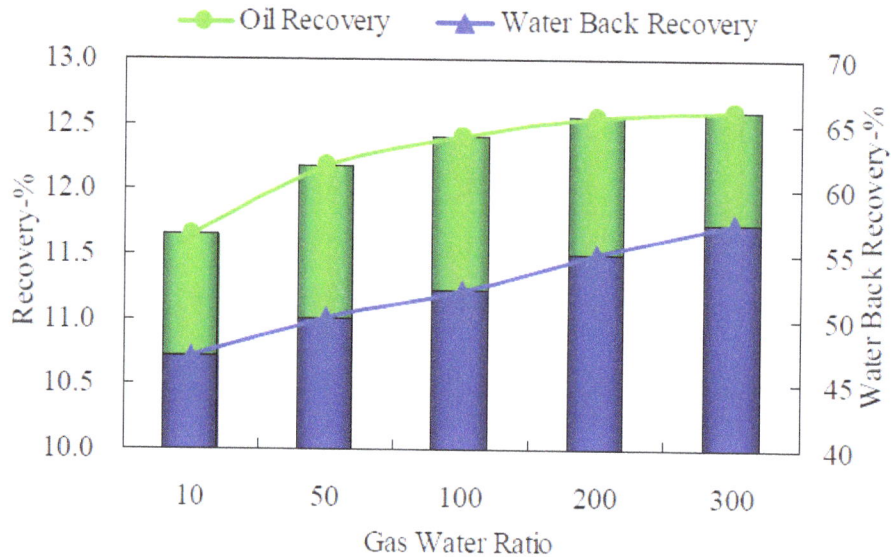

Figure 17. Impact of gas water ratio.

CO₂ content

Multi-thermal fluid stimulation effect achieves better recovery with the increment of CO_2 content in non-condensate gas (Figure 18). Recovery can be increased 1% by injection pure CO_2 than pure N_2.

CONDITIONS SUIT FOR MULTI-THERMAL FLUID

The injection of N_2 and flue gas is a new technology development. Though in some oil and gas field it has achieved good results, but increasing the ultimate recovery only limits to laboratory research. Thus how to choose reservoir that suit for this technology lacks an adequate practical basis. A responding displacement mechanism could be chosen by considering the reservoir geology, reservoir fluid composition and properties, reservoir pressure and temperature conditions. Based on this displacement mechanism, injection ways, injection layer and the injection pressure and other parameters also can be determined.

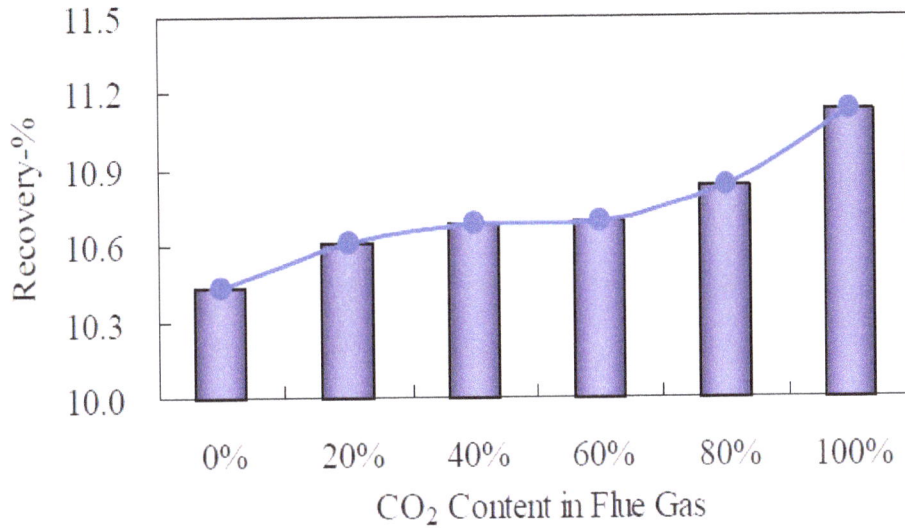

Figure 18. Impact of CO_2 content.

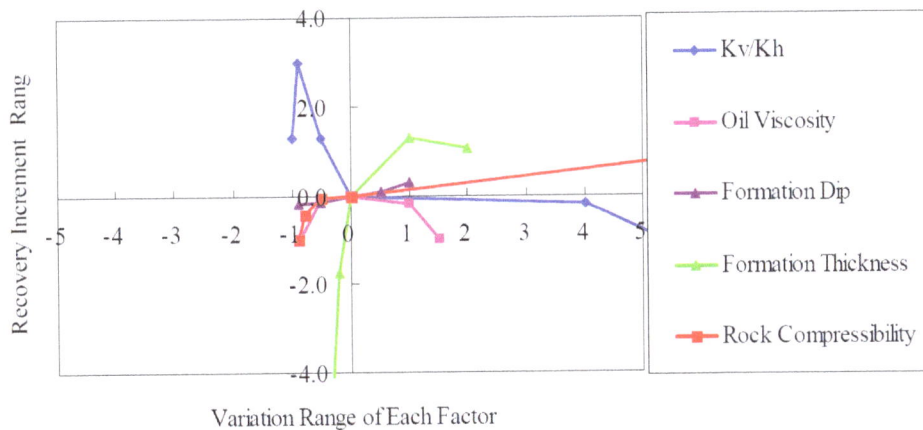

Figure 19. Sensitivity analysis of oil recovery increment.

Oil recovery increment

Sensitivity analysis of oil recovery increment is shown in Figure 19, the sensitivity degree for each parameter from big to small order successively is: reservoir thickness > vertical and horizontal permeability ratio>formation oil viscosity > formation dip > rock compressibility coefficient.

Oil recovery rate

The oil recovery rate and each factor is shown in Figure 20, the sensitivity degree for each parameter about the recovery rate from big to small order successively is:

formation oil viscosity > rock compressibility coefficient > vertical and horizontal permeability ratio > reservoir thickness >steam injection volume each cycle > carbon dioxide content > gas water ratio.

Oil recovery

The oil recovery and each factor chart is shown in Figure 21, the sensitivity degree for each parameters from big to small order successively is: reservoir thickness > rock compressibility coefficient > steam injection volume each cycle > formation oil viscosity > formation dip > gas water ratio > carbon dioxide content and vertical and horizontal permeability ratio.

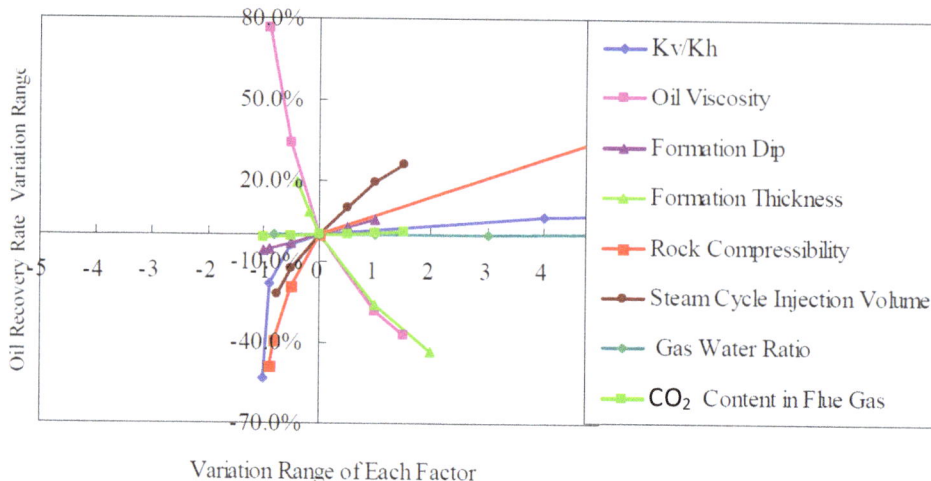

Figure 20. Sensitivity analysis of oil recovery rate.

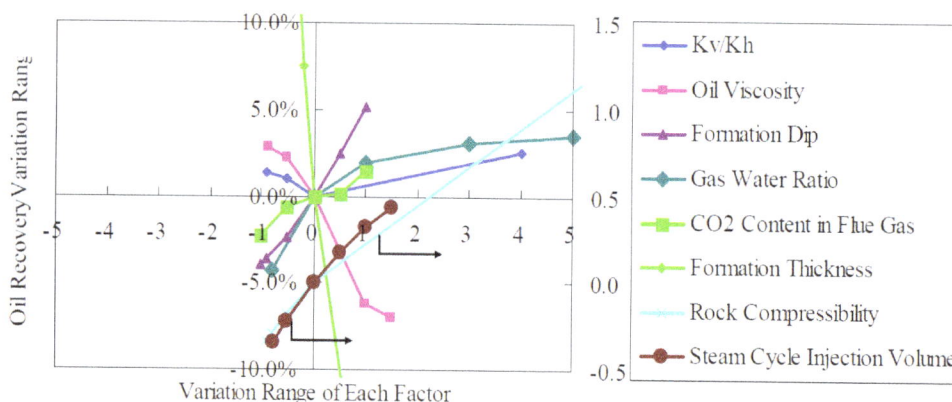

Figure 21. Sensitivity analysis of oil recovery.

MULTI-THERMAL FLUID STIMULATION PILOT TEST

Multi-thermal fluid stimulation pilot test was carried out at M oilfield with compact thermal recovery equipment since 2008, and it shows positive results (Figure 22).

From January 11th to February 13th in 2010, B28 injected thermal fluids for 25 days. During this period, 20 days were spent to inject multi-thermal fluid with wellhead temperature of 240℃, and 5 days to inject fluid with wellhead temperature of 120℃. The cumulative volume of injected multi-thermal medium as follows: 4,904 tons of hot water, 1,232 tons of N_2, 264 tons of CO_2, N_2 injected into the annular insulation with 249,700 m³ in standard state (Figure 23). Then it shut well for 3 days. Peak oil production was 134 m³/day, and B28h well cold production capacity was 38 m³/day by prediction, that is, tripled production capacity can be achieved by multi-thermal fluids stimulation, this is expected to increase oil 8,800m3 in first cycle (Figure 24).

The second thermal recovery test well B29m injected multi-thermal hot fluid from May 5th to May 28th in 2010. It injected multi-thermal fluid with wellhead temperature of 240℃ for 24 days. The cumulative amount of injected multi-thermal medium is as follows: 4,003 tons of hot water, 284 tons of N_2, 224 tons of CO_2, N_2 injected into the annular insulation with 222,310 m³ in standard state (see Figure 23). Then the well was shut down for 3 days. Peak oil production was 112 m³/d, and B29m well cold production capacity was 45m³/d by prediction, that is, tripled production capacity can be got by multi-thermal fluids stimulation, this is expected to increase oil 9,600 m³ in first cycle (Figure 24).

CONCLUSIONS

1. Flue gas assisted steam stimulation can obviously improve recovery efficiency. It is an efficient way to

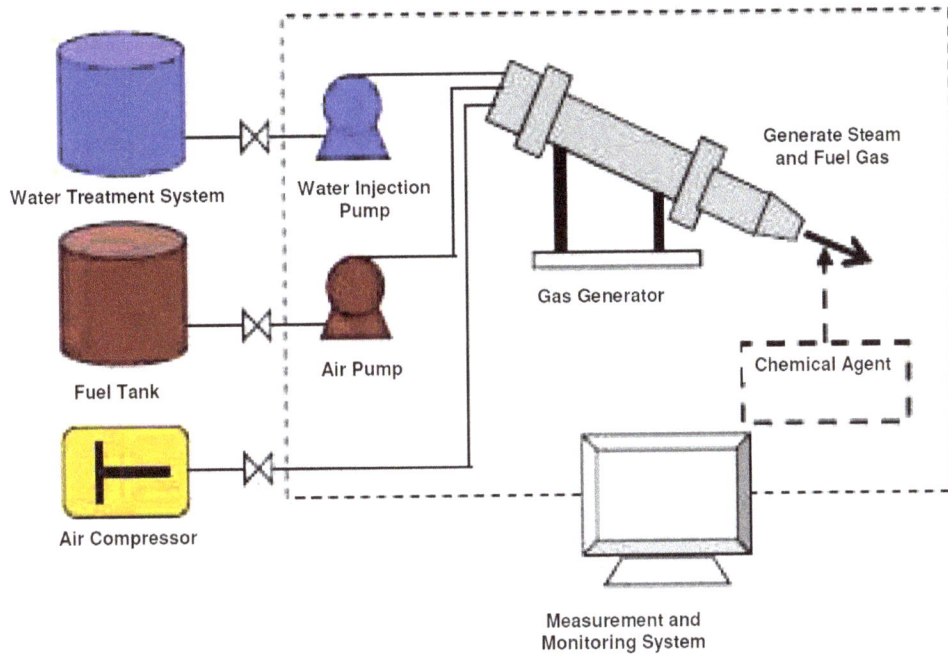

Figure 22. Schematic Diagram for Multi-thermal fluid Injection

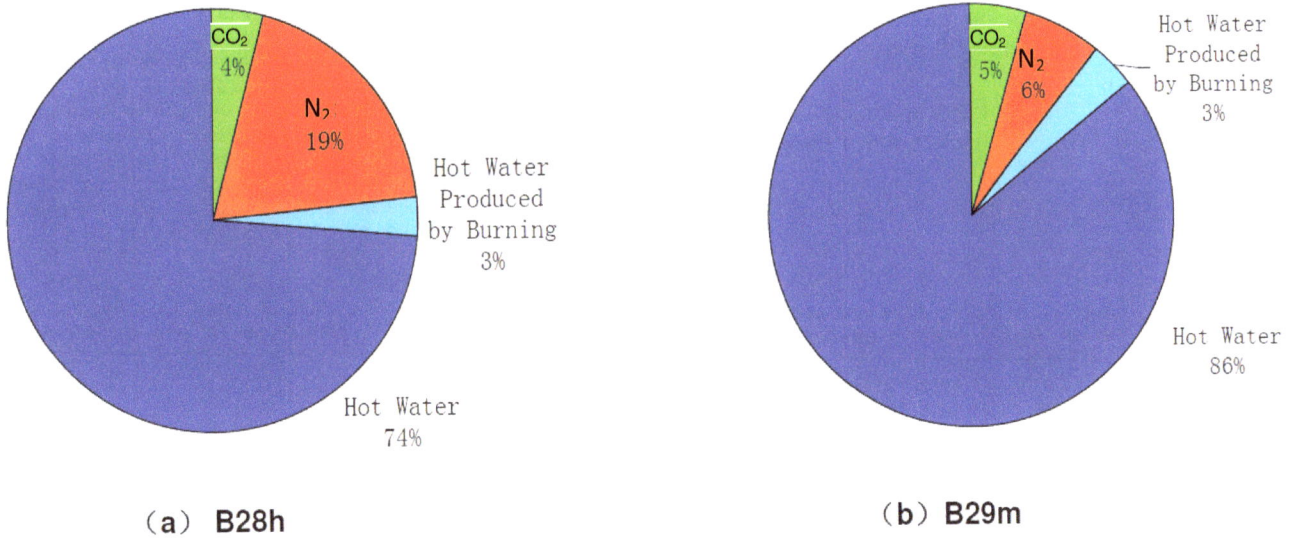

（a）**B28h**

（b）**B29m**

Figure 23. Mass fraction of multi-thermal fluid injected for well B28h and B29 m.

enhance steam sweep zone and slow down the production decline for heavy oil.

2. The mechanisms of multi-thermal fluid stimulation are reducing oil viscosity by heating and dissolving gas, increasing pressure by injecting gas, and expanding the heating range and reducing heat loss and interfacial tension, and gas assisting gravity drive etc.

3. The favorable geological conditions for multi-thermal fluid stimulation are high dip angle, high reservoir permeability, positive rhythm layers, higher rock compressibility and low oil viscosity.

4. The sensitive degree of each parameter to oil increment by multi-thermal fluids from big to small order successively is: reservoir thickness > vertical and horizontal permeability ratio > formation oil viscosity > formation dip > rock compressibility coefficient.

（a） B28h

（b） B29m

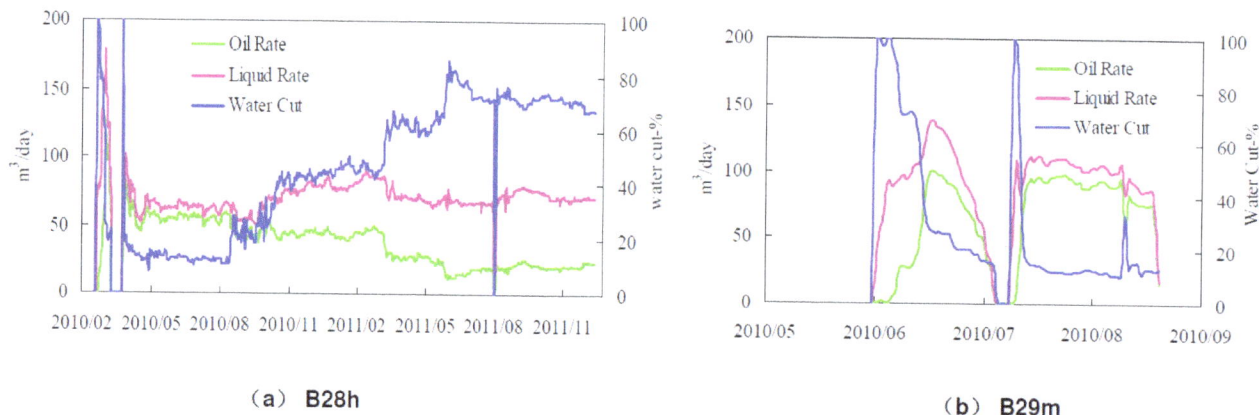

Figure 24. Production Curve for well B28h and B29m.

ACKNOWLEDGEMENTS

Some laboratory test data in this paper was obtained from many other departments of CNOOC in Tianjin. We would like to thank them all for their contributions to this work.

Nomenclature: Go, Driving force of the gravity under the oil-water two-phase flow conditions (Pa); Gog, Driving force of the gravity under the gas-water two-phase flow conditions (Pa); ρ_o, Oil density (kg/m^3); ρ_w, Water density (kg/m); g, Acceleration of gravity (m/s^2); h_{st}, Steam Chamber Height (m); θ, Formation Dip; K_v, Vertical Permeability (10 μm); K_h, Horizontal Permeability (10 μm).

REFERENCES

Chung F, Jones R, Nguyen H (1988). Measurements and correlations of the physical properties of CO_2-heavy crude oil mixtures. SPE Reservoir Eng. 3(3), 822-828.

Feng Y, Ji B, Gao P, Li Y (2010). An Improved Grey Relation Analysis Method and Its Application in Dynamic Description for a Polymer Flooding Pilot of Xingshugang Field, Daqing. North Africa Technical Conference and Exhibition.

Fu M (2001). Numerical Simulation Study of CO_2, N_2 and Flue gas Stimulation for Oil Production from Block Du 84. J. Oil Gas Technol. 30(5):328-337.

Gao P, Towler B (2011). Investigation of polymer and surfactant-polymer injections in South Slattery Minnelusa Reservoir, Wyoming. J. Pet. Explor. Prod. T. 1(1):23-31.

Li X, Wu Y, Zhao X, Gao P, Qin Q, Yu M (2012). A Cyclic-Steam Injection in Fuyv Reservoir, Daqing. SPE EOR Conference at Oil and Gas West Asia.

Liu H (2001). Research on mixed flooding of flue gas with steam in cores for Block Gao23 oil field. Pet. Explor. Dev. 28(5):79-81.

Liu H, Fan Y, Zhao D (2000). Principles and methods of thermal oil recovery technology. Dongying: China University of Petroleum Press, pp. 138-141.

Lookeren JV (1983). Calculation methods for linear and radial steam flow in oil reservoirs. Old SPE J. Res. 23(3):427-439.

Peng T (2009). Study and Application on EOR by Gaseous Nitrogen - Assistant Cyclic Steam Injection in Heavy Oil Reservoir of Shallow Buried Formation in Xingjiang. Xingjiang Oil Gas 5(3):45-47.

Zhao X, Wu Y, Baiming Z, Yu S, Xianbao Z, Gao P (2011). Strategies to Conduct Steam Injection in Water flooded Light Oil Reservoir and a Case in Fuyu Reservoir, Jilin. SPE Middle East Oil and Gas Show and Conference.

Zhao X, Zhong H, Gao P, Wu Y, Zhong H, Hasi B, Qian Y (2010). Investigation of Steam Injection at Critical Condition and a Pilot Test in 119-52, Chao-601. International Oil and Gas Conference and Exhibition in China.

14

Development of a new surfactant-polymer system and its implementation in Dagang oilfield

Hongyan Wang[1] and Brian Miller[2]*

[1]Geological and Scientific Research Institute, Shengli Oilfield, SINOPEC, Dongying 257015, PR China.
[2]EnProTech, Houston, USA.

In this paper, a new surfactant–polymer enhanced oil recovery (EOR) system is investigated for the feasibility of injection in Dagang. The newSP flooding system has been designed and developed for Dagang oilfield with Dagang petroleum sulfonate (SLPS) as the primary ingredient. The dynamic behavior of the system and the interactions of the system components have been investigated through various methods, including dissipative particle dynamics (DPD) molecular modeling technology and dynamic interfacial-tension analysis. The results have shown a significant synergistic effect between sulfonate and nonionic surfactant. The interfacial tension (IFT) and its time to reach equilibrium could be dramatically decreased, suggesting a fast diffusion–adsorption characteristic of ionic surfactants as well as the high surface activity of nonionic surfactants. The SP flooding formulation was optimized. A pilot test has been carried out. The field trial provides useful information for the further large-scale application of the SP system in Dagang oilfield.

Key words: Surfactant-polymer flooding, sulfonate, interfacial tension, dynamic behavior, pilot field trial.

INTRODUCTION

After being developed for more than thirty years, Dagang oilfield becomes increasingly expensive to exploit due to high water cut in the main oilfield reservoir. Meanwhile, it is getting extremely difficult to explore new oil reservoirs. Therefore, it is of great importance to improve oil production of current oilfields by using new technologies.

Alkaline-surfactant-polymer (ASP) flooding invented in 1980s has been regarded as a potential enhanced-oil-recovery (EOR) technology which is more powerful than polymer flooding. Extensive studies on ASP technology have been carried out in the U.S., Germany, and the North Sea (Hernandez et al., 2003; Carrero et al., 2007; Yin et al., 2010). In China, a number of bench-scale and on-site experiments on ASP technology were carried out in Daqing and Dagang oilfields, which had indicated satisfactory capability of ASP systems to increase oil production (Baoyu et al., 1994; Xulong et al., 2002; Kang, 2001). Meanwhile, disadvantages identified during the implementation process of the ASP flooding technology, such as, severe scaling in the injection lines and strong emulsification of the produced fluid, (Wang et al., 2005: 2009; Zhang and Xiao, 2007; Gao et al., 2010) also limited its further application in the field.

In order to overcome the drawbacks associated with ASP flooding, alkali-free surfactant–polymer (SP) flooding technology was developed, and its application in Dagang oilfield has been extensively investigated by authors' research institute. The pilot field trial of our S–P flooding system in southwest Dagang 7th region G-2 was the first field experiment of the S–P flooding technology in China. The objective of the work focused on demonstrating the feasibility of the S–P flooding technique for further

Table 1. List of chemicals.

No.	Name	Company	Purity grade
1	Dodecyl polyoxyethylene polyoxypropylene ether $C_{12}H_{25}(EO)4(PO)5H$ (LS45) and $C_{12}H_{25}(EO)5(PO)4H$ (LS54)	Henkel company, Germany	Greater than 99.95%
2	Sodium hexadecyl sulfonate (AS);	Company, China	Purity grade
3	Sodium dodecyl sulfonate (SLS);		
4	Nonylphenol polyethylene oxide ether TX, n = 8 – 9		
5	Shengli petroleum sulfonate (SLPS)	Shengli zhongsheng	31.4% active

enhancement of oil recovery after polymer flooding. Moreover, the field trials provided useful information on the S–P flooding technology, such as, the formulation design and optimization. This could eventually lead to wider implementation of the S–P flooding system in Dagang oilfield.

Austad and Fjelde had previously indicated that significant improvements can be obtained by coinjecting surfactant and polymer at a rather low chemical concentration. Furthermore, the key factor inselecting chemicals is to avoid S–P complex formation in order to still maintain a very low IFT at low surfactant concentration (Austad et al., 1994; Gao et al., 2011). Among all the efforts to selecting good chemicals for S–Pflooding, petroleum sulfonate, a good oil displacement agent, has gained more and more attention as used in chemical enhanced oil recovery technology. Zhang et al. (2004) studied the effect of different acidic fractions in crude oil on dynamic interfacial tensions in surfactant/alkali/model oil systems (Van der and Joos, 1980; Gao et al., 2009). A study by Al-Hashim had focused on the adsorption and precipitation behaviour of petroleum sulfonates on Saudi Arabian limestone (Zhang et al., 2004). DeBons and Whittington compared performance of the petroleum sulfonate with lignin in Berea sandstone cores (Al-Hashim et al., 1988; Gao et al., 2012). All these studies were limited on the laboratory development stage, none petroleum sulfonate–polymer pilot test have been report using Dagang petroleum sulfonate (SLPS) as the primary component, this effort studied the effect of the secondary surfactant and polymer in the formulation, as well as their capability to enhance the overall oil-recovery performance.

MATERIALS AND METHODS

Surfactant design for the S–P flooding system

Dodecylpolyoxyethylenepolyoxypropyleneether,$C_{12}H_{25}(EO)4(PO)5H$ (LS_{54}), or $C_{12}H_{25}(EO)5(PO)4H$(LS_{54}), (both purchased from Henkel company, Germany) are colorless, viscous liquid with purity greater than 99.95%; sodium dodecyl benzene sulfonate (SDBS), sodium hexadecylsulfonate (AS), sodium dodecyl sulfonate (SLS), and nonylphenol polyethylene oxide ether (TX, n =8 to 9) are all of analytical purity grade, purchased from Shanghai Reagent

Company, China; Dagang petroleum sulfonate (SLPS) contains 31.4% active ingredient (Table 1).

Droplet volume method was utilized to measure dynamic surface tension; TX-500C spin drop apparatus from Bowing Industry Corporation, USA, was used to measure IFT. Viscosity of the polymers was examined using DVIII viscometer in reservoir conditions. Chromatographic separation tests were conducted for the surfactant flooding system under reservoir conditions. The tests were performed in a tube model with inner diameter of 1.5 cm and length of 50 cm. In the tests, unconsolidated model porous media composed of fine silica sands of different mesh was used and its permeability was about 1.5×10^{-3} μm^2 which was close to the reservoir conditions. The tube was prevacuumized and then saturated by water before injection of the surfactant flooding fluid of 0.3 PV. Afterwards, the tube was flooded by water until the outgoing concentration of surfactants became zero.

In oil displacement test, the formation water and re-injection water were prepared at salinity of 4876 and 6188 mg/L, respectively. The oil mixture was formulated by kerosene and dehydrated crude oil from Dagang 16 to 011 well to simulate underground crude oil at 50 mPa.s viscosity. Testing temperature was 70°C. The tests were performed in a core with inner diameter of 2.5 cm and length of 30 cm. The heterogeneousness of the formation was simulated by dual-core system of different permeability: 1500×10^{-3} and 4500×10^{-3} μm^2. The core-flooding procedure is as follow:

Vacuum-pump test core for 2 h, then saturated with reservoir water. Afterwards the core was flushed with (1) oil until a state corresponding to S_{wi} was reached (2) water until a state corresponding to S_{or} was reached and (3) a slug of surfactant formula until water cut to 100%. Injection speed was 0.23 mL/min.

RESULTS AND DISCUSSION

Dynamic surface tension of the sulfonate–nonionic surfactant system

SLPS was been selected as the primary surfactant of the S–P flooding system for the pilot field trial in Dagang oilfield due to its compatibility with oil reservoir as SLPS is produced directly from Dagang crude oil. Meanwhile Employing SLPS as the main component of the S–P flooding formula also lowers the reliance on the outside chemical sources. As an anionic surfactant, SLPS shows strong electric repulsion among the polar heads. A variety of hydrophobic chains structures is expected as SLPS is produced from Dagang crude oil. This results in a loose

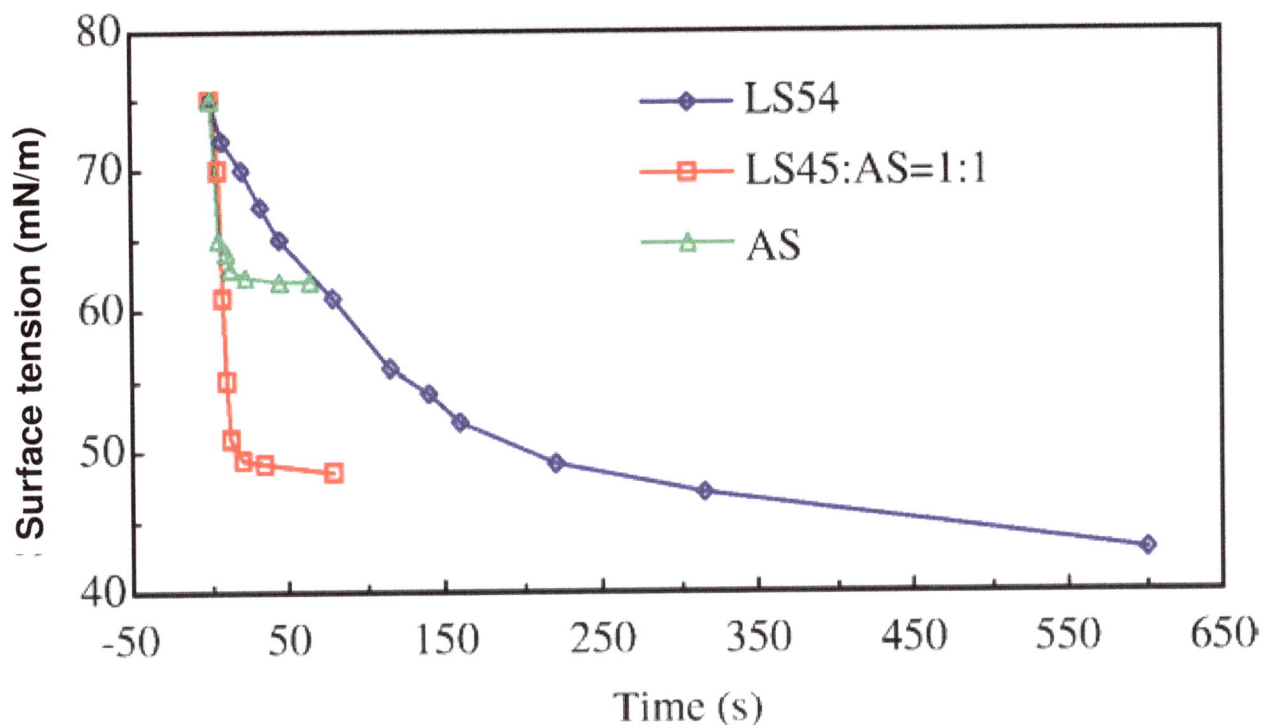

Figure 1. Dynamic surface tension of the sulfonate–nonionic surfactant system.

arrangement of chains in the interfacial membrane and thus the low interfacial activity. As a consequence, SLPS itself is incapable of decreasing the oil/water IFT to an extremely low level of 10^{-3} mN/m. Nevertheless, previous studies had indicated that co-adsorption of different types of surfactants in oil/ water interface could generate interfacial membranes with tight and ordered arrangement of molecules due to weaker steric and electric interactions in the system (Salager and Mongan, 1979; Myers, 2009; Sheng and Wang, 2001; Gao et al., 2010). In this case, the oil/water IFT can be further reduced as a result of the synergistic effects between different surfactants.

The dynamic surface tension of AS-LS54 was measured, as shown in Figure 1, to investigate the activating behavior of the sulfonate–nonionic surfactant system and the synergistic interactions between the two types of surfactants.

Figure 1 illustrates that the surface tension for AS equilibrates immediately in water while it takes a much longer time for LS54. The equilibrium surface tension are determined to be 62 mN/m for AS and 42 mN/m for LS54, respectively, which is 20 mN/m lower than that for AS. In contrast, the surface tension of AS-LS54 (in 1:1 ratio) reaches the steady state as efficient as AS, and possesses a low value similar to that for LS54. Since AS-LS54 shows augmentation of diffusion coefficient as well as surface activity compared to the single surfactant system, sulfonate–nonionic surfactant has been

determined to be the basic formulation of the S–P flooding system (Zhao et al, 2010).

Synergistic effect between sulfonate and nonionic surfactant

Structure–function relationships have been investigated for combinations of sulfonate and various nonionic surfactants through measurement of oil/water IFT. Total concentration of each surfactant system studied is 0.05% by weight in 0.7% NaCl brine.

Table 2 demonstrates a remarkable decrease of 1-octane/water IFT upon the addition of a small amount of nonionic surfactant (TX-100 or Tween-80) in SDBS, suggesting a certain synergistic effect between SDBS and nonionic surfactants. With toluene as the oil phase, combination of SDBS and Tween gives the lowest oil/water IFT among all the surfactant systems. It has been found that the oil/water IFT gradually increases with the content of saturated alkanes in oil phase. Lower oil/water IFT by introducing toluene into saturated alkanes shows that structure similarity between surfactant hydrophobic chains and oil phase molecules favors the IFT reduction. Clearly, the most prominent synergistic effect for SDBS comes from the combination of nonionic surfactants that contain aromatic rings in the hydrophobic chains. Since the ratio of aromatic hydrocarbon and alkane in Dagang crude oil is near 1:2

Table 2. IFT (mN/m) for sulfonate and sulfonate–nonionic surfactant systems.

| Oil | Oil phase surfactant | | | | | | |
	SDBS	SDBS: LS45=9:1	SDBS:TX-100=8:2	SDBS:TW-80=8:2	SDBS: AES=9:1	SLS	TX
1-Octane	1.209	/	0.528	0.983	/	>3	0.5
Tolune:1-Octane=1:2	0.456	0.505	0.142	0.321	0.418	2.46	/
Tolune:	0.333	/	0.227	0.082	/	/	>3

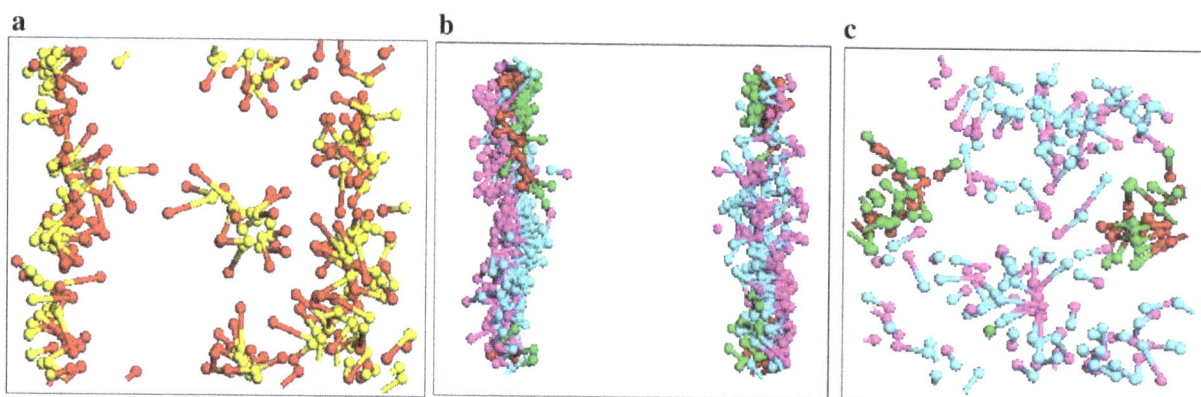

Figure 2. Arrangement of sulfonate and sulfonate–nonionic surfactant in the interface through molecular modeling. (a) SDBS array in the interface. Red and yellow colors represent the head and tail of SDBS respectively. (b), (c) SDBS-TX array in the interface. Pink and blue colors represent the head and tail of SDBS respectively, while red and green colors correspond to the head and tail of TX respectively. (For interpretation of the references to color in this figure legend, the reader is referred to the web version of this article).

the combination of the right molecule chain and the number of EO with SLPS gain the lowest interfacial tension in surfactant–polymer flooding system.

Molecular modeling of the synergistic effect between sulfonate and nonionic surfactants

Dissipative particle dynamics (DPD), one of the molecular modeling techniques, was used in this study to simulate the synergistic effect between sulfonate and nonionic surfactants (Dong et al., 2004). DPD originally proposed by Hoogerbrugge and Koelman in 1992 is a state-of-the-art mesoscale simulation method for the study of complex fluids, such as polymeric or colloidal suspensions. In DPD, molecular cluster in complex fluid system is denoted as 'bead', which is taken to be an effective soft sphere that acts as a center of mass. Based on Newton's motion equation, each bead interacts with the remaining beads through soft potentials, subjected to dissipative and fluctuating forces. In current simulation, each surfactant molecule was represented by two beads, hydrophilic head bead and hydrophobic tail bead, linked by elastic springs with elastic constant to be 4.0 KT.

Simulation was carried out in a 20 × 10 × 10 simulation box at density of 3.0 using 20,000 time steps and a time step interval of 0.05. The mass of bead and system temperature were all set to be 1.0 DPD unit.

Simulation (Figure 2a) shows loose arrangement of SDBS in the interface with cavities that can not be inserted by other free SDBS molecules no matter how large the concentration of SDBS is. However, the TX clusters can enter the cavities (Figure 2b, c) because of the weaker repulsion between the nonelectric polar head of TX and electric polar head of SDBS than that between two electric polar heads of SDBS. Therefore, combination of TX with SDBS offers a synergistic effect to tremendously diminish the IFT by significantly increasing the surfactant density in the interface.

Surfactant formulation design for the S–P flooding system in Dagang oilfield

The crude viscosity was determined to be 45 mPa.s and the reservoir temperature to be 68 °C in the southwest Dagang 7th oilfield Ng54-61. The estimated salinity is 6188 mg/L for injection water and 8207 mg/L for produced

Table 3. IFT of SLPS and its combination with various surfactants.

Number	Surfactant system	IFT (mN/m)
1	0.3%SLPS	7.62×10^{-2}
2	0.3%SLPS-01+0.1%JDQ-1	8.61×10^{-3}
3	0.3%SLPS-01+0.1%JDQ-2	5.65×10^{-3}
4	0.3%SLPS-01+0.1%1#	2.95×10^{-3}
5	0.3%SLPS-01+0.1%4#	6.03×10^{-3}
6	0.3%SLPS-01+0.1%T1501	9.81×10^{-3}
7	0.3%SLPS-01+0.1%T1402	6.00×10^{-3}
8	0.3%SLPS-01+0.1%4-02	5.10×10^{-3}

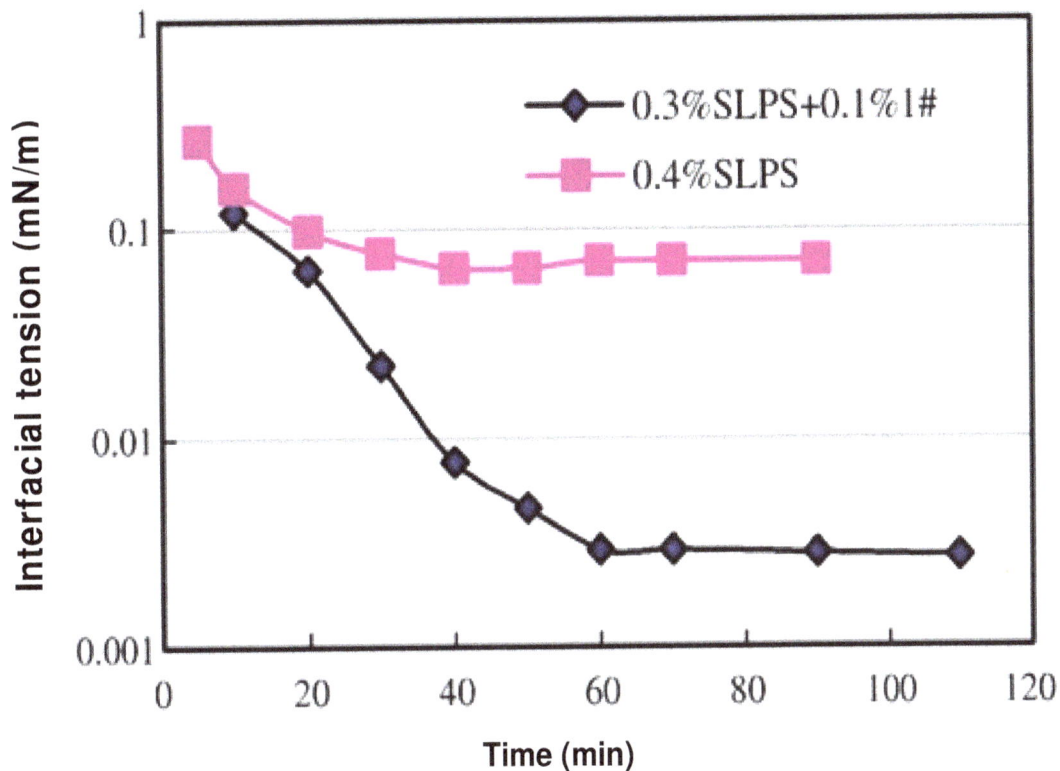

Figure 3. Dynamic IFT of the surfactant flooding system.

water. The bivalent ion (Ca^{2+} and Mg^{2+}) in injection water is 189 mg/L.

Based on aforementioned synergistic studies and reservoir conditions, SLPS was formulated as the primary ingredient together with a variety of complementary surfactants of different types and structures. Table 3 shows the oil/water IFT for these formulations. The formulation with the lowest IFT (2.95×10^{-3} mN/m) in Table 3 corresponds to the combination of SLPS and the secondary surfactant, 1#, which is a nonionic surfactant with TX-100 as the basic ingredient. Furthermore, the dynamic oil/water IFT for both SLPS and 1# has been illustrated in Figure 3.

Chromatographic separation of surfactants in the S–P flooding system in Dagang oilfield

An important consideration of S–P flooding system is to avoidpossible chromatographic separation of surfactants, which occurs during the movement of the flooding chemicals in the oilfield formation since the flooding system is composed of surfactants of various structures. Undoubtedly, chromatographic separation would dramatically decrease the flooding efficacy and oil recovery (Austad et al., 1994; Wang et al., 2005; Sui Xihua et al., 2000).

The dynamic adsorption of the injected SLPS-1#

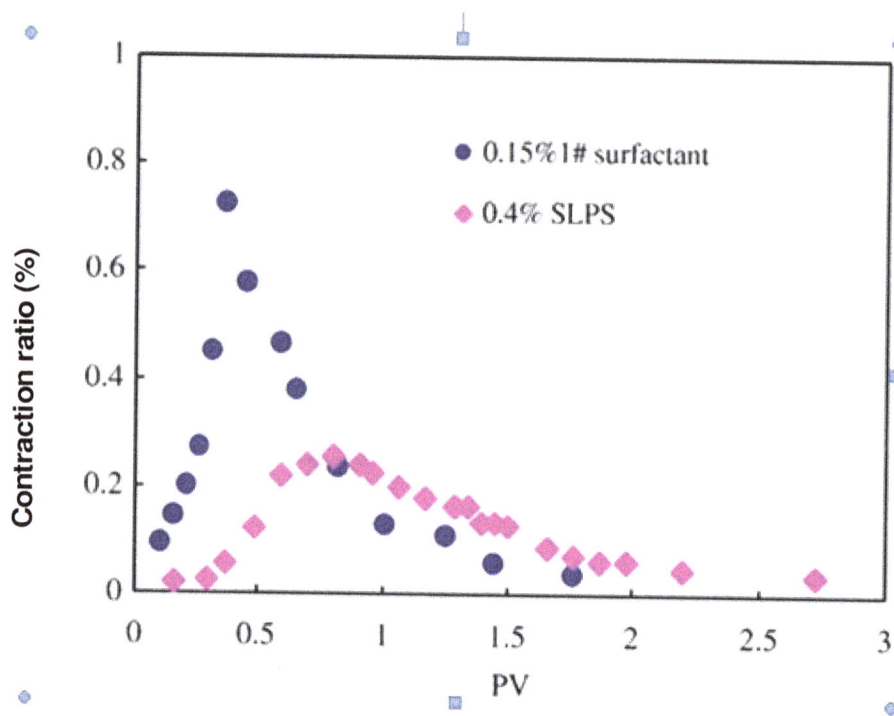

Figure 4. Dynamic adsorption of SLPS and 1#.

mixture in Figure 4 suggests that there exists a chromatographic separation phenomenon between SLPS and 1#. The time difference of 0.5 PV between the outgoing concentration peaks of these two surfactants was observed. To decrease the possible surfactant adsorption and chromatographic separation in the oilfield formation, it has been suggested to increase the total concentration of the injected surfactants to be higher than the threshold value of 0.15%. As noted before, the IFT for SLPS-1# is able to achieve the super low level (10^{-3} mN/m) as long as the total concentration of two surfactants is above 0.15% and below 0.65%. Synergism between SLPS and 1# might be weakened at the front edge of the flooding fluid because of the dilution effect by underground water and the adsorption of surfactants in the formation. Based on above studies, it is recommended to start with 0.45% SLPS + 0.15% 1# in the field trial.

Polymer design for the S–P flooding system

Polymer viscosity swept volume by polymer (Cao et al., 2002). As a result, it is recommended to incorporate polymer into the surfactant flooding system to maximize oil recovery. Four polymers have been carefully selected for the S–P flooding system based on the oil recovery effects in the previous field applications. Viscosity of the polymers was examined using DVIII viscometer in reservoir conditions, that is, brine salinity of 6188 mg/L

and temperature of 68°C. The experiment shows high viscosity for all of the four polymers at concentration of 1500 mg/L, as shown in Figure 5.

Influence of polymer on the interfacial tension

Various polymers (0.15%) have been added into the surfactant flooding system consisting of 0.3% SLPS and 0.1% 1#, and their influence on the IFT has been evaluated accordingly. Because of the elevation of system viscosity upon the addition of polymers, the diffusion of surfactant from water phase towards oil/water interface slows down, extending the time for IFT to reach the super low level (Figure 6). Nevertheless, there is no difference on the order of the lowest IFT for both the S–P flooding system and polymer-free surfactant system, indicating that addition of polymer does not affect surfactants' ability to reduce the oil/water IFT.

Oil displacement test

Oil displacement tests were performed to investigate the EOR performance of different S-P flooding systems under reservoir conditions: actual formation temperature, pressure, permeability, and the degree of oil saturation. The results are shown in Table 4.

It has been found that an oil recovery enhancement of

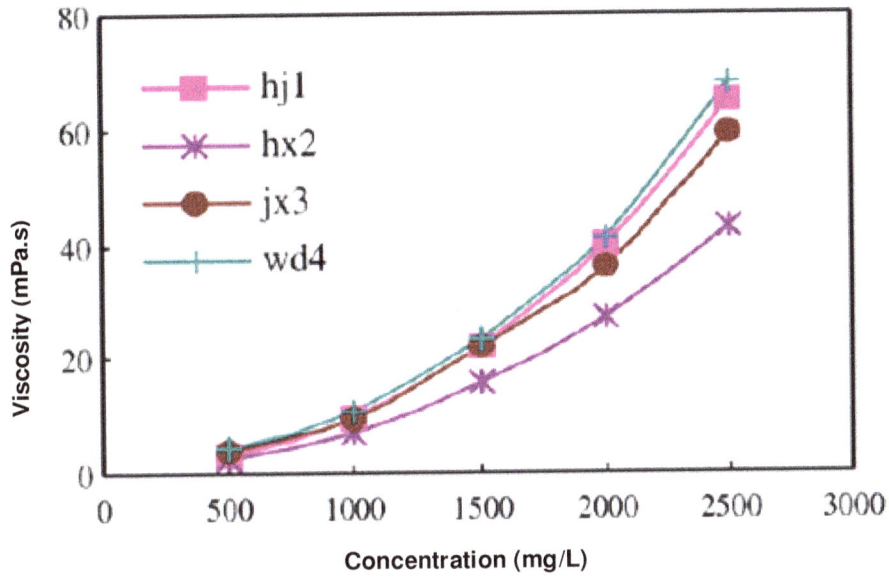

Figure 5. Polymer viscosity–concentration relationship.

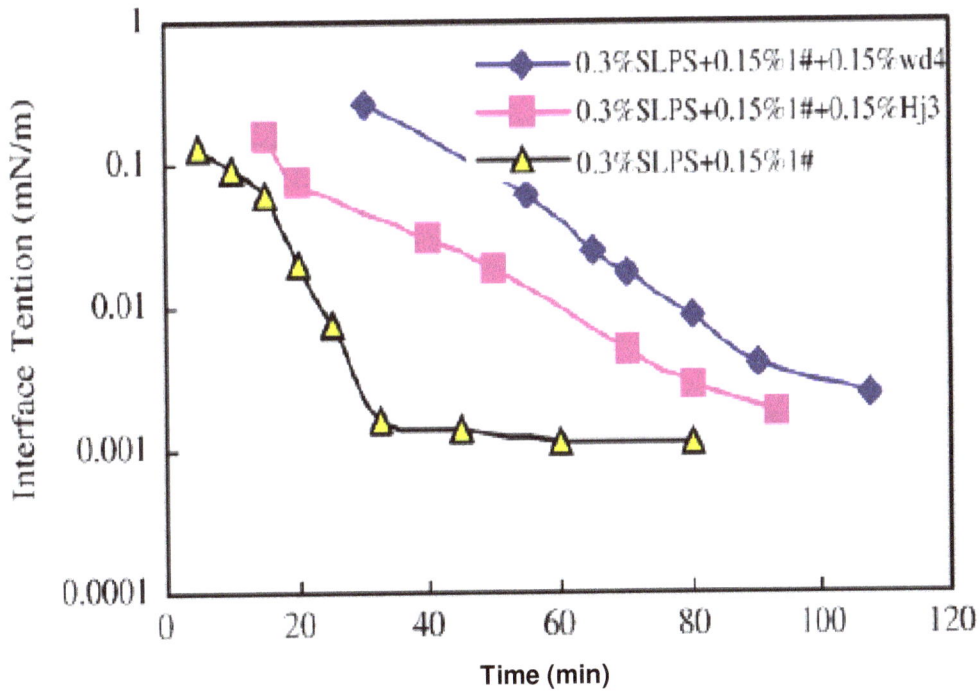

Figure 6. Influence of polymer on the IFT of surfactants in the flooding system.

Table 4. EOR comparison between S–P flooding and polymer flooding.

No.	Formulation	Injection (PV) (%)	OOIP %
Model-11#	0.3%SLPS+0.1%1#+0.15%P	0.3	18.1
Model-18#	0.15%P	0.54	15.2
Model-6#	0.15%P	0.3	11.7

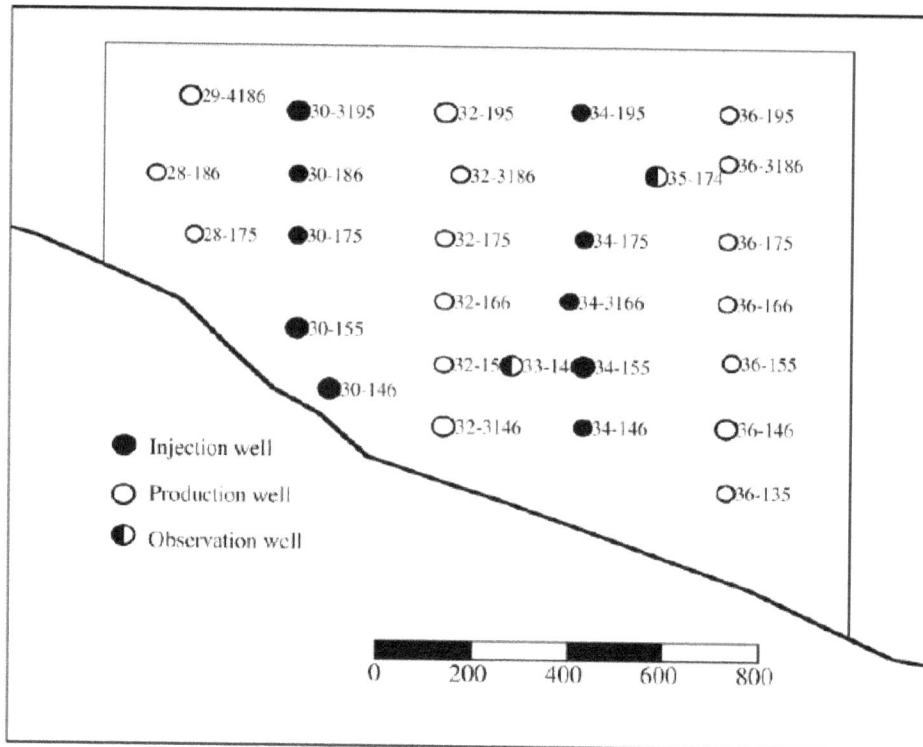

Figure 7. Well distribution for the pilot field trial of the S-P flooding system in G-2 oilfield.

18.1% can be achieved by injecting 0.3% SLPS + 0.1% 1# + 0.15% hj1 of 0.3 PV. Tests also showed that S-P flood outperformed polymer flood under the same core conditions and economical cost. Considering the adsorption consumption and costs, the polymer concentration is suggested to be0.15 to 0.2% in the S-P flooding system.

Application of the S–P flooding system in the pilot field trial in Dagang oilfield

The pilot experiment of the S–P flooding system was carried out in the southeast zone of west Dagang 7th oilfield Ng54-61, which is characteristic of oil containing area of 0.94 km², and oil reservoir of 277.5 × 104 t in depth of 1261~1294 m. The target zone consists of three oil containing strata (54, 55, and 61) with approximately 34% porosity.

The field trial involved twenty six wells, including sixteen production wells, ten injection wells, and three observation wells, as shown in Figure 7. The water cut of the field was determined to be 98.2%, and the oil recovery to be 34.4% before the trial. In order to alleviate "fingering" and "crossflow" of the flooding system in the formation, the three-slug injection methodology was established in the field trial. The first slug was the polymer pre-protection slug of 2000 mg/L polymer solution of 0.05 PV. The second slug was the main slug

of 0.3 PV solution consisting of 1700 mg/L polymer, 0.45% SLPS, and 0.15%1# and the third one was the polymer post-protection slug of 1500 mg/L polymer solution of 0.05 PV.

The pilot field trial has showed significant water-cut reduction and oil- production enhancement since the injection of the main slug of the S–P flooding system in June 2004. Most current field data show that the water cut has been continuously decreased by 13%, that is, from 98.2% in year 2004 to 85.2% in year 2007. Until July 2008, the oil production had increased dramatically by 159 t/day, that is, from 34 t/day to 193 t/day, The single well's oil production has reached to 2.0 × 104 t and the cumulative oil production had risen by 11.5 × 104 t with the oil recovery enhancement of4.15%. Fourteen out of sixteen production wells in the field trial have manifested water-cut reduction and oil-production improvement at various degrees.The increase rates from field trials are clearly higher than that of exclusive polymer flooding.

Conclusion

The SP formulation in this work exhibits the efficient diffusion– adsorption properties of ionic surfactants as well as the high surface activity of nonionic surfactants.

An excellent synergistic effect has been obtained between the primary surfactant (SLPS) and secondary nonionic surfactant.

The finalized S–P flooding formulation used in the pilot field trial is capable to reduce the IFT to 2.95×10^{-3} mN/m, and improve the oil recovery by 18.1% in laboratory oil displacement tests. Since the injection of the main slug of the S–P flooding system in June 2004, the field trial has demonstrated tremendous decrease of water cut and enhancement of oil production. It has been reported that the accumulative oil-production increase had reached 17.8×10^4 t in July 2008. The increase rate of oil recovery and decrease rate of water cut from S–P flooding are clearly higher than those using single polymer flooding.

Current studies of surfactant dynamic activities and synergistic effects for the S–P flooding system will provide theoretical guidance for the future design and development of combination flooding systems. Meanwhile, the success of the pilot field trial of the S–P flooding system in Dagang oilfield will build up a solid foundation for the further large-scale application of the system.

REFERENCES

Al-Hashim HS, Celik MS, Oskay MM, Al-Yousef HY (1988). Adsorption and precipitation behavior of petroleum sulfonates from Saudi Arabian limestone. J. Pet. Sci. Eng. pp. 335–344.

Austad T, Fjelde I, Veggeland K, Taugbol K (1994). Physicochemical principles of low tension polymer flood. J. Pet. Sci. Eng. 10:255-269.

Baoyu W, Xulong C, Qiwei W, Shengwen Z, Xiaohong C (1994). Phase transition and composition concentration of the produced fluid in the alkaline– surfactant–polymer flooding field trial in Dagang oilfield with short-well spacing. Oilfield Chem. 11(4):327–330.

Cao X, Jiang S, Sun H, Jiang X, Li F (2002). Interactions of polymers and surfactants. Appl. Chem. 19(9):866–869.

Carrero E, Queipo NV, Pintos S, Zerpa LE (2007). Global sensitivity analysis of alkali–surfactant–polymer enhanced oil recovery processes. J. Pet. Sci. Eng. 58(1-2):30–42.

Dong FL, Li Y, Zhang P (2004). Mesoscopic simulation study on the orientation of surfactants adsorbed at the liquid/liquid interface. Chem. Phys. Lett. 399:215.

Gao P, Towler B (2011). Investigation of polymer and surfactant-polymer injections in South Slattery Minnelusa Reservoir, Wyoming. J. Pet. Explor. Prod. Technol. 1(1):23-31.

Gao P, Towler B, Li Y, Zhang X (2010). Integrated Evaluation of Surfactant-Polymer Floods. In SPE EOR Conference at Oil & Gas West Asia.

Gao P, Towler B, Wang L (2012). An Integrated Investigation of Enhanced Oil Recovery in South Slattery Minnelusa Reservoir, Part 1: Polymer Injection. Pet. Sci. Technol. 30(21):2208-2217.

Hernandez C, Chacon LJ, Anselmi L (2003). ASP system design for an offshore application in La Salina Field, Lake Maracaibo. SPE Reserv. Eval. Eng. 6:147–156.

Hongyan W, Benyan Z, Jichao Z, Wenli T, Lenyan Z, Wen Z (2005). Scale inhibitors for the alkaline–surfactant–polymer flooding system inZhengli oilfield. Oilfield Chem. 22(3):252–254.

Kang W (2001). Mechanism of the Alkaline–Surfactant–Polymer Flooding Chemicals in Daqing Oilfield, Vol. 1. Petroleum Industry Publishers, Inc., Beijing, pp. 4–15.

Myers D (2009). Surfactant Science and Technology. VCH Publishers, Inc., New York, pp. 6–24.

Salager JL, Mongan JC (1979). Optimum formulation of surfactant/water/oil system for minimum interfacial tension or phase behavior. Soc. Pet. Eng. J. pp. 107–115.

Sheng Z, Wang G (2001). Colloid and Surface Chemistry [M]. Chemical Industry Publishers, Inc., Beijing, pp. 347–351.

Sui X, Cao X, Wang D, Wang H (2000). Chromatographic separation effects of the alkaline-surfactant-polymer flooding system in the west Gudao oilfield. Oil Gas Recov. Technol. 7(4):1-3.

Van der BR, Joos P (1980). Diffusion-controlled adsorption kinetics for a mixture of surface active agents at the solution–air interface. J. Phys. Chem. 84:190–194.

Wang H, Cao X, Zhang J, Tian Z (2005). Dynamic adsorption of the Dagang petroleum sulfonateflooding system. Dev. Specialty Petrochem. 6(3):28–29.

Wang Hongyan, Cao X, Zhang J, Zhang A (2009). "Development and application of dilute surfactant–polymer flooding system for Shengli oilfield." J. Pet. Sci. Eng. 65.1:45-50.

Xulong C, Huanquan S, Yanbo J, Xiansong Z, Lanlei G (2002). Combination flooding field trial in the west Gudao oilfield. Oilfield Chem. 19(4):350–353.

Yin D, Gao P, Zhao X (2010). Investigation of a New Simulator for Surfactant Floods in Low Permeability Reservoirs and Its Application in Chao-522 Field, 2010 Update. In SPE EOR Conference at Oil & Gas West Asia.

Zhang L, Xiao H (2007). Optimal design of a novel oil–water separator for raw oil produced from ASP flooding. J. Pet. Sci. Eng. pp. 213–218.

Zhang Lu, Luo L, Zhao S (2004). Effect of different acidic fractions in crude oil on dynamic interfacial tensions in surfactant/alkali/model oil systems. J. Pet. Sci. Eng. pp. 189–198.

Zhao X, Zhong H, Gao P, Wu Y, Zhong H, Hasi B, Qian Y (2010). Investigation of Steam Injection at Critical Condition and a Pilot Test in 119-52, Chao-601. In International Oil and Gas Conference and Exhibition in China.

15

Effect of pH on interfacial tension and crude oil–water emulsion resolution in the Niger Delta

Olanisebe, E. B., and Isehunwa, S. O.*

Department of Petroleum Engineering, University of Ibadan, Nigeria.

It is important to know the produced volumes and to effectively separate the oil and water phases in oil well effluents to the surface during production. While it is known that the effectiveness of separating the phases and hence the accuracy of measurements are often affected by the nature of the fluids, the specific contributions of the physical parameters of the fluids are little known. In this study, the effect of crude oil PH on emulsion resolution was investigated. Crude oil samples from the Niger Delta with known amounts of water present were obtained. After obtaining the pH and interfacial tension, the oil and water phases were separated by centrifuge at different speeds. The results were used to establish a relationship between interfacial tension, pH and the basic sediments and water (BSW). The results show that without the use of de-emulsifiers, separation of between 66 and 90% was achieved in the light crude oil samples, but less than 30% in heavy crudes. Oil pH was found to affect emulsion resolution and has implications on the choice of de-emulsifiers. On the other hand, interfacial tension was independent of the volume of water and hence BSW of the crude oil samples.

Key words: Oil production, interfacial tension, basic sediments and water (BSW), Niger Delta, bottle test.

INTRODUCTION

Most oil wells produce petroleum with some basic sediments and water. High flow rates and agitation along the production tubings and flow lines could lead to formation of emulsions. Interfacial tension is created at the oil-water interface as a result of imbalance of molecular forces of the dissimilar molecules (McCain Jnr, 1990). Interfacial tension is an indication of the degree of surface activity of a given crude-brine system (McCaffery, 1972) and it is a function of the dispersion forces of the oil and water molecules (Fowkes, 1964). Interfacial tension has been defined as the work required creating a unit area of interface at a constant temperature, pressure, and chemical potential (Drelich et al., 2002; Abhijit, 2006). According to Gong et al. (2001), interfacial tension can be static or dynamic. Static interfacial tension measures the excess energy associated with unsaturated inter-molecular interactions which tend to drive the interface to adopt geometries that minimize the interfacial

area while dynamic interfacial tension measures freshly created interfaces (Buckley and Tianguang, 2005).

Over the years, several methods have been employed in the measurement of interfacial tension. These have been classified as direct and indirect measurement methods (Drelich et al., 2002). Direct measurement involves the use of Wilhemy plate or the du Nuoy ring to measure the excess energy per unit length either by static measurement or in detachment mode (Isehunwa and Olanisebe, 2012).

A wide range of methods have been used to measure the BSW of crude oil during production. (EESFLOW, 2007). Warren (1962) developed a technique of measuring BSW using dielectric-constant. It works on the basis of variation of impedance and circuit phase angle with cell capacitance which changes when cell is filled with clean oil and increases when filled with wet oil. The sensitivity to other contaminants other than water and influence of viscosity at lower temperatures on the dielectric constant are major issues of Warren's technique despite its widespread use.

The installation of water-cut meters which measures oil and water production at time intervals is another widely used method. The meter measures the properties of the production fluids which is mainly oil and water and analyses the differences in their properties translating the results to volumetric concentrations. The major disadvantages of this method are its expensive installation process, inefficiency at elevated temperatures, unsuitability for saline crudes and incorrect field calibration. Manual estimation of BSW has also been done by the grab sample analysis and the well known bottle test method. The bottle test method is laborious in that it involves visual observation of phase separation, difficult task of efficient demulsifier selection, time consuming and costly. Furthermore, data obtained from this method could be unreliable because samples used are often taken in batches and may not be representative of the total flow line and makes data reconciliation a necessity.

Obtaining accurate estimates of oil and water produced in simple and cost effective ways over a wide range of operational conditions is a continued challenge to the oil petroleum industry and has led to studies on measurement of BSW based on parameters such as critical electric field and or the interfacial tension of crude oil-water systems. This work investigated the relationship between BSW, pH and interfacial tension of crude oil systems.

THEORETICAL FRAMEWORK

The relationship between capillary pressure and centrifuge data has been noted by several researchers such as Slobod (1951), Bentsen (1977), Melrose (1988) and Abhijit (2006). It should be possible to extend the relationship to interfacial tension and BS&W.

In general, the pressure difference between two immiscible fluids is given by:

$$\Delta P = (\rho_1 - \rho_2)gh \tag{1}$$

Using the method of Hassler and Brunner (1945), the relation between capillary pressure and average saturation of cores in centrifugal field can be written as:

$$P_{c,r_1} = \frac{1}{2} \Delta\rho \, \omega^2 \, (r_2^2 - r_1^2) \tag{2}$$

Using Equation (2):

$$r = r_2 \sqrt{1 - \frac{P_c}{\frac{1}{2}\Delta\rho w^2 r_2^2}} \tag{3}$$

$$a = \frac{v^2}{r} \tag{4}$$

$$v = \frac{2\pi r w}{60} \tag{5}$$

$$a = \frac{4\pi^2 r w^2}{3600}$$

Acceleration due to gravity g in Equation (1) is converted to centrifugal gravity g' by dividing centrifugal acceleration by acceleration due to gravity g;

$$g' = \frac{4\pi^2 r w^2}{3600 \, g} = 0.00001119 \, r \, w^2 \tag{7}$$

Using Slobod et al. (1951) relations for acceleration in the centrifuge [Equations (4) to (7)], Equation (1) can be expressed as:

$$\Delta P = (\rho_1 - \rho_2) \, g' \, h \quad \left(\frac{g}{cm^2}\right)$$

Substituting values and converting $\left(\frac{g}{cm^2}\right)$ to psig, Equation (8) becomes:

$$\Delta P = 1.5912 * 10^{-7} \, (\rho_1 - \rho_2) \, w^2 \, r \, h$$

Given that

$$h = r_2 - r_1 \ and \ r = \frac{r_2 + r_1}{2}$$

and assuming capillary pressure at the outer face of the centrifuge tube is zero, then for a centrifuge with r_1 and r_2 of 5.0 and 5.7 cm respectively, the capillary pressure at any particular rpm can be expressed as:

$$P_c(r) = 7.956 * 10^{-8}(\rho_1 - \rho_2) \, w^2 \, (r_2^2 - r_1^2)$$

Or, in general:

$$P_c(r) = Y * w^2 \, Psig \tag{11}$$

Where, Y is a constant that varies with the centrifuge speed, and crude oil type. From the well-known relationship between capillary pressure and interfacial tension of two immiscible fluids, the interfacial tension can be expressed as:

$$\sigma_{ow} = \frac{P_c(r)}{2(Cos \, \theta)} \tag{12}$$

Where,

Table 1. Average properties of crude oil samples.

Crude oil	Density (g/cm³)	pH	API	Static interfacial tension (dynes/cm)	Y (Equation 11)
Sample A	0.82	6.5	40	11.8	4.32×10^4
Sample B	0.83	5.2	56	8.4	10.0×10^4
Sample C	0.88	6.2	29	8.4	7.17×10^4
Sample D	0.87	7.3	32	13.5	8.60×10^4
Sample E	0.93	7.8	21	22.8	2.18×10^4

$$\cos\theta = \frac{r_1}{r_2} \tag{13}$$

Thus, combining Equations (11) to (13), dynamic interfacial tension can be determined.

MATERIALS AND METHODS

Crude oil samples obtained from five different reservoirs were collected and emulsified by rigorous mixing with known quantities of water. pH was measured using an analytical pH meter while interfacial tension was measured using the CSC-DuNouy Tensiometer. The Hermle Centrifuge model Z323 was used to separate the water and crude at 2000, 3000 and 4000 rpm over a duration of 5 min. Results were analyzed by adapting the derivations of Hassler and Brunner (1945).

RESULTS AND DISCUSSION

The crude oil samples used varied from light to heavy crudes; with density ranging from 0.82 to 0.93 g/cc. Detailed average physical properties of the samples at room temperature of 29°C are presented in Table 1. Table 2 shows the estimated dynamic interfacial tension using Equations (10) to (13).

Table 3 shows that the percentage of water recovered from the oil centrifuged at 4000 rpm varied with the initial BSW for each crude oil sample. The recovery of samples A, B and C increased with increasing initial percentage BSW giving an average of 70% recovery while samples D and E gave an average recovery of 20%. This shows that crude oil samples D and E did not readily separate at higher water content and centrifugal revolution per minute. The incomplete recovery of water in the water-in-oil emulsions in all the crude oil samples shows that full resolution cannot be achieved by only centrifugation without use of de-emulsifiers.

Furthermore, it was observed that the static interfacial tension values obtained by the Du Nouy Ring method in Table 1 were quite different from the calculated dynamic interfacial tension using capillary pressure and shown in Figures 3 and 4. This is as a result of continuous structural changes which occur on both sides of the physical interface of the two fluids in contact. According to Drelich et al. (2002), the creation of fresh surfaces is accompanied by constantly refreshed surfaces of which

composition has not reached equilibrium. The dynamic interfacial tension at these interfaces is therefore low because solute redistribution has not yet occurred between the water and crude oil phases. The process of agitation while using the centrifuge continuously creates fresh interface in the crude oil-water system.

Figures 3 and 4 show that the behavior of dynamic interfacial tension and BSW at different revolution per minute of centrifuge. The plots show that while interfacial tension increases with the rpm of the centrifuge, the interfacial tension does not vary with BSW. This result is perhaps due to the fact that since interfacial tension involves adhesive forces between the two liquid phases, interaction occurs at the interface of the fluids involved and does not therefore depend on the relative amounts of the immiscible fluids present. The practical implication of this is that BSW may not be accurately determined from the measurement of interfacial tension.

Furthermore, it was observed that pH has a strong effect on the Interfacial tension and on the volume of water recovered after centrifuge. The relationships were established by Equations (14) and (15) as:

$$y = 3.0553x^2 - 34.722x + 106.8 \tag{14}$$

While,

$$P = 26.254x^2 - 313.91x + 957.13 \tag{15}$$

Where, $y = static\ interfacial\ tension,$ and $P = BSW\ (\%)\ (after\ \text{centrifuge})$, and $x = pH\ (5.2 < x < 7.6)$

The coefficients of correlation (R^2) were estimated as 0.92 and 0.96 for Equations (14) and (15) respectively (Figures 1 and 2).

Conclusion

This study revealed the following:

1. Interfacial tension is an interface phenomenon and is not affected by the volume of water present in crude. The practical implication is that in an oil-water system, interfacial

Table 2. Dynamic Interfacial tension of emulsified crude samples during centrifuge.

Rotary speed (rpm)	Interfacial Tension (dynes/cm)				
	A	B	C	D	E
2000	0.56	1.34	0.93	1.12	0.28
3000	1.26	2.93	2.09	2.52	0.64
4000	2.24	5.21	3.73	4.47	1.13

Table 3. Percentage water recovery.

Initial BSW (%)	% Water recovered after centrifuge at 4000 rpm				
	A	B	C	D	E
9.1	77	83	88	12	33
16.7	51	78	90	24	24
23.1	69	65	91	12	20
28.6	67	64	88	14.7	18
19.4*	66.0*	72.5*	89.3*	18.2*	23.8*

*Average.

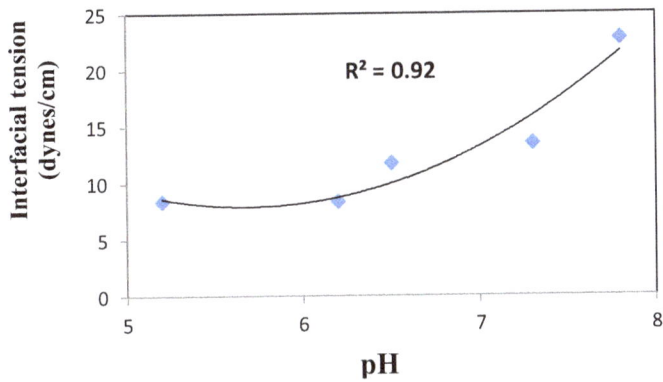

Figure 1. Interfacial tension (dynes/cm) versus pH.

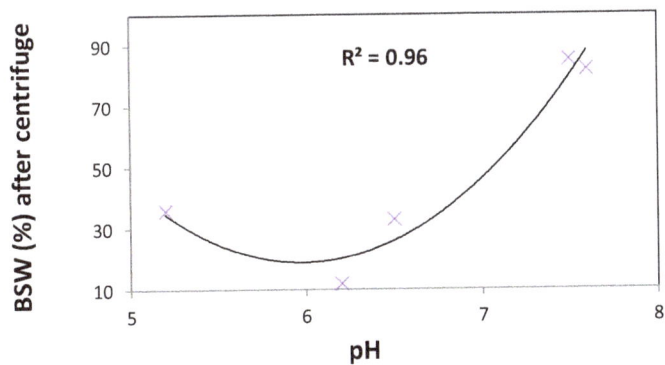

Figure 2. BSW (%) after centrifuging against pH.

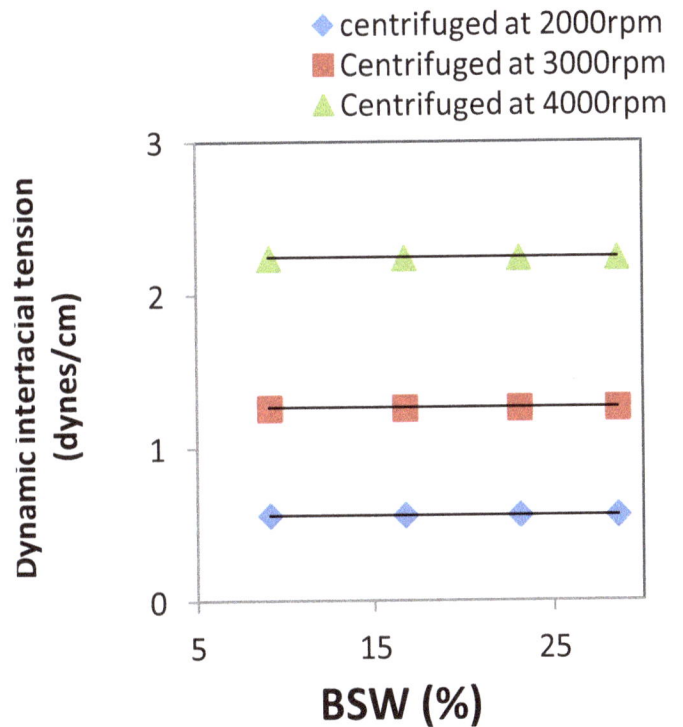

Figure 3. Interfacial tension (dynes/cm) versus (dynes/cm) versus BSW (%) of sample A at different rpm.

tension may not be a good correlator of the relative volumes of the fluids present.
2. pH of crude samples affect the volume of water recovered and also affects interfacial tension. This observation has implications on the choice of de-emulsifiers and their effectiveness in breaking oil-water emulsions.
3. Static interfacial tension of crude oil-water system is higher than dynamic interfacial tension.

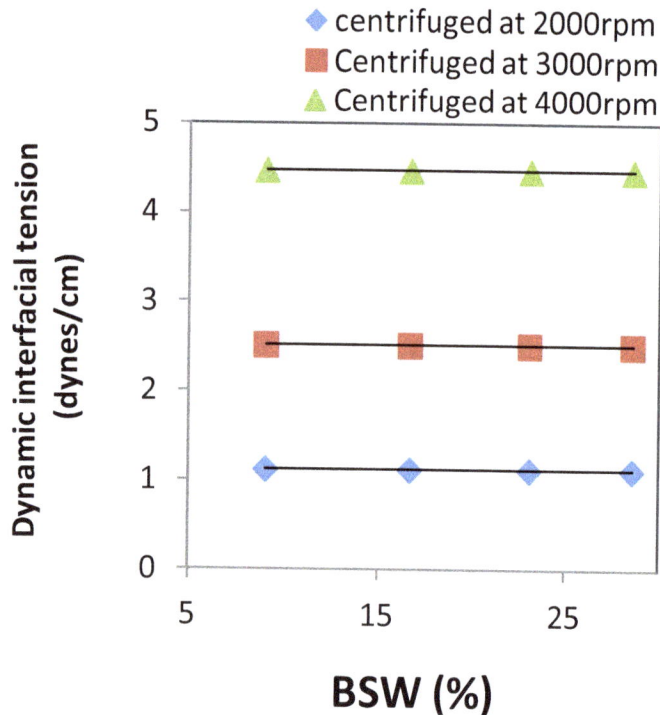

Figure 4. Interfacial tension BSW (%) of sample D at different rpm.

ACKNOWLEDGMENT

This study was supported through research grant from the Shell Petroleum Development Company to the Shell Professorial Chair in Petroleum Engineering, University of Ibadan.

Nomenclature: ΔP, Pressure difference (psia); $\square$, height above the reference level (cm); ρ_1, ρ_2, densities of the two fluids present (g/cc); g, acceleration due to gravity (cm/s^2); w, centrifuge angular speed (rpm); σ_{ow}, oil-water interfacial tension; θ, contact angle; **Pc,** capillary pressure, psig; **Y,** constant that depends on centrifuge speed.

REFERENCES

Abhijit DY (2006). Petroleum Reservoir Rock and Fluid Properties, Taylor and Francis Group, LLC. CRC Press, Boca Raton.

Bentsen RG, Anli J (1977). Using parameter estimation techniques to convert centrifuge data into a capillary-pressure curve. SPEJ 17(1):57-64.

Buckley JS, Tianguang F (2005). Crude Oil/Brine Interfacial Tensions, Paper SCA 2005-01, Proceeding of International Symposium of the Society of Core Analysts, Toronto, Canada pp. 1-12.

Drelich J, Fang C, White CL (2002). Measurement of Interfacial Tension in Fluid-Fluid Systems. Encyclopedia of Surface and Colloid Science. Marcel Dekker.

EESIFLOW News (2007). BS&W in Crude Oil. http://www.eesiflo.com/bs&w-in-crude-oil.html Retrieved from on 20 March, 2011.

Fowkes FM (1964). Attractive Forces at Interfaces, Industrial Engineering Chemistry. http://garfield.library.upenn.edu/classic1980/A1980JP30400001.pdf retrieved on 10 March, 2011.

Gong C, James TW, Baker P. (2001) Study of Dynamic Interfacial Tension for Demulsification of Crude Oil Emulsion, Paper SPE 65012 presented at the SPE International Symposium on Oilfield Chemistry Houston, Texas.

Hassler GL, Brunner E (1945). Measurement of Capillary Pressure in Small Core Samples, Trans AIME, 160:114.

Isehunwa SO, Olanisebe EO (2012). Interfacial Tension of Crude Oil-Brine Systems in the Niger Delta. IJRRAS. 10(3):460-465.

McCaffery FG (1972). Measurement of Interfacial Tension and Contact Angles at High Temperature and Pressure, Paper JCPT 72-03-03 presented at 23rd Annual Technical Meeting of the Petroleum Society of CIM, Calgary, Alberta.

McCain Jr W (1990). The Properties of Petroleum Fluids. PennWell Publishing Company, Tulsa, Oklahoma, USA.

Melrose JC (1988). Interpretation of Centrifuge Capillary Pressure Data. Log Analyst 29(1):40-47.

Slobod RL (1951). Use of Centrifuge for Determining Connate water, Residual Oil, and Capillary Pressure Curves of Small Core Samples. Trans AIME 192:127.

Warren WJ (1962). BS&W Measurement - Principle and Practices, Paper SPE 345 presented at the SPE Production Automation Symposium, Hobbs.

16

Saturation and pressure distributions in a two-phase flow through semi-infinite porous media

C. O. C Oko* and O. E. Diemuodeke

Department of Mechanical Engineering, University of Port Harcourt, Rivers State, Nigeria.

This paper presents models and solutions for the saturation and pressure distributions in a two-phase flow through semi-infinite porous media. The mathematical model, which was developed from the popular Darcy's equation and the continuity equation, contains a substance source/sink term. The implicit finite difference formulation of the model was solved for saturation and pressure using the modified Gaussian elimination method, which was implemented in Microsoft Excel Visual Basic for Application. The solution was tested using hypothetical oil/water well production data.

Key words: Two-phase flow, semi-infinite porous media, finite difference method.

INTRODUCTION

The modelling of the flow of fluids through porous media is very important in many spheres of human endeavours, such as reservoir engineering, soil science and environmental engineering, where the working fluid is usually in multiphase. The ability to predict the fluid flow behaviour in various processes is central to the efficiency and effectiveness of the design and operation of most industrial flow processes (Brenem, 2005; Kolev, 2005). Since the 1960s, significant progress has been made in mathematical modelling of flow and transport processes in porous media. Research efforts, driven by the increasing need to model petroleum and geothermal reservoirs have developed many numerical modelling approaches and techniques (Pruess and Narassimhan, 1985). An accurate reservoir modelling capability is required for reservoir engineering calculations for maximized production.

Aziz and Settari (1979) modelled two-phase (oil and water) immiscible flow through porous media by extending Darcy's law for single-phase flow through porous media. Methods used to determine porosity and/or absolute permeability are presented in Seto et al. (2001). The relative permeabilities and capillary pressure are functions of the fluid saturation. Knut-Andreas et al. (2004) presented a hierarchical multiscale method for the numerical solution of two-phase flow in strongly hetero-geneous porous media. The method is based upon a finite-element formulation, where basic multiphase functions are computed numerically on a coarse grid to correctly account for subscale permeability variations from the underlying finite-scale geomodel. The multiphase flow equations were simplified by neglecting effects of compressibility, gravity and capillary forces. Reuben et al. (2004) presented a front-tracking method for hyperbolic multiphase models. Krogstad and Durlofsky (2007) presented a coupled wellbore-reservoir flow model that is based on a multiscale mixed finite element formulation for reservoir flow linked to a drift-flux wellbore flow representation. Douglas and Bartley (2005) provided a one-dimensional parallel tube perfect cross-flow model (PCFM) for a two-phase flow in porous media. The model assumed that fluid can flow without resistance between tubes that contain the same phase (oil or water) at a given location. By comparing results from their model with the results from the conventional two-phase numerical simulator (based on the modified Darcy's law), they showed that the PCFM is an exact analogy of the modified Darcy's law. Although each of these mathematical models of the two-phase flows approximate real flows to varying degrees, none of them contains a substance source or sink term.

The two main dependent variables of interest in two-phase flow in porous media are saturation and pressure. Saturation is the ratio of the volume that a fluid occupies to the pore volume of a porous media. The relative amounts of oil, gas or water that will flow when more than one phase is present in a porous medium are dependent

---
*Corresponding author. E-mail:chimaoko@yahoo.com.

on the individual phase saturation. On the other hand, reservoir pressure is used for characterizing a reservoir, estimating its oil capacity and predicting its future behaviour. The production of oil and water in a well is a function of the reservoir pressure, which depends on the amount of oil and water in the reservoir, which, on its part, is described by the saturation (Craft and Hawkins, 1991). There are many computer software packages available for reservoir simulation. However, none of them accurately simulates existing reservoirs, because the mathematical models upon which they are based are only approximate representation of reality. This is why the mathematical modelling of multiphase flow is ongoing judging by the nature of journal papers on the subject yearly. An example of popular simulators is the ECLIPSE reservoir simulators, which have been the benchmark for commercial reservoir simulation for many years. It covers the entire spectrum of reservoir simulation, specializing in blackoil, compositional and thermal finite-difference reservoir simulation, and streamline reservoir simulation (Onyekonwu, 1997).

This paper, therefore, models a two-phase flow in semi-infinite porous media with a source/sink term, for the determination of the saturation and pressure distributions. The governing equations are formulated in one-dimension and two-dimensions, for the saturation and pressure, respectively, which are then discretised based on the finite difference method. The resulting numerical model was solved by the modified Gaussian elimination scheme (Oko, 2008) using Visual Basic for Application (VBA) in Microsoft Excel. Two-dimensions are used for the pressure distribution, because petroleum reservoirs are usually more permeable in the horizontal direction than in the vertical direction, and many reservoirs are modelled as two-dimensional space coordinates (Blunt, 2001).

THE GOVERNING EQUATIONS

In absence of hydrodynamic dispersion, the Darcy's equation is written to relate the superficial velocity of the simultaneous flow of each phase to the pressure gradient of the phase:

$$\hat{q}_w = -\frac{\hat{k}_w}{\mu_w}\left(\nabla p_w + \rho_w g\right)$$

(1)

$$\hat{q}_{nw} = \frac{\hat{k}_{nw}}{\mu_{nw}}\left(\nabla p_{nw} + \rho_{nw}\right)$$

(2)

where $\hat{q}$ [m/s] is the production by cross sectional area of flow; p [Pa] is the fluid pressure; g [m/s^2] is the acceleration due to gravity; $\hat{k}$ [m^2], μ [Pas] and

ρ [kg/m^3] are the permeability, viscosity and density, respectively; and the indices "w" and "nw" refer to the wetting and non-wetting

fluid, respectively; $\nabla \equiv \dfrac{\partial}{\partial n}$ [1/m] is the nabular operator; and n [m] is the unit vector.

Due to the surface tension and curvature of the interphase between the two phases, one phase referred to as the wetting phase "w" tends to wet the porous medium more than the other phase, referred to as the non-wetting phase "nw" (Xue, 2004). Since the void volume is completely occupied by the two fluid phases, the following equation applies for the saturation, S :

$$S_w + S_{nw} = 1$$

(3)

The pressures of the two phases are related to each other through the capillary pressure, p_c (Xue, 2004);

$$p_c\left(S_w\right) = p_{nw} - p_w$$

(4)

By relating the effective permeabilities $\hat{k}_w$ and $\hat{k}_{nw}$ to the single phase permeability, $\hat{k}$, the relative permeabilities k_{rw} and k_{rnw} can be defined:

$$\hat{k}_w = k_{rw}\hat{k}$$

(5)

$$\hat{k}_{nw} = k_{rnw}\hat{k}$$

(6)

The relative permeabilities are empirically taken to be functions of saturation and are assumed to be independent of direction (Aziz and Settari, 1979).

Introducing equation (5) into equation (1), and also equation (6) into equation (2), we have, respectively;

$$\hat{q}_w = -\frac{\hat{k}_{rw}\hat{k}}{\mu_w}\left(\nabla p_w + \rho_w g\right)$$

(7a)

$$\hat{q}_w = -\frac{\hat{k}_{rnw}\hat{k}}{\mu_{nw}}\left(\nabla p_{nw} + \rho_{nw}g\right)$$

(7b)

The continuity equation is written for each phase (Xue, 2004) as:

$$\frac{\partial\left(\phi\rho_w S_w\right)}{\partial t} = -\nabla.\left(\rho_w v_w\right) + q_{v,w}$$

(8a)

$$\frac{\partial\left(\phi\rho_{nw} S_{nw}\right)}{\partial t} = -\nabla.\left(\rho_{nw} v_{nw}\right) + q_{v,nw}$$

(8b)

where ϕ [-] is the porosity, or void fraction of the porous media; and $q_{v,w}$ [kg/m^{3}s] and $q_{v,nw}$ [kg/m^{3}s] represent the sink or source capacities for the wetting and non-wetting phases, respectively; and t [s] is production time.

Equations (7) and (8) represent the mathematical model for the flow of two immiscible phases in porous media. In order to solve it for the transient saturation and pressure of each phase, the following additional information is to be provided:
(i)   Capillary pressure and relative permeabilities as functions of saturation;
(ii)   Appropriate boundary and initial conditions;
(iii)   The porosity and fluid properties (densities and viscosities).

For the two-phase flow considered here, we shall mainly be concerned with the oil and water reservoir in which water is the wetting fluid (water flooding). The two phases, oil and water, will be denoted by subscripts "o" and "w", respectively. The effects of capillary pressure and buoyancy force are neglected. We shall also neglect the influence from other neighbouring wells and assume a one point well. With these assumptions, equations (7), (8) and (3), respectively, become

$$q_w = -\frac{\hat{k}_{rw}\hat{k}}{\mu_w}\nabla p_w$$

(9a)

$$q_o = -\frac{\hat{k}_{ro}\hat{k}}{\mu_o}\nabla p_o$$

(9b)

$$\frac{\partial(\phi\rho_w S_w)}{\partial t} = -\nabla(\rho_w \hat{q}_w) + q_{v,w}$$

(10a)

$$\frac{\partial(\phi\rho_o S_o)}{\partial t} = -\nabla(\rho_o \hat{q}_o) + q_{v,o}$$

(10b)

$$S_w + S_o = 1$$

(11)

It should be noted that a positive value of q_w or q_o connotes fluid injection, while a negative value connotes production.

Substituting Equations (9) into Equations (10) and using Equation (11) to eliminate S_o from (10b), we obtain

a set of coupled partial differential equations.

$$\frac{\partial(\phi\rho_w S_w)}{\partial t} = \nabla\left(\frac{\rho_w k_{rw}\hat{k}}{\mu_w}\nabla p_w\right) + q_{v,w}$$

(12a)

$$\frac{\partial[\phi\rho_o(1-S_w)]}{\partial t} = \nabla\left(\frac{\rho_o k_{ro}\hat{k}}{\mu_o}\nabla p_o\right) + q_{v,o}$$

(12b)

The coupled differential equations, equations (12) have to be decoupled for pressure and saturation.

Pressure equation

The compressibilities for water, β_w [Pa^{-1}], oil, β_o , and the porous media, β_r , can be related as:

$$\beta_i = \frac{1}{\rho_i}\frac{d\rho_i}{dp}, i = w, o$$

(13a, b)

$$\beta_r = \frac{1}{\phi}\frac{d\phi}{dp}$$

(13c)

Considering the compressibilities, equations (13), introducing some identities and assuming the pressure gradient, $\Delta p = \Delta p_w = \Delta p_o$, the coupled differential equations, Equations (12), are decoupled for pressure by expanding each of the left-hand side (lhs) terms in Equations (12); using Equations (13) to obtain some identities and substituting them into the expanded equations to eliminate the porosity; the resulting equations for oil and water are then added to obtain

$$\phi[\beta_r + \beta_o(1-S_w) + \beta_w S_w]\frac{\partial p}{\partial t} = \frac{1}{\rho_w}\nabla\left[\frac{\rho_w k_{rw}\hat{k}}{\mu_w}\nabla p\right] + \frac{1}{\rho_o}\nabla\cdot\left[\frac{\rho_o k_{ro}\hat{k}}{\mu_o}\nabla p\right] + q_{v,\rho}$$

(14a)

where $q_{v,\rho}$ [1/s] is the total source or sink rate and is given as:

$$q_{v,\rho} = \frac{q_{v,w}}{\rho_w} + \frac{q_{v,o}}{\rho_o}$$

(14b)

The densities on the right hand side (rhs) of equation (14a) are eliminated by neglecting the spatial variation of density to obtain:

$$\phi\beta_T\frac{\partial p}{\partial t} = \left(\frac{k_{rw}}{\mu_w}\beta_w + \frac{k_{ro}}{\mu_o}\beta_o\right)k\nabla p.\nabla p + \nabla M_T\nabla p + q_{v,\rho}$$

(15)

where,

$$\beta_T = \beta_r + \beta_o(1-S_w) + \beta_w S_w \text{ [Pa}^{-1}\text{] is the total}$$

compressibility; $M_T = \left(\frac{k_{rw}}{\mu_w} + \frac{k_{ro}}{\mu_o}\right)\hat{k}$

[kg^{-1}sm^3] is the total mobility.

By applying the order of magnitude, one neglects the non-linear term in equation (15) and if we assume that β_T, ϕ and M_T to be constant parameters, equation (15) becomes:

$$\beta_T\phi\frac{\partial p}{\partial t} = \nabla(M_T\nabla p) + q_{v,\rho}$$

(16)

Considering the circular bounded reservoir, equation (16) becomes:

$$\frac{1}{\alpha}\frac{\partial p}{\partial t} = \frac{\partial^2 p}{\partial r^2} + \frac{1}{r}\frac{\partial p}{\partial r} + \frac{1}{r^2}\frac{\partial^2 p}{\partial \theta^2} + q^*_{v,\rho}, \quad r_b < r < r_\infty,$$
$$0 < \theta < 2\pi \tag{16}$$

where,

$\alpha = \dfrac{M_T}{\beta_T \phi}$ [m²/s] is the momentum diffusivity and $q^*_{v,\rho} = \dfrac{q_{v,\rho}}{M_T}$

[Pa/m⁻²] is the sink or source term.

Let the sink or source capacity vary with the radial distance, angle and time then the last equation becomes:

$$\frac{\partial p}{\partial t} = \alpha\left(\frac{\partial^2 p}{\partial r^2} + \frac{1}{r}\frac{\partial p}{\partial r} + \frac{1}{r^2}\frac{\partial^2 p}{\partial \theta^2}\right) + \frac{\xi}{r^2}e^{-\varsigma t}\left(1 - \cos\omega\theta t\right),$$
$$r_b < r < r_\infty, \quad 0 < \theta < 2\pi \tag{17}$$

where the last term on rhs of the last equation represents the

source/sink terms, $\alpha q^*_{v,\rho}$; $\xi \equiv \dfrac{\alpha^2}{M_T}$ [kgs⁻³m]; ς [1/s] and ω [1/s]

are constant coefficients and θ [rad] is the angular distance. The boundary conditions at the bore, infinity, and angular distances

$(0, \dfrac{\pi}{2}, \pi, \dfrac{3\pi}{2}, 2\pi)$ are, respectively.

$$\eta p + \zeta r_w \frac{\partial p}{\partial r} = \frac{Q_T \mu_T}{2\pi h k} = p_\rho, \quad r = r_b,$$
$$0 < \theta < 2\pi, t > 0 \tag{18a}$$
$$\frac{\partial p}{\partial r} = 0, \quad r = r_\infty, \ 0 < \theta < 2\pi, \ t > 0 \tag{18b}$$
$$\frac{\partial p}{\partial \theta} = 0, \quad \theta = 0, \frac{\pi}{2}, \pi, \frac{3\pi}{2}, 2\pi, \ t > 0 \tag{18c}$$

where Q_T [m³/s] is the constant total well production/injection rate; μ_T [Pas] is the total viscosity; h[m] is the well thickness and η [-] and ζ [-] are dimensionless constants intended to be adjusted to meet desired results from field data.

The initial condition is taken as the pressure of the well measured before commencement of production.

$$p = p_{in}, \ t = 0, \ r_b \le r \le r_\infty, \ 0 \le \theta \le 2\pi \tag{19}$$

By taking the dimensionless form of the model, we have:

$$\frac{\partial\Pi}{\partial\tau} = \frac{\partial^2\Pi}{\partial R^2} + \frac{1}{R}\frac{\partial\Pi}{\partial R} + \left(\frac{1}{\theta_{max}}\right)\frac{1}{R^2}\frac{\partial^2\Pi}{\partial\Psi^2} + \frac{\xi}{R^2\alpha\wp}e^{\frac{-\varsigma r_\infty^2}{\alpha}\tau}\left(1 - \cos\frac{\omega\theta_{max}\Psi r_\infty^2}{\alpha}\tau\right)$$
$$\tag{20}$$

Subject to the following dimensionless boundary/initial conditions:

$$\gamma\Pi + \beta R_0 \frac{\partial\Pi}{\partial R} = C, \ R = R_b, \tau > 0 \tag{21a}$$

$$\frac{\partial\Pi}{\partial R} = 0, \ R = 1, \tau > 0 \tag{21b}$$

$$\frac{\partial\Pi}{\partial\psi} = 0, \ \psi = 0, \frac{1}{4}, \frac{1}{2}, \frac{3}{4}, 1; \tau > 0 \tag{21c}$$

$$\Pi = 1, \ \tau = 0 \tag{21d}$$

The dimensionless parameters are defined as:

$$\Pi \equiv \frac{\wp'}{\wp} = \frac{p - p_{ref}S_w}{p_{in} - p_{ref}S_w}; R \equiv \frac{r}{r_\infty}; R_b \equiv \frac{r_b}{r_\infty};$$

$$\Psi \equiv \frac{\theta}{2\pi}; \tau \equiv \frac{\alpha t}{r_\infty^2} \text{ and } C \equiv \frac{p_\rho - p_{ref}S_w}{p_{in} - p_{ref}S_w} \tag{22}$$

where $\Pi, R, R_0, \Psi, \text{ and } \tau$ stand for the dimensionless pressure, dimensionless radius and dimensionless wellbore radius, dimensionless angle and dimensionless time, respectively; $p, p_{ref}, p_{in}, r_\infty[m], r[m],$ and $r_b[m]$ stand for the unknown pressure, reference pressure, initial pressure, semi-infinite radius, radius from the well bore and wellbore radius, respectively.

Saturation equation

The velocities v_i [m/s] in the continuity equation, equations (8), are related for water and oil as:

$$v_i = -\frac{k_{ri}k}{\mu_i}\nabla p, \ i = w, o \tag{23}$$

To derive a single equation for saturation, we eliminated the pressure gradients from equations (8a) and (8b) by dividing equation (8a) by equation (8b) and introducing fractional flow of water, f_w [-], to obtain:

$$\frac{\partial(\phi\rho_w S_w)}{\partial t} = -\nabla.(\upsilon\rho_w f_w v_t) + q_{v,w} \tag{24}$$

where,

$$f_w = \frac{x_w/x_o}{1 + x_w/x_o} \text{ and } x_i = \frac{k_{ri}k}{\mu_i}, \ i = w, o$$

Dividing equation (24) by ρ_w and expanding the rhs, one obtains.

$$\phi \frac{\partial S_w}{\partial t} = -f_w \nabla v_t - v_t \nabla f_w + \frac{q_{v,w}}{\rho_w} \qquad (25)$$

Considering the identity, equation (26).

$$\nabla f_w = \frac{df_w}{dS_w} \nabla S_w \qquad (26)$$

and substituting into equation (25) one obtains:

$$\phi \frac{\partial S_w}{\partial t} = -f_w \nabla v_t - v_t \frac{df_w}{dS_w} . \nabla S_w + \frac{q_{v,w}}{\rho_w} \qquad (27)$$

If one is concerned with flow away from the exterior boundary and neglecting the source or sink capacity for the saturation, then $q_{v,w} = 0$ and $\nabla . v_t = 0$. Therefore, equation (27) becomes:

$$\phi \frac{\partial S_w}{\partial t} = -v_t \frac{df_w}{dS_w} \nabla S_w \qquad (28)$$

Equation (28) is a first order hyperbolic equation and if we assume that variation of saturation with angle is negligible then the model is subject to the following initial and boundary conditions.

$$S_w(r,0) = S_{in}; \quad r_b \le r \le r_\infty , t = 0 \qquad (29)$$

$$S_w(r_b,t) = S_o; \quad t \ge 0 \qquad (30)$$

$$\frac{\partial S_w(r_\infty,t)}{\partial r} = 0; \quad t \ge 0 \qquad (31)$$

By transforming equation (28) into dimensionless form we obtained:

$$\frac{\partial S_w}{\partial \tau} = -\frac{df_w}{dS_w} \frac{\partial S_w}{\partial R} \qquad (32)$$

where the dimensionless parameters are define as:

$$\tau = \frac{v_t t}{r_\infty \phi} \equiv \frac{\alpha t}{r_\infty^2}, \quad R = \frac{r}{r_\infty}$$

With the dimensionless initial and boundary conditions as:

$$S_w(R,0) = S_{in}; \quad R_b \le R \le 1 , \tau = 0 \qquad (33a)$$

$$S_w(R_b,\tau) = S_o; \quad \tau \ge 0 \qquad (33b)$$

$$\frac{\partial S_w(1,\tau)}{\partial R} = 0; \quad \tau \ge 0 \qquad (33c)$$

In simple form, equation (32) can be written as:

$$\frac{\partial S_w}{\partial \tau} = -f_w' \frac{\partial S_w}{\partial R} \qquad (34)$$

Differentiating data presented by Foroozesh (2008) for oil and water one obtains the value of f_w'.

Discretized equations

We employ the central difference of the Crank-Nicholson finite difference method to discretize equations (32), (33), (20) and (21).

Discretized saturation equations

We denote $s_w(R,\tau)$ as $s_{wi}^\tau, i = 0,1,2,...M$, hence, the implicit finite difference method is formulated for Equations (32) and (33) as:

$$\xi_s'\left(S_w^{\tau+1}\right) = f_s'\left(S_w^\tau\right) \qquad 1 < i < M , \tau > 0 \qquad (35)$$

where,

$$\xi_s'\left(S_w^{\tau+1}\right) = S_{wi}^{\tau+1} + \frac{\Delta \tau f_w'}{4\Delta R} S_{wi+1}^{\tau+1} - \frac{\Delta \tau f_w'}{4\Delta R} S_{wi-1}^{\tau+1} \qquad (36a)$$

$$f_s'\left(S_w^\tau\right) = S_{wi}^\tau - \frac{\Delta \tau f_w'}{4\Delta R} S_{wi+1}^\tau + \frac{\Delta \tau f_w'}{4\Delta R} S_{wi-1}^\tau \qquad (36b)$$

Considering the boundary condition of equation (33b) we obtain for $R = R_b (i = 1)$.

$$S_{wi}^{\tau+1} + \frac{\Delta \tau f_w'}{4\Delta R} S_{wi+1}^{\tau+1} = S_{wi}^\tau + \frac{\Delta \tau f_w'}{4\Delta R} S_{wi+1}^\tau + 2\frac{\Delta \tau f_w'}{4\Delta R} S_o \qquad (37)$$

Also, considering the boundary condition of equation (33c) we obtain for $R = 1 (i = M)$:

$$S_{wi}^{\tau+1} = S_{wi}^\tau \qquad (38)$$

Discretised pressure equations

We denote $\Pi(R,\psi,\tau)$ as $\Pi_{i,j}^\tau$ for $i = 0,1,2,...M$; and $j = 0,1,2,...N$. The implicit finite difference is formulated for Equations (20) and (21) as:

$$\xi'_p\left(\Pi^{\tau+1}\right)=f'_p\left(\Pi^{\tau}\right)\qquad 0<i<M;$$
$$0<j<N \tag{39}$$

where,

$$\xi'_p\left(\Pi^{\tau+1}\right)\equiv\gamma_i\Pi^{\tau+1}_{i,j}-\delta_i\Pi^{\tau+1}_{i+1,j}+\sigma_i\Pi^{\tau+1}_{i-1,j}-\varepsilon_i\Pi^{\tau+1}_{i,j+1}-\varepsilon_i\Pi^{\tau+1}_{i,j-1}$$
$$\tag{40a}$$
$$f'_p\left(\Pi^{\tau}\right)\equiv\gamma'_i\Pi^{\tau}_{i,j}+\delta_i\Pi^{\tau}_{i+1,j}-\sigma_i\Pi^{\tau}_{i-1,j}+\varepsilon_i\Pi^{\tau}_{i,j+1}+\varepsilon_i\Pi^{\tau}_{i,j-1}+0.5\Delta\tau\left(\Lambda^{\tau}_{i,j}+\Lambda^{\tau}_{i,j}\right)$$
$$\tag{40b}$$

where,

$$\gamma_i=\left(1+\frac{\Delta\tau}{\left(\Delta R\right)^2}+\frac{\Delta\tau}{\theta_{max}\left[\left(R_b+i\Delta R\right)\left(\Delta\psi\right)\right]^2}\right)\tag{41a}$$

$$\gamma'_i=\left(1-\frac{\Delta\tau}{\left(\Delta R\right)^2}-\frac{\Delta\tau}{\theta_{max}\left[\left(R_b+i\Delta R\right)\left(\Delta\psi\right)\right]^2}\right)\tag{41b}$$

$$\delta_i=\left(\frac{\Delta\tau}{4\left(R_b+i\Delta R\right)\Delta R}+\frac{\Delta\tau}{2\left(\Delta R\right)^2}\right)\tag{41c}$$

$$\sigma_i=\left(\frac{\Delta\tau}{2\left(\Delta R\right)^2}-\frac{\Delta\tau}{4\left(R_b+i\Delta R\right)\Delta R}\right)\tag{41d}$$

$$\varepsilon_i=\left(\frac{\Delta\tau}{2\theta_{max}\left[\left(R_b+i\Delta R\right)\left(\Delta\psi\right)\right]^2}\right)\tag{41e}$$

$$\Lambda^{\tau+1}_{i,j}=\frac{\xi e^{\frac{\wp^2_{\infty}((k+l)\Delta\tau)}{\alpha}}}{\left(R_b+i\Delta R\right)\alpha\wp}\left[1-\cos\left(\frac{\omega\theta_{max}j(\Delta\psi)r^2_{\infty}}{\alpha}(k+1)\Delta\tau\right)\right]\tag{41f}$$

$$\Lambda^{\tau}_{i,j}=\frac{\xi e^{\frac{\wp^2_{\infty}(k\Delta\tau)}{\alpha}}}{\left(R_b+i\Delta R\right)\alpha\wp}\left[1-\cos\left(\frac{\omega\theta_{max}j(\Delta\psi)r^2_{\infty}}{\alpha}k\Delta\tau\right)\right]\tag{41g}$$

Considering the boundary conditions at $R=R_b\left(i=0\right)$,

$R=1\left(i=M\right)$, $\psi=0\left(j=0\right)$ and $\psi=1\left(j=N\right)$, then, equation (39) becomes:

$$\left(\gamma_i+\frac{2(\Delta R)\eta\sigma_i}{R_b\zeta}\right)\Pi^{\tau+1}_{i,j}+\left(\sigma_i-\delta_i\right)\Pi^{\tau+1}_{i+1,j}-\varepsilon_i\Pi^{\tau+1}_{i,j-1}-\varepsilon_i\Pi^{\tau+1}_{i,j+1}=\tag{42}$$
$$\left(\gamma_i-\frac{2(\Delta R)\eta\sigma_i}{R_b\zeta}\right)\Pi^{\tau+1}_{i,j}+\left(\delta_i-\sigma_i\right)\Pi^{\tau+1}_{i+1,j}+\varepsilon_i\Pi^{\tau+1}_{i,j-1}+\varepsilon_i\Pi^{\tau+1}_{i,j+1}+0.5\Delta\tau\left(\Lambda^{\tau}_{i,j}+\Lambda^{\tau+1}_{i,j}\right)+2D_C\sigma_i$$

(i)   for $R=R_b(i=0)$, $\psi\neq 0(j\neq 0)$ and $\psi\neq 1(j\neq N)$:

(ii)   for $R=1(i=M)$, $\psi\neq 0(j\neq 0)$ and $\psi\neq 1(j\neq N)$:

$$\left(\sigma_i-\delta_i\right)\Pi^{\tau+1}_{i-1,j}+\gamma_i\Pi^{\tau+1}_{i,j}-\varepsilon_i\Pi^{\tau+1}_{i,j-1}-\varepsilon_i\Pi^{\tau+1}_{i,j+1}=$$
$$\left(\delta_i-\sigma_i\right)\Pi^{\tau}_{i-1,j}+\gamma'_i\Pi^{\tau+1}_{i,j}+\varepsilon_i\Pi^{\tau+1}_{i,j-1}+\varepsilon_i\Pi^{\tau+1}_{i,j+1}+0.5\Delta\tau\left(\Lambda^{\tau}_{i,j}+\Lambda^{\tau+1}_{i,j}\right)\tag{43}$$

(iii)   for $\psi=0(j=0)$, $R\neq R_b(i\neq 0)$ and $R\neq i(i\neq M)$:

$$\sigma_i\Pi^{\tau+1}_{i-1,j}+\gamma_i\Pi^{\tau+1}_{i,j}-\delta_i\Pi^{\tau+1}_{i+1,j}-2\varepsilon_i\Pi^{\tau+1}_{i,j+1}=$$
$$-\sigma_i\Pi^{\tau}_{i-1,j}+\gamma'_i\Pi^{\tau}_{i,j}+\delta_i\Pi^{\tau}_{i+1,j}+2\varepsilon_i\Pi^{\tau+1}_{i,j+1}+0.5\Delta\tau\left(\Lambda^{\tau}_{i,j}+\Lambda^{\tau+1}_{i,j}\right)\tag{44}$$

(iv)   For $\psi=1(j=N)$, $R\neq R_b(i\neq 0)$ and $R\neq i(i\neq M)$:

$$\sigma_i\Pi^{\tau+1}_{i-1,j}+\gamma_i\Pi^{\tau+1}_{i,j}-\delta_i\Pi^{\tau+1}_{i+1,j}-2\varepsilon_i\Pi^{\tau+1}_{i,j-1}=$$
$$-\sigma_i\Pi^{\tau}_{i-1,j}+\gamma'_i\Pi^{\tau}_{i,j}+\delta_i\Pi^{\tau}_{i+1,j}+2\varepsilon_i\Pi^{\tau+1}_{i,j-1}+0.5\Delta\tau\left(\Lambda^{\tau}_{i,j}+\Lambda^{\tau+1}_{i,j}\right)\tag{45}$$

Computational algorithm

The computation of the saturation and pressure distributions was carried out using the algorithm (Chart 1)

RESULTS AND DISCUSSION

The system of linear equations, equations (35) to (39) and (42) to (45) was solved using the algorithm of the Modified Gaussian Elimination Method (MGEM) (Oko, 2008).The algorithm was implemented in Visual Basic for Application in Microsoft Excel. The results obtained based on hypothetical well production data are shown in Figures 1 to 6.

Figures 1 and 2 show that significant variation of saturation occurs close to the wellbore and the variation approaches steady-state condition as one moves far away from the wellbore. The figures show the practical phenomena of water conning.

Figure 3 shows the dimensionless pressure distribution at water saturation of 0.7[-] and dimensionless time of 0.0005 [-], while Figure 4 shows dimensionless pressure distribution at corresponding saturation at the same dimensionless time. One will observes from Figures 3 and 4, that in Figure 3 with a water saturation of 0.7 [-] has a lesser magnitude of pressure distribution, which can be seen at the dimensionless radius 1.0[-] as compared to Figures 4 with varied saturation at the same

start

  saturation distribution;

      input parameter;

        differentiate water fractional flow with respect to water saturation numerically;

          compute:

            initialize time and radius increment counter

            repeat

              generate coefficient matrix from the set of systems of linear equation;

              generate constant matrix from the set of systems of linear equation;

              use the modified Gaussian elimination scheme to solve for saturation distribution;

              increase time and dimensionless radius with their respective steps;

            until (maximum dimensionless time and radius are reached)

            put the computed saturation in cells;

            plot appropriate data from the computed results;

  pressure distribution;

        input parameter;

       computation;

          initialize time and radius increment counter;

          repeat

            compute total compressibility and mobility from the computed water saturation;

            generate coefficient matrix from the set of systems of linear equation;

            generate constant matrix from the set of systems of linear equation;

            use the modified Gaussian elimination method to solve for saturation distribution;

            increase time and dimensionless radius with their respective steps;

          until (maximum dimensionless time and radius are reached)

     put the computed dimensionless pressure in cells;

    convert data to dimensional values; **Optional step**

    plot appropriate from the computed results;

stop

Chart 1. Algorithm used in the computation of saturation and pressure distributions.

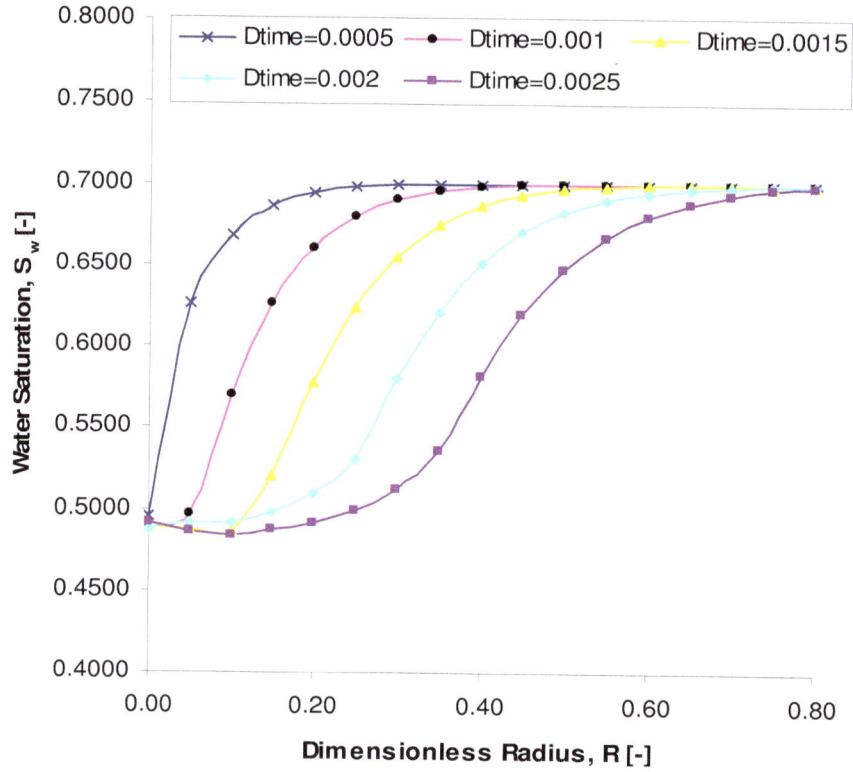

Figure 1. Variation of Water Saturation with Dimensionless Radius and Time (Dtime).

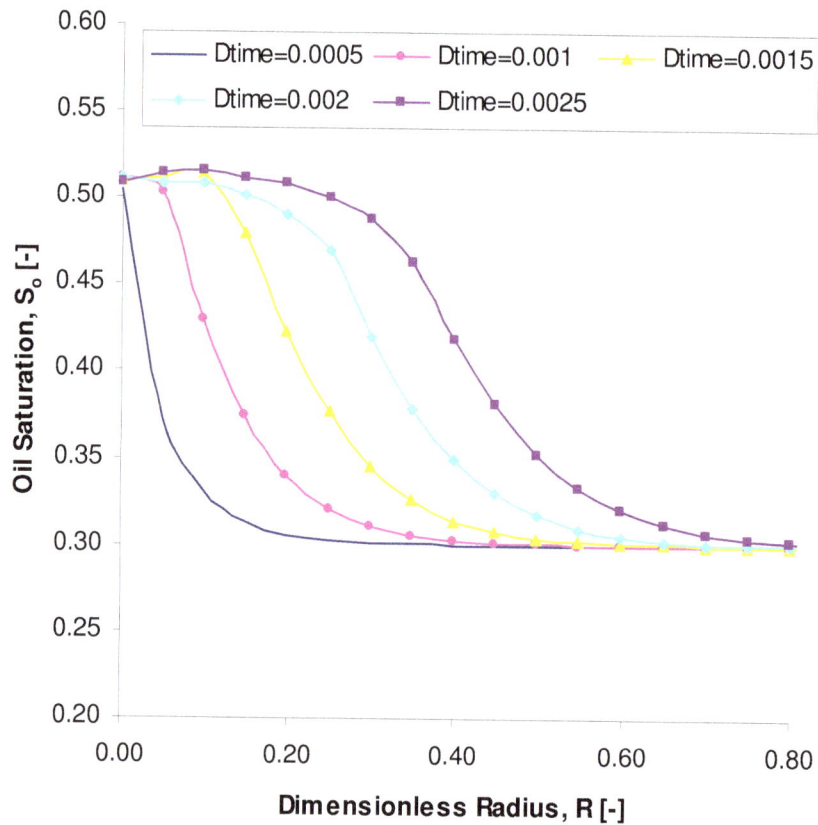

Figure 2. Variation of Oil Saturation with Dimensionless Radius and Time (Dtime).

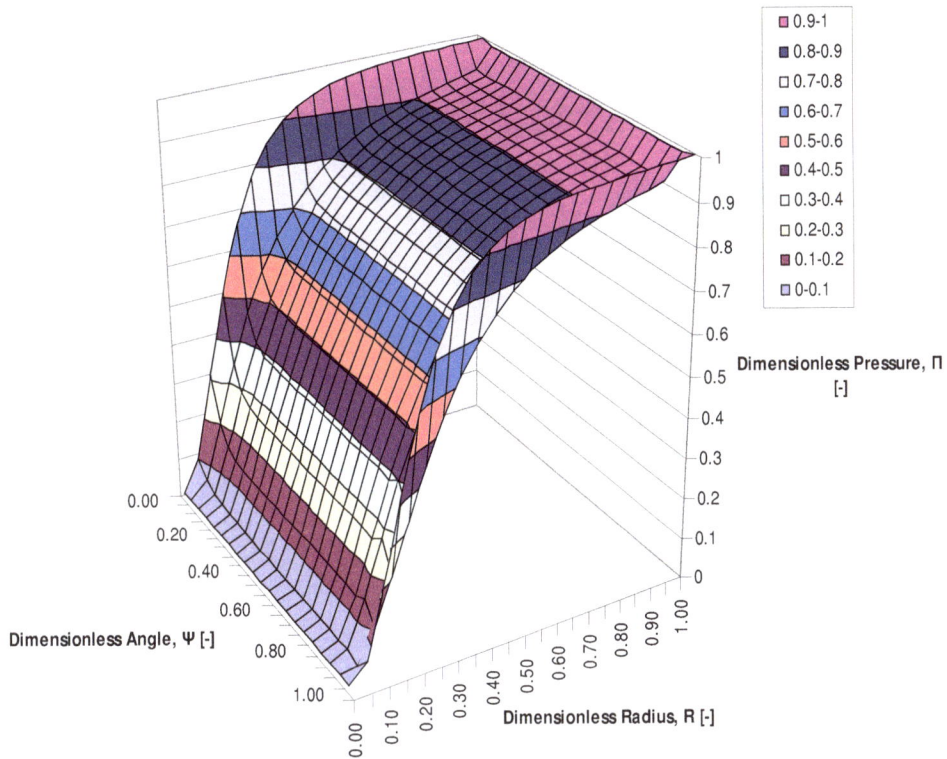

Figure 3. Dimensionless pressure distribution at dimensionless time = 0.0005 and constant water saturation, S_w = 0.7.

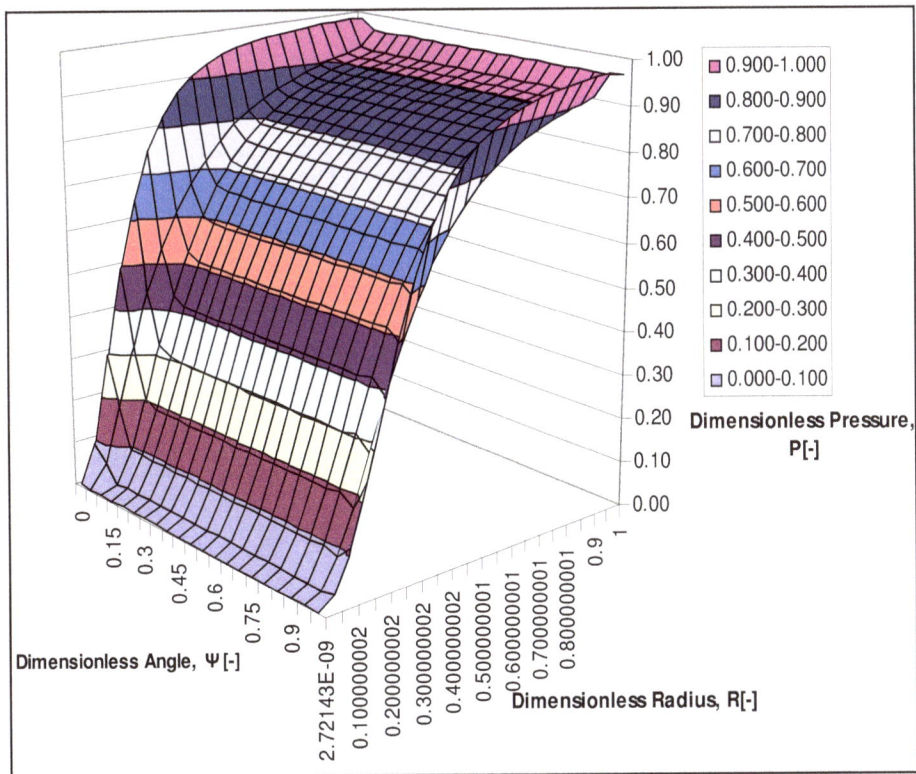

Figure 4. Dimensionless pressure distribution at dimensionless Time= 0.0005 at varied water saturation.

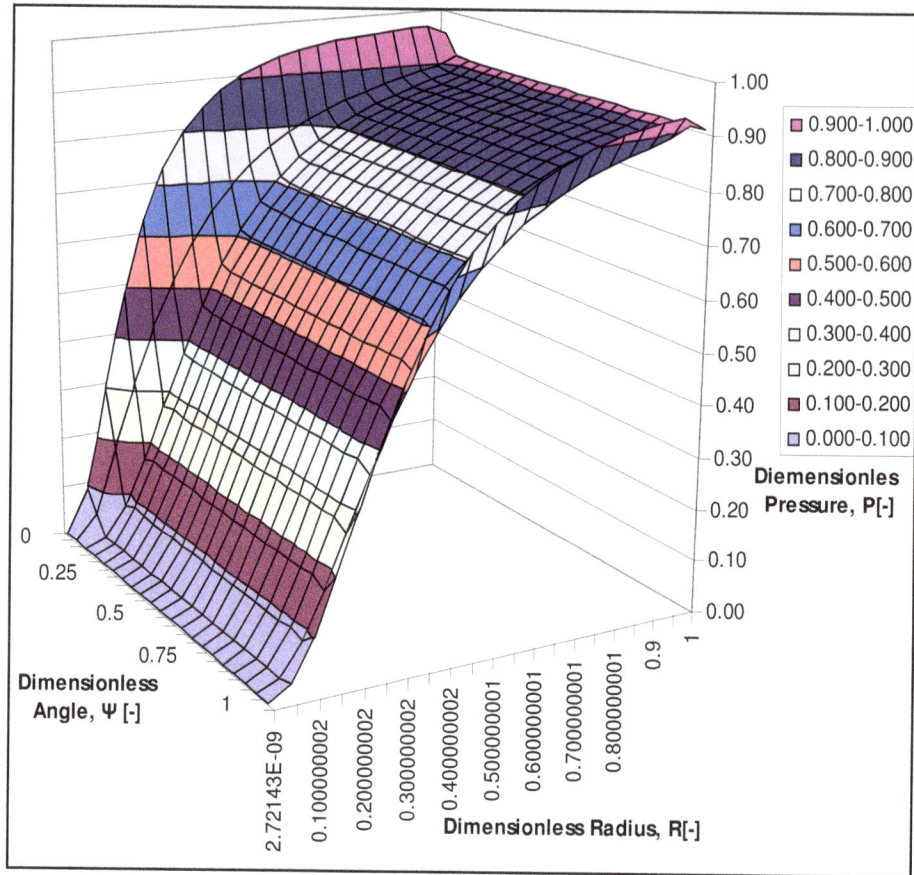

Figure 5. Dimensionless pressure distribution at dimensionless Time = 0.0025 at varied water saturation.

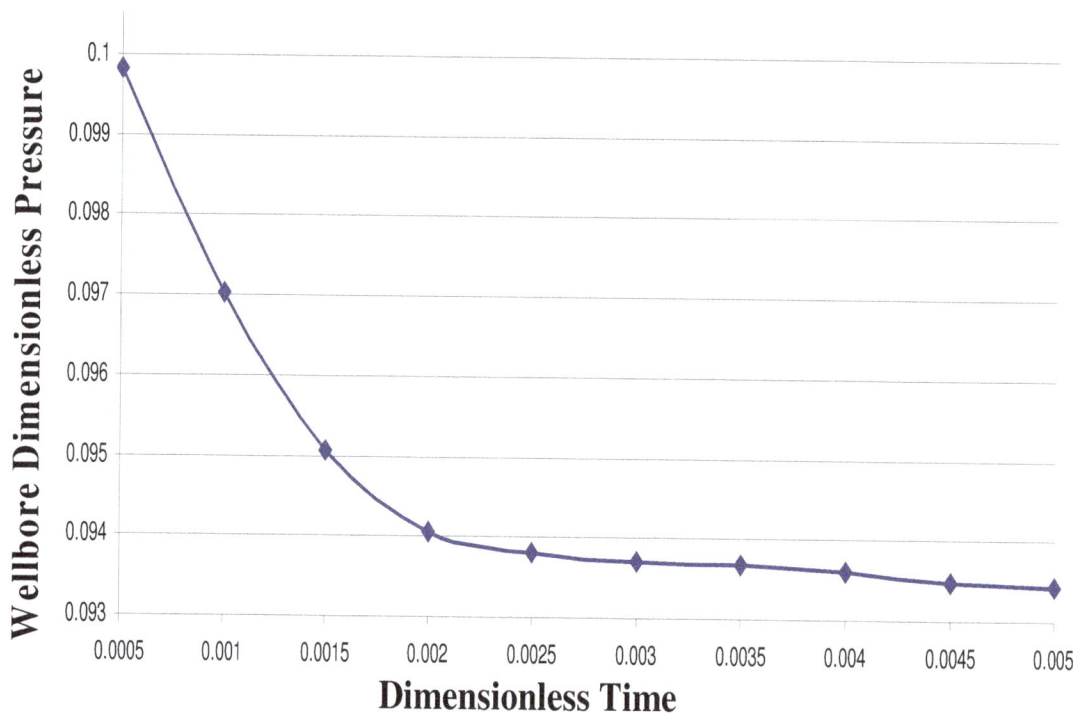

Figure 6. Wellbore pressure against Dimensionless time.

dimensionless time of 0.0005 [-]. Therefore, Figures 3 and 4 shows that pressure distribution in porous media is a strong function of saturation.

Figures 4 and 5 show a decrease in the magnitude of the pressure distribution as the dimensionless time increases from 0.0005 to 0.0025 [-] at the corresponding saturation distribution and at constant media properties and fluid properties, which is an indication of the transient pressure and saturation distribution. Figures 3, 4 and 5 also indicate variation of the pressure with the angle, which may be due to presence of aquifer and/or the topography of the reservoir.

The wellbore pressure (Figure 6) is decaying with time (pressure depletion). As this happens, then we shall expect that a time will come when the wellbore pressure will be equal to the atmospheric pressure. When this happens, natural production in the well is going to stop and hence further attempt to get oil from the well can only be by artificial means in which case there has to be a gas injection or water flooding or other means of recovery (Onyekonwu, 1997).

Conclusion

This paper has succeeded in presenting the saturation and pressure distributions in oil and water flow through semi-infinite porous media. The generalized (dimensionless) results can easily be inverted to obtain dimensional values for concrete cases by using the appropriate inversion formula, Equation (22). The solution scheme and computer language adopted in this work are easy to apply and use as opposed to the sophisticated and expensive computer software used by most researchers in the area. Since the results presented were based on hypothetical well production data, the immediate future work should, therefore, be the validation of the model and solution presented in this work with typical oil well production data.

REFERENCES

Aziz K, Settari A (1979). Petroleum Reservoir Simulation, Applied Science Publishers, London.

Blunt MJ (2001). Flow in Porous Media-Pore-Network Models and Multiphase Flow, Current Opinion in Colloid & Interface Science, 6: 197-207.

Brenem CE (2003). Fundamentals of Multiphase Flow, California Institute of Technology, Pasadana.

Douglas R, Bartley J (2005). A Perfect Cross-Flow Model for Two-Phase Flow in Porous Media, University of Manitoba, Manitoba.

Foroozesh J, Barzegeri D, Ayatollahi SH, Jahanniri BN (2008). Simulation of water coning in Oil Reservoir Using a corrected IMPES Method. Iranian J. Chem. Eng., 5(4):1-11.

Knut-Andreas L, Jorg EA, Vegard K, Stein K (2004). Multiscale Methods for Flow in Porous Media. Multiscale Model Simulation, 2(3): 421-439.

Kolev NI (2005). Multiphase Flow Dynamics 1: Fundamentals, Springer-Verlag, Berlin.

Krogstal S, Durlofsky LJ (2007). Multiscale Mixed Finite Element Modelling of Coupled Wellbore/Near-Well Flow, SPE Reservoir Simulation Symposium, 26th -28th February, 2007, Houston.

Oko COC (2008). Engineering Computational Methods: An Algorithmic Approach, University of Port Harcourt Press, Port Harcourt.

Onyekonwu MO (1997). General Principles of Bottom-Hole Pressure Tests, Laser Engineering Consultants, Port Harcourt.

Pruess K, Narasimhan TN (1985). A Practical Method for Modeling Fluid and Heat Flow in Fractured Porous Media, Soc. Petroleum Eng. J., 25 :14–26.

Reuben J, Knut-Andreas L, Vegard K (2004). A Front-tracking Method of Hyperbolic Three-phase Models, European Conference on the Mathematics of Oil Recovery, 30 August – 2 September, 2004, Cannes.

Seto K, Hollenshead JT, Watson AT, Chang CTP, Slattery JC (2001). Determining Permeability using NMR Velocity Imaging, Transaction in Porous Media, 42: 351- 388.

Xue S (2004). Towards Improved Methods for Determining Porous Media Multiphase Flow Functions, M.Sc Thesis, Texas A & M University, Texas.

Craft BC, Hawkins MF (1991). Applied Petroleum Reservoir Engineering, 2ed/ Revised by Ronald E T, Prentice Hall Inc NY.

17

Microbial enhanced oil recovery using potent biosurfactant produced by *Pseudomonas* sp. from Arabian Sea, Mumbai

Seema Dhail

Department of Biotechnology, Manipal University, Jaipur-302004, India.

Increase in oil pollution due to accidental leakages during various ship operations and human activities make these hydrocarbons the most common global environmental pollutants. Current understanding in degradation of these harmful oils involves the isolation of biosurfactants from various isolated microbial strains from contaminated sites of Marine Sea. Biosurfactants are the surface active molecules synthesized by microorganisms. Due to various side effects of chemical surfactants the demand for biosurfactant production and its synthesis has been steadily increasing and may eventually replace their chemically synthesized counterparts. In this study, isolation and identification of biosurfactant producing bacteria were assessed from oil-spilled area of Arabian Sea. To confirm the ability of isolates in biosurfactant production, various biosurfactant activity assay tests were performed. Marine biosurfactants produced by some marine microorganisms have been paid more attention, particularly for the bioremediation of the sea polluted by crude oil. Among all of the isolated strains *Pseudomonas* sp. showed the highest biosurfactant activity. The biosurfactant component was detected as glycolipid or other anionic surfactants in analysis of a phenotypic assay test using CTAB. The isolated culture filtrate was found to be highly effective in microbial enhanced oil recovery (MEOR) using sand pack method.

Key words: Biosurfactant, *Pseudomonas* sp., microbial enhanced oil recovery, oil spilled area.

INTRODUCTION

Petroleum products are the major source of energy for industry, transportation as well as day today life; it also poses major concern over hydrocarbon release during its production, transportation and accidental leakages. On 7th of August, 2010, 800 tonnes of oil being spilled into the Arabian Sea, due to the collision of Panamanian vessels MSC Chitra and MV Khalijia III had severe impact around 1,273 ha of mangroves. The oil spills from marine water are treated using bioremediation methods using biosurfactant as it is one of the promising technologies in future. Microbial compounds, which exhibit pronounced surface activity are classified as biosurfactants. Biosurfactants or surface-active compounds are a heterogenous group of surface active molecules produced by microorganisms, which either adhere to cell surface or are excreted extracellularly in the growth medium (Fietcher, 1992; Zajic and Stiffens, 1994; Makker and Cameotra, 1998). Biosurfactants have unique amphiphatic properties due to hydrophilic and hydrophobic portion in it. As a result, they can partition preferentially at the interfaces (Desai and Banat, 1997) and are environmentally compatible (Georgiou et al., 1990). Biosurfactants have various advantages, such as high biodegradability, low toxicity, environmental compatibility, high selectivity, and specific activity at extreme temperatures, pH, and salinity (Desai and Banat, 1997; Lang and Wullbrandt, 1999). The use of biosurfactants to protect the marine environment seems possible since a number of marine bacterial strains can produce biosurfactants during growth on hydrocarbons (Bertrand et al., 1993). Microbial enhanced oil recovery (MEOR) is a good alternative in improving the recovery of crude oil from reservoir rocks by using microorganisms and their metabolic by-products. Recently many investigations on MEOR have used whole cells and their biosurfactants to improve the efficiency of oil recovery (Joshi et al., 2008; Toledo et al., 2008; Jinfeng et al., 2005; Rashedi et al., 2005; Mei et al., 2003). There are three mechanisms by which microorganisms can contribute to increased oil production: i) microorganisms can produce biosurfactants and biopolymers on the cell surface, ii) microorganisms produce gases and acids to recover trapped oil and iii) microorganisms can selectively plug high permeability channels into the reservoir (Bryant, 1987). The main objective of this paper is to isolate biosurfactant producing bacteria from oil spilled area of Marine Sea for use in MEOR.

MATERIALS AND METHODS

Isolation and enrichment of biosurfactant producing microorganisms

Soil samples were collected from different oil spilled areas of Marine sediment. Each sample was loaded into sterile 250 ml flasks. The sediments were collected through scuba diving to the depth of 1 to 100 m. The samples were collected in plastic bags and immediately transported to the Microbiology lab, IIS University, Jaipur, India, and stored in a refrigerator at 4°C up to further processing. 1 g of soil sample was taken and serially diluted in 0.85% sterile saline water. All dilutions were performed in triplicates and then the samples were spread on nutrient agar plates and incubated at room temperature for 1 to 2 days. After incubation, plates were enumerated and morphologically different bacteria were selected for biosurfactant screening (approximately 5 to 6 isolates per plate) and purified by re-streaking twice. Isolated colonies were inoculated into 100 ml of Marine Broth 2216 (Difco) containing 2 to 3 drops of petrol+kerosene+diesel (P+K+D) in 1:1:1 ratio, and incubated with continuous shaking (200 rpm) for 24 to 48 h at room temperature using a shaker. Colonies possessing biosurfactant-producing activity, as evidenced by emulsification of oil, were chosen for further experimentation. In addition, the cell suspensions of isolated strains were tested for presence of surfactant by using haemolytic activity, the qualitative drop collapsing test, quantitative oil displacement test and emulsification activity.

Screening for biosurfactant producers

Haemolytic activity

Biosurfactant producing capacity was found to be associated with haemolytic activity. Haemolytic activity therefore appears to be a good screening criterion for surfactant-producing strains. Isolated strains were screened on blood agar plates containing 5% (v/v) goat blood and incubated at room temperature for 24 h. Haemolytic activity was detected as the occurrence of a define clear zone around a colony (Carrillo et al., 1996).

Drop collapsing test

Two micro-liter of mineral oil was added to each well of a 96-well micro-liter plate lid. The lid was equilibrated for 1 h at room temperature, and then 5 µl of the cultural supernatant was added to the surface of oil. The shape of the drop on the oil surface was inspected after 1 min. Biosurfactant-producing cultures giving flat drops were scored as positive '+'. Those cultures that gave rounded drops were scored as negative '-', indicative of the lack of biosurfactant production (Youssef et al., 2004).

Oil activity assay test

The isolates present in the clearing zone on Haemolytic activity test were re-screened by hydrolyzing oil activity assay on oil agar plates (Morikawa et al., 1993). Potential biosurfactant-producing strains were inoculated into the Minimal salt medium (MSM) (Himedia), and the cultured filter was further analyzed by improved degreasing effect assay test and emulsification activity measurement. Improved degreasing effect assay was carried out following Bi et al. (2009) work. The emulsification activity was determined by adding 5 ml of kerosene and an equal volume of cell-free supernatant to a 20 ml tube. The sample was homogenized in a vortex at high speed for 2 min and allowed to settle for 24 h. The emulsification index was then calculated by given formula.

Emulsification Activity = Height of emulsion layer/Total height

Identification of potential strain

The potential strain that was showing good biosurfactant activity was identified using Bergey's manual of determinative bacteriology (Buchanan and Gibbons, 1974).

Production, isolation and identification of rhamnolipids

The identified strain was initially maintained on *Pseudomonas* isolation agar at 30°C for 24 h. Random single colonies were transferred into the MSM with the addition of 200 g/ml cetyltrimethylammonium bromide (CTAB, Sigma), 5 g/ml methylene blue, and 1.5% (w/v) agar, as described by Siegmund and Wagner (1991). A colony showing a dark blue halo was selected and grown in Kay's minimal medium (0.3% $NH_4H_2PO_4$, 0.2% K_2HPO_4, 0.2% glucose, 0.05 mg% $FeSO_4$, 0.1% $MgSO_4$) at 30°C for 5 days with shaking at 250 rpm. The culture filtrate was centrifuged at 8,000 × g for 10 min at 4°C to remove the cells and the debris. The pH of the supernatant obtained was adjusted to 2.0 using 12 M HCl, and then stored overnight at 4°C. The precipitates were collected by centrifugation at 8,000 × g for 20 min, and then extracted three times with a chloroform–methanol (2:1, v/v) mixture. The mixture was evaporated, leaving behind an oil-like appearance as the crude biosurfactant (Liu et al., 2011).

Microbial enhanced oil recovery (MEOR)

The MEOR process was carried out by the sand pack method described by Abu-Ruwaida et al. (1991). Hydrocarbon saturated

Table 1. Oil displacement activity, emulsification activity and drop collapsing test of cultural supernatant from different-strains.

Strain	Emulsification activity (%)	Oil displacement test (cm^3)	Drop collapsing test	Haemolytic activity	Degreasing effect assay test
B1	65.2 ± 0.26	2.25 ± 0.02	+	+	+
B2	40.0 ± 0.18	1.11 ± 0.02	+	+	+
B3	60.0 ± 0.30	2.30 ± 0.03	+	+	+
B4	50.0 ± 0.30	1.14 ± 0.02	+	+	+
B5	25.0 ± 0.13	1.05 ± 0.01	+	+	+
B6	50.0 ± 0.27	1.13 ± 0.03	+	+	+
B7	28.0 ± 0.15	1.09 ± 0.02	+	+	+

Table 2. Characterization of bacterial strains.

Bacterial isolates	Gram's stains	Cell shape	Spore	Motility	Penicillin sensitivity	Oxidase	Catalase	Urease	Gelatinase	Flurescent pigment
B1 (*Pseudomonas* Sp.)	-	Rods	-	+	-	+	+	-	+	+

sand pack column was treated with the culture filtrate and cell-free supernatant at 30°C. Distilled water served as the control. The oil displacement rate is calculated by given formula.

Oil displacement (%) = $(M_2/M_1) \times 100$

M_1: Oil content in the sand (g)
M_2: Wash out oil content (g).

The assay was repeated three times with three replications for each treatment.

RESULTS

This study revealed that B1 strains detected as *Pseudomonas* sp. out of 15 isolated strains had shown the higher biosurfactant activity. Among the 15 isolates, 7 formed a hydrolyzing oil spot on the oil agar plate, and were considered potential biosurfactant-producing isolates. Moreover, as determined by improved degreasing effect assay, B1 strain showed the highest emulsification activity (65.2 ± 0.26) and oil displacement test (2.25 ± 0.02) as shown in Table 1. This isolate was characterized as Gram negative having a slender rod with rounded ends. The morphology of the colonies, as well as the physiological and biochemical characteristics of the strain, is shown in Table 2. It was therefore found as *Pseudomonas* sp. Bacterium can be screened for rhamnolipid production using CTAB–methylene blue indicator plates. The results of the current work showed a dark blue halo after 48 h of incubation at 30°C, whereas the other strains did not show a positive reaction, demonstrating that the isolate can produce biosurfactant rhamnolipids. The oil displacement rates caused by the fermentation broth and culture filtrate (except for the cell) at 30°C were 68% and 60.2% respectively, whereas those caused by bacterial cell suspensions and distilled

water were 11 and 10%, respectively. The results showed that both the fermentation broth and culture filtrate (except for the cell) produced by *Pseudomonas* sp. was highly effective in recovering crude oil from the sand pack column.

DISCUSSION

Present study revealed that biosurfactant produced by various microbes can reduce pollution of Sea that occurred due to various human activities. Chemical treatment has several disadvantages. Thus biological treatment may be preferred due to their eco-friendly nature, low toxicity, biodegradable and biocompatible and selective (Desai and Banat, 1997). In present study the initial isolation of suspected biosurfactant producers was done on blood agar plates, utilizing the ability of many biosurfactants to lyse erythrocytes, which results in a band of beta hemolysis surrounding biosurfactant-producing bacterial colonies (Bernheimer and Avigad, 1970; Banat, 1995a, b; Lin, 1996). Single screening method is unsuitable for identifying all types of biosurfactants, and recommended that more than one screening method should be included during primary screening to identify potential biosurfactant producers (Kiran et al., 2010). Therefore, drop collapsing test, oil activity assay test and emulsification activity measurement were used to screen the biosurfactant producer. Strain *Pseudomonas* sp. showed positive results in all the screening methods used. Thus, we confirm that this bacterium can produce biosurfactants with positive responses. *Pseudomonas* spp. are known to produce different types of rhamnolipids. For example, production of polymeric biosurfactant by *Pseudomonas nautica* was achieved (Husain et al., 1997). The major constituents were proteins, carbohydrates and lipid at the ratio of 36:63:2, respectivey. A simple method using CTAB-methylene blue indicator plates can be used to screen rhamnolipids produced by a wide range of *Pseudomonas* species, as well as other types of bacteria. We conclude that *Pseudomonas* sp. B1 may produce rhamnolipids, and first assayed the strain by this method. The main advantages of microbiological method of bioremediation of hydrocarbon polluted sites are use of biosurfactant producing bacteria without necessarily characterization of the chemical structure of the surface active compounds. The cell free culture broth containing the biosurfactants can be applied directly or by diluting it appropriately to the contaminated site. The other benefit of this approach is that the biosurfactants are very stable and effective in the culture medium that was used for their synthesis (Płociniczak et al., 2011).

Conclusion

Biosurfactants have been proven as one of the promising agent for controlling the oil pollution by hydrocarbons, particularly oil polluted in marine environment. Synthetic detergents used to clean up these spillages have often led to more destruction of the environment. From the environmental view point it is important that all substances released into the environment are degradable, firstly to assess their potential for causing environmental damage and secondly to safeguard against the possibility of future harm due to build up in the environment. Therefore, potent biosurfactant producing microorganisms should be intensively isolated and screened for bioremediation without causing adverse effects to the environment.

ACKNOWLEDGEMENT

The authors are grateful to Dr. Shruti Mathur, HOD of Mirobiology Department, ICG Jaipur, India for cooperating while doing this work.

REFERENCES

Abu-Ruwaida AS, Banat IM, Hadirto S, Saleem A, Kadri M (1991). Isolation of biosurfactant producing bacteria-product characterization and evaluation. Acta. Biotechnol. 4:315-324.

Banat IM (1995a). Biosurfactants Production and Possible Uses in Microbial-Enhanced Oil Recovery and Oil Pollution Remediation: A Review. Bioresour. Technol. 51:1-12.

Banat IM (1995b). Characterization of Biosurfactants and Their Use in Pollution Removal – State of the Art (Review). Acta Biotechnol. 3:251-267.

Bernheimer AW, Avigad LS (1970). Nature and Properties of a Cytolytic Agent Produced by *Bacillus subtilis*. J. Gen. Microbiol. 61:361-369.

Bertrand JC, Bonin P, Goutx M, Mille G (1993). Biosurfactant production by marine microorganisms: Potential application to fighting hydrocarbon marine pollution. J. Mar. Biotechnol. 1:125-129.

Bi SN, Wang YJ, Zuo YH (2009). Improvement and application of oil spreading to detect biosurfactant. J. Heilongjiang. Bayi. Agric. Univ. 6:58-60.

Bryant RS (1987). Potential uses of microorganisms in petroleum recovery technology. Proc. Okla. Acad. Sci. 67:97-104.

Buchanan RE, Gibbons NE (1974). Bergey Manual of Determinative Baeteriology (8th ed) Baltimore: The Wolliams & Wilkins Company, pp. 529-545.

Cameotra SS, Makkar RS (1998). Synthesis of biosurfactants in extreme conditions. Appl. Microbiol. Biotechnol. 50:520-529.

Carrilo PG, Mardaraz C, Pitta – Alvarez SI, Giulietti AM (1996). Isolation and selection biosurfactant–producing bacteria. World J. Microbiol. Biotechnol. 12:82-84.

Desai JD, Banat IM (1997). Microbial production of surfactants and their commercial potential. Microbiol. Mol. Biol. Rev. 61:47–64.

Fietcher A (1992). Biosurfactant moving toward industrial application. Tibtech. 10:208-217.

Georgiou G, Lin SC, Sharma MM (1990). Surface active compounds from microorganisms. Bio/Technol. 10:60-65.

Husain DR, Goutx M, Acquaviva M, Gilewicz M, Bertrand JC (1997). The effect of temperature on eicosane substrate uptake modes by a marine bacterium *Pseudomonas nautical* strain 617: Relationship with the biochemical content of cells and supernatant. World J. Microbiol. Biotechnol. 13:587-590.

Jinfeng L, Lijun M, Bozhong M, Rulin L, Fangtian N, Jiaxi Z (2005). The field pilot of microbial enhanced oil recovery in a high temperature petroleum reservoir. J. Pet Sci. Eng. 48:265-271.

Joshi S, Bharucha C, Jha S, Yadav S, Nerurkar A, Desai AJ (2008). Biosurfactant production using molasses and whey under thermophilic conditions. Biores. Technol. 99:195-199.

Kiran GS, Thomas TA, Selvin J, Sabarathnam B, Lipton AP (2010). Optimization and characterization of a new lipopeptide biosurfactant produced by marine *Brevibacterium aureum* MSA13 in solid state culture. Bioresour. Technol. 101:2389–2396.

Lang S, Wullbrandt D (1999). Rhamnolipids-biosynthesis, microbial production and application potential. Appl. Microbiol. Biotechnol. 51:22–23.

Lin S (1996). Biosurfactants: Recent advances. J. Chem. Tech. Biotechnol. 66(2):109-120.

Liu T, Hou J, Zuo Y, Bi S, Jing J (2011). Isolation and characterization of a biosurfactant producing bacterium from Daqing oil-contaminated site. Afr. J. Microb. Res. 5(21):3509-3514.

Mei S, Wei L, Guangzhi L, Peihui H, Zhaowei H, Xinghong C, Ying W (2003). Laboratory study on MEOR after polymer flooding. SPE84865. In: Proceedings of the SPE International Improved Oil Recovery Conference in Asia Pacific, October 20-21, Kuala Lumpur.

Morikawa M, Daido H, Takao T, Marato S, Shimonishi Y, Imanaka T (1993). A new lipopeptide biosurfactant produced by Arthrobacter sp. strain MIS 38. J. Bacteriol. 175:6459–6466.

Płociniczak MP, Płaza GA, Seget ZP, Cameotra SS (2011). Environmental Applications of Biosurfactants: Recent Advances. Int. J. Mol. Sci. 12:633-654.

Rashedi H, Jamshidi E, Mazaheri Assadi M, Bonakdarpour B (2005). Isolation and production of biosurfactant from *Pseudomonas aeruginosa* isolated from Iranian southern wells oil. Environ. Sci. Technol. 2:121-127.

Siegmund I, Wagner F (1991). New method for detecting rhamnolipids excreted by *Pseudomonas* species during growth on mineral agar. Biotechnol. Technol. 4:265–268.

Toledo FL, Gonzalez J, Calvo C (2008). Production of bioemulsifier by *Bacillus subtilis*, *Alcaligenes faecalis* and *Enterobacter* species in liquid culture. Bioresour. Technol. 99:8470-8475.

Youssef NH, Dunacn KE, Nagle DP, Savage KN, Knapp RM, McInerney MJ (2004). Comparison of methods to detect biosurfactant production by diverse microorganism. J. Microbiol. Methods 56:339–347.

Zajic JE, Stiffens W (1994). Biosurfactants. CRC Rev. Biotechnol. 1:87-106.

18

Removal of polyaromatic hydrocarbons from waste water by electrocoagulation

Ehssan Mohamed Reda Nassef

Petrochemical Engineering Department, Faculty of Engineering, Pharos University, Canal El Mahmoudia Street, Beside Green Plaza Complex Alexandria, Egypt.

In the present study, two important groups of factors were studied; the first one corresponds to the Electrocoagulation (EC) process parameters such as current density, pH, electrolysis time and electrolyte concentration that were investigated in terms of their effects on the removal efficiencies of polyaromatic hydrocarbons, specially β-naphthols. The second group of factors analyses the effect of three different anode geometries on the percentage removal of β-naphthols. The present study achieved the following results: The optimum conditions for the removal of β-naphthols for cell No. 1 and 2 were achieved at current density 20 mA/cm², initial pH 7 treated volumes 1.5 L, NaCl concentration 1 g/L and temperature 25°C. Meanwhile, the optimum conditions for cell No. 3 achieved at 106 mA/cm², initial pH = 7, NaCl concentration = 1 g/L and temperature 25°C, treated volume = 0.5 L. The maximum separation efficiency was achieved for cell No. 1 and the lowest for cell No. 2.

Key words: PolyaromaticHydrocarbons, β-naphthols, electrocoagulation, waste water treatment.

INTRODUCTION

Polycyclic aromatic hydrocarbons (PAHs) are composed of "wo or more fused aromatic (benzene) rings". However, "fused aromatic rings" is probably the best definition (Do and Chen, 1994). PAHs are to a certain degree resistant to biodegradation and are sometimes included in a class of persistent organic pollutants (POPs) (Sengil et al., 2004). PAHs can be present in both particulate and gaseous phases, depending upon their volatility. Light molecular weight PAHs (LMW PAHs) that have two or three aromatic rings are emitted in the gaseous phase, while high molecular weight PAHs (HMW PAHs), with five or more rings are emitted in the particulate phase (Aakinson et al., 1990), α- and β-naphtols

can be considered as the derivatives of naphthalene which is the simplest polycyclic aromatic hydrocarbon (PAH).

PAHs can be formed from both natural and anthropogenic sources, though the anthropogenic sources contribute most to the hazards associated with PAHs (Marian, 2010; Baek et al., 1991). They are usually found as a mixture containing two or more of these compounds and commercially available pure PAHs are usually colourless, white or pale yellow - green solids which are odorless or have a faintly pleasant odor (Busetti et al., 2006). Some PAHs are used in medicines, dyes, plastics and pesticides. Others are contained in asphalt used in road construction as well as found in substances such as crude oil, coal, coal tar pitch, creosote and roofing tar (ATSDR, 1995). Due to the widespread release of PAHs which consequently results in considerable health and environmental hazards (Nazimek and Ćwikła, 2004), the European Union as well as the United States Environmental Protection Agency (USEPA) have specified permissible limits for the 16 priority PAHs (Fang et al., 2002).

The major sources of PAH emissions may be divided into four classes: stationary sources including domestic (Chen et al., 2007; Ravindra and Sokh, 2008; Fabbri and Vassura, 2006) and industrial sources (Gerhard et al., 2009; Fan et al., 1995), mobile emission (Guo et al., 2003), agriculture activities (Biswas and Lazarescu, 1991; Lin and Peng, 1994), and natural sources (Pakpahan et al., 2009; Lui et al., 2001). There are different methods for the separation of PAHS: the thermal treatment of polycyclic aromatic hydrocarbons under controlled conditions is one of the more effective methods of degradation (Eriksson et al., 2003; Summers and Snoeyink, 1999; Rajeshwar et al., 1994; Golder et al., 2007), adsorption or sorption (Arlette et al., 2012; Drogui et al., 2007; Ebubekir et al., 2013), direct photolysis (Khandegar and Anil, 2013), and electrocoagulation (Pons et al., 2005; Modirshahla et al., 2008; Emamjomeh and Sivakumar, 2009; Mohd Salleh et al., 2011; Kliaugaitė et al., 2013).

Electrocoagulation has the potential to extensively eliminate the disadvantages of the classical treatment techniques. Moreover, the mechanisms of EC are yet to be clearly understood and there has been very little consideration of the factors that influence the effective removal of ionic species, particularly metal ions from wastewater by this technique. The mechanism of coagulation has been the subject of continual review (Modirshahla et al., 2008; El-Ashtoukhy et al., 2013). The theory of EC has been discussed by many authors (Daneshvar et al., 2006).

ELECTROCOAGULATION THEORY

Electrocoagulation (EC) is a technology in which the coagulant is generated *in situ* by oxidation of a metal anode material when applying electrical current (Golder et al., 1994). As well known, Al^{3+} ions dissolve and combine with hydroxyl ions in water (Daneshvar et al., 2003), when direct current passes through the Al anodes, they form metal hydroxides which are partly soluble in the water under definite pH values.

Additionally, electrolytic reactions evolve gas (usually as hydrogen bubbles) at the cathode that can enhance the process; this effect is known as electroflotation which results in better removal of contaminants (Linares-Hernández et al., 2009; Sengil, 2008). Electrocoagulation has occurred in three steps. In first step, coagulant has

formed because of oxidation of anode. In second step, pollutants have destabilized. In last step, destabilized matters have united (Kobya and Delipinar, 2008). The most common electrode materials for electrocoagulation are aluminum and iron. They are cheap, readily available, and proven effective (Un and Aytac, 2013). When aluminum is used as electrode material, the reactions are as follows:

At the anode: $Al \rightarrow Al^{3+} + 3e$
(Equation 1)

At the cathode: $2H_2O + 2e^- \rightarrow H_{2\,(g)} + 2OH^-$
(Equation 2)

When the anode potential is sufficiently high, secondary reactions may occur, especially oxygen evolution (Equation 3):

$$2H_2O \rightarrow O_2 + 4H^+ + 4e-$$
(Equation 3)

Aluminum ions (Al^{3+}) produced by electrolytic dissolution of the anode (Equation 1) immediately undergoes spontaneous hydrolysis reactions which generate various monomeric species according to the following sequence (Pulkka et al., 2014):

$$Al^{3+} + H_2O \rightarrow Al\,(OH)_2^+ + H^+$$
(Equation 4)

$$Al(OH)_2^+ + H_2O \rightarrow Al(OH)_2^+ + H^+$$
(Equation 5)

$$Al(OH)_2^+ + H_2O \rightarrow Al(OH)_3 + H^+$$
(Equation 6)

The overall reaction in the solution:

$$Al^{3+}_{\,(aq)} + 3H_2O \rightarrow Al(OH)_3 + 3H^+_{\,(aq)}$$
(Equation 7)

EC process like other treatment methods has some advantages and disadvantages. Main disadvantages of this method are lack of a systematic approach to EC reactor design and operation, replacement of electrodes at regular intervals, high cost of electricity and anode passivation (Behbahani et al., 2011).

On the other hand, main advantages of EC are: the simplicity of the equipment (Linares-Hernandez et al., 2009), no need of additional chemical matter after or before treatment, relatively low area demand, also sludge from this process is intensive and has low water (El-Naas et al., 2009).

For these reasons the present work is focused on the EC

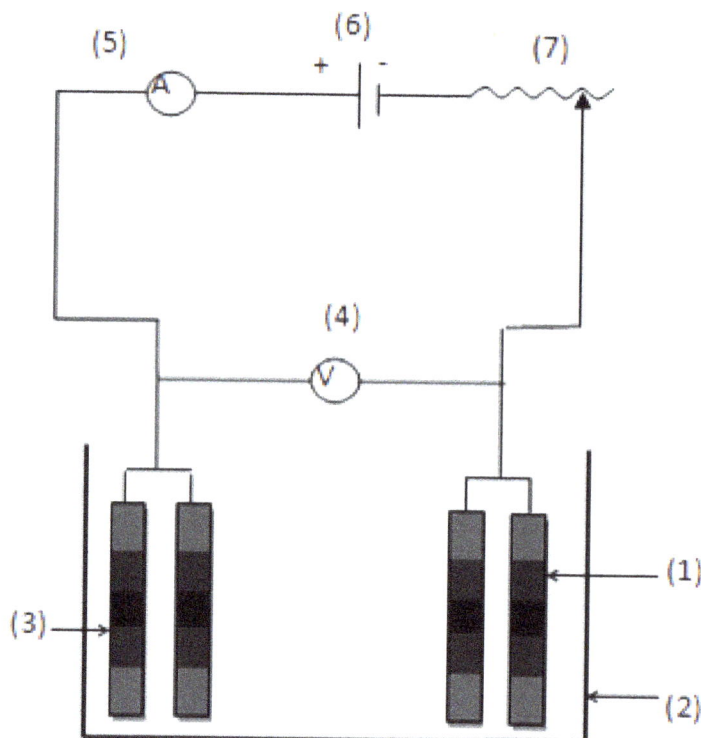

Figure 1. A schematically. Diagram for Cell No. 1 (1) Al anode, (2) Plexiglas stank, (3) Al cathode, (4) Voltammeter, (5) Ampere meter, (6) Power supply and (7) Resistance.

processes using a sacrificial aluminum anode to remove PAH from waste water by studying variables like current density, initial concentration, time, temperature, pH, and electrolyte concentration (NaCL). The objective of this study is to remove β-naphthol which is very dangerous to people and environmental. The removal technique done in this work is electrocoagulation method. The main objective is to examine the different parameters which affect the PAH removal. These parameters are the initial concentration of the solution, the pH of the solution, the contact time and the current density.

EXPERIMENTAL PART

Experimental set-up

The three experimental set-ups used in the present work are schematically shown in Figures 1, 2 and 3. The EC Cell No. 1 consisted mainly of a rectangular vessel made of plexi glass with dimensions 10×10 cm^2 base and a height of 20 cm (Figure 1). A two perforated aluminum sheets (10×20 cm) which placed vertically in the vessel and operate as an anode, the anode-cathode distance was kept at 4 cm. The other cell (No. 2) made of the same rectangular vessel with the same dimensions, but with different anode geometry which consists of two smooth rods of aluminum sheets (10×20 cm). The anode-cathode distance was kept also at 4 cm as shown in (Figure 2). Meanwhile, the third cell consists of a cylinder vessel made of plexi glass with diameter 6 cm and a height of 10 cm (Figure 3). A perforated aluminum sheet with diameter (5.5

cm) was isolated with epoxy from its back and then placed in plexi glass cylinder and acts as the cathode, while the anode in the bottom consists of a perforated cylindrical connected together with a thin wire of aluminum to make sure that the layer of anode was fed with the electrical current. The anode-cathode distance was kept at 2 cm. The electrical circuit of each cell consists of power supply (10V, 5A) with a voltage regulator and multi range ammeter, all connected in series with the cell; a voltmeter was connected in parallel with the cell to measure its voltage.

Reagents and analytical procedures

Stock solution of β-naphthols compound was prepared by using an analytical grade of chemicals and dissolving it in distilled water. Experimental solutions of desired concentrations were obtained by successive dilution with distilled water. The pH of the solution was adjusted by means of HCl and/or NaCl solutions. A digital calibrated pH-meter (Hanna, Model pH 211) was used to measure the pH of waste solutions. The analytical determination of β-naphthols was carried out with the standard spectrophotometric procedure using U. V. spectrophotometer (UNICO, Model U.V 2100) and measured at wave length 320 μm. At the end of each experiment, the treated solutions were filtered by using Whitman No. 40 filter paper and centrifuged to remove any traces before analysis.

Electrocoagulation procedure

For each run, a known volume of synthetic waste solution was mixed with the appropriate amount of sodium chloride which was used as a supporting electrolyte. The solutions were placed into the

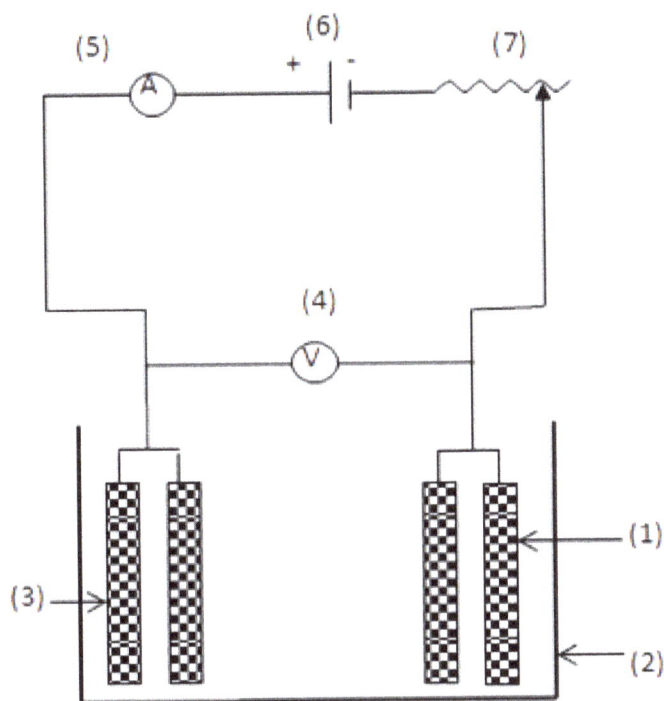

Figure 2. A schematically. Diagram for Cell No. 2 (1) Al anode, (2) Plexiglas stank, (3) Al cathode, (4) Voltammeter, (5) Ampere meter, (6) Power supply and (7) Resistance.

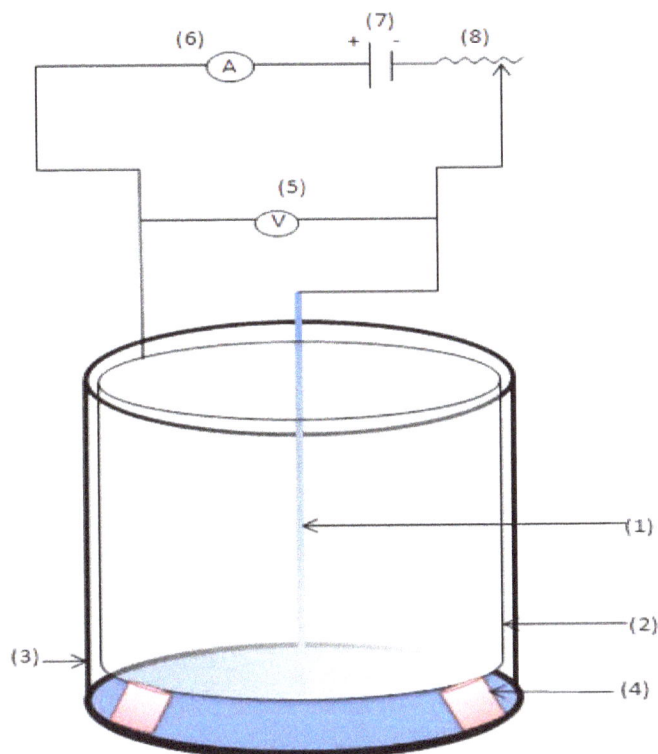

Figure 3. A schematically. Diagram for Cell No. 3 (1) Al anode, (2) Aluminum Cathode, (3) Plexi-glass tank, (4) Epoxy from plexi-glass, (5) Voltammeter, (6) Ammeter and (7) Power supply (8) Resistance.

electrolytic cell. The pH was adjusted by the addition of NaOH and/or HCl solutions. Direct current from the D. C. power supply was passed through the solution via the two electrodes during 120 min electrolysis; 10 ml of the solution was drawn at specified interval time during the experiment. The location of the drawn samples was kept constant for each run. Samples were filtered and centrifuged, and then taken for absorbance measurements at an appropriate wavelength of the maximum absorption for β-naphthols. The measured absorbance was then converted to the residual concentration of the compound using a calibration curve obtained from a plot between the absorbance versus the concentration. The electrodes were washed with HCl solution (15% w/v) before each run in order to remove any adhering scales or oxides. Following each run, the electrodes were washed with distilled water, dried until the second use. The efficiency of β-naphthol removal (% Removal) was calculated as:

$$\% \ Removal = \frac{Ci - Cf}{Ci} * 100 \qquad (1)$$

Where C_i is the initial β-naphthols concentration (mg/l) and C_f is the final β-naphthol concentration (mg/l).

RESULTS AND DISCUSSION

Effect of electrolysis time

Figures 4 to 6 show the effect of electrolysis time on the removal efficiency of β-naphthols at a current density of 15 mA/cm² for cells No. 1 and 2), and current density

of 106 mA/cm² for cell No. 3. As the duration of the electrolysis treatment increased, a comparable increase in the removal efficiency of β-naphthols was observed. It was also observed that as the reaction time increased from 30 to 180 min, the removal efficiency increased from 27 to 77%, from 20.8 to 58.3% and from 25.0 to 72.9% for Cells 1, 2 and 3, respectively. This is due to that, as the time of electrolysis increased, the time of mixing and reaction also increased (Golder et al., 1994). This is ascribed also to the fact as the time increased, more hydrogen bubbles were generated at the cathode; these bubbles improved the degree of mixing and enhanced the flotation ability of the cell with a consequent increase in the percentage removal (Mouedhen et al., 2008; Zaroual et al., 2009).

Effect of current density

Another important parameter influencing the performance and the economy of the electrocoagulation process is the density of current applied at the electrodes (Lacasa et al., 2011). To study the effect of current density on the efficiency of electrocoagulation in removal of β-naphthols, the experiments carried out from 2.5 to 20 mA/cm². From

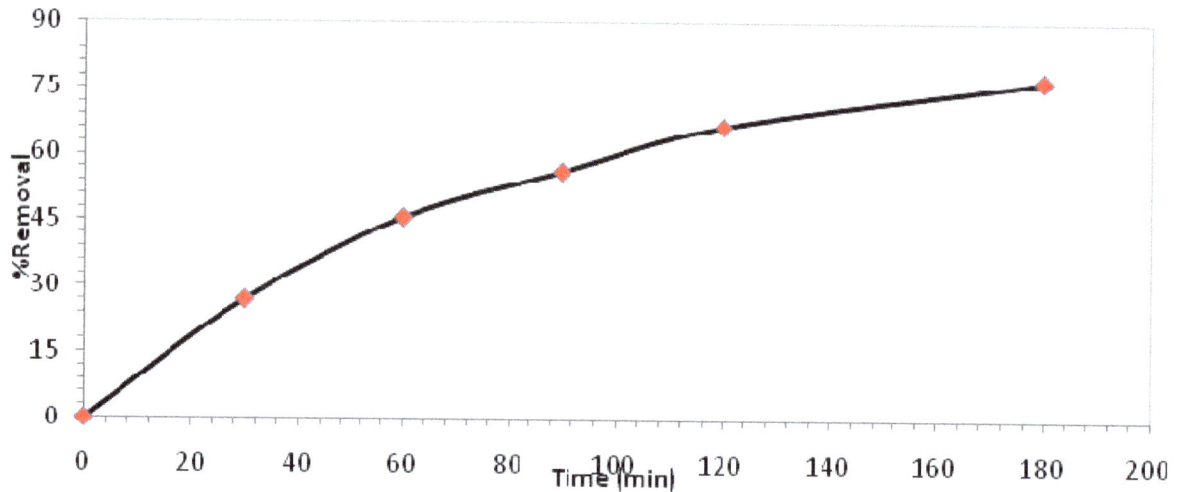

Figure 4. Effect of electrolysis time on the removal efficiency (Current density 20 mA/cm^2, Initial pH7, initial concentration 40 ppm, NaCl concentration = 1 g/l, temperature 25°C, Cell No. 1).

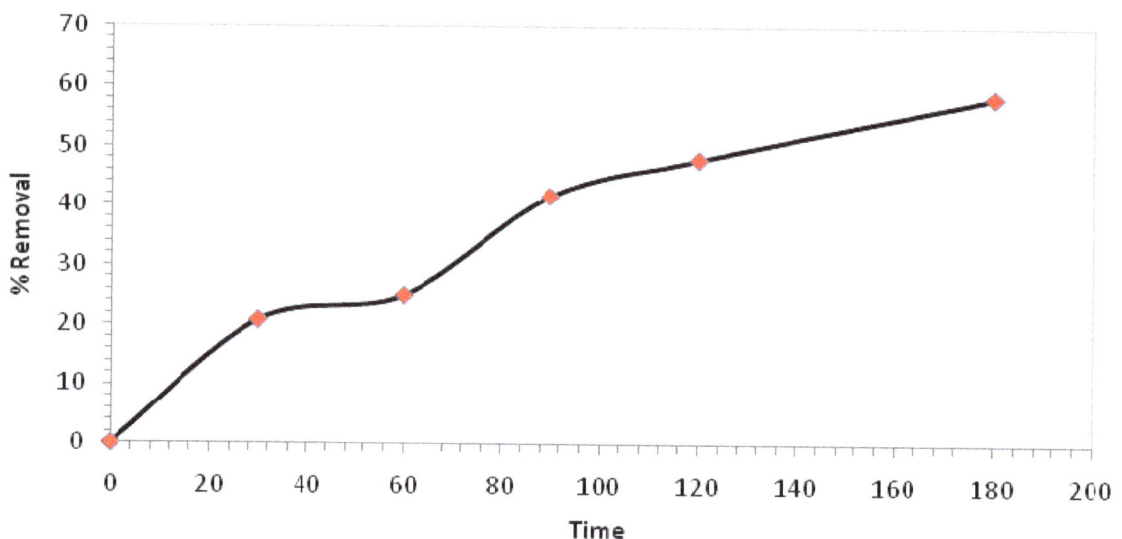

Figure 5. Effect of electrolysis time on the removal efficiency (At Current density 20 mA/cm^2, initial pH7, Initial concentration 40 ppm, NaCl Concentration = 1 g/l, temperature 25°C, cell No. 2).

Figures 7 and 8, it was observed that as the current density increased from 2.5 to 20 mA/cm^2, the efficiency of electrocoagulation increased from 25 to 79% and from 16.7 to 66.6% for Cells 1 and 2, respectively. But for Cell No. 3, the same trend was observed, but with higher values of current densities due to the difference in the area of the cathode as shown in Figure 9. This trend was observed for the three cells due to the fact that as the current density increased, the formation of flocs of aluminum hydroxide at the anode of the cell increased and this enhanced the rate of adsorption of β-naphthols on the flocs of aluminum hydroxide. Also, by increasing the current density of the cell, the amount of hydrogen bubbles at the cathode increased, resulting in a greater upward flux and a faster removal of β-naphthols and sludge flotation (Kurt et al., 2008; Khandegar et al., 2013). At high current density, the quantity of Al^{3+} from anode dissolution increases according to Faraday's low. An increase in current density above the optimum current density does not result in an increase in the pollutant removal efficiency as sufficient number of metal hydroxide flocs are available for the sedimentation of the pollutant.

Effect of pH

Solution pH is one of main factors affecting electrochemical

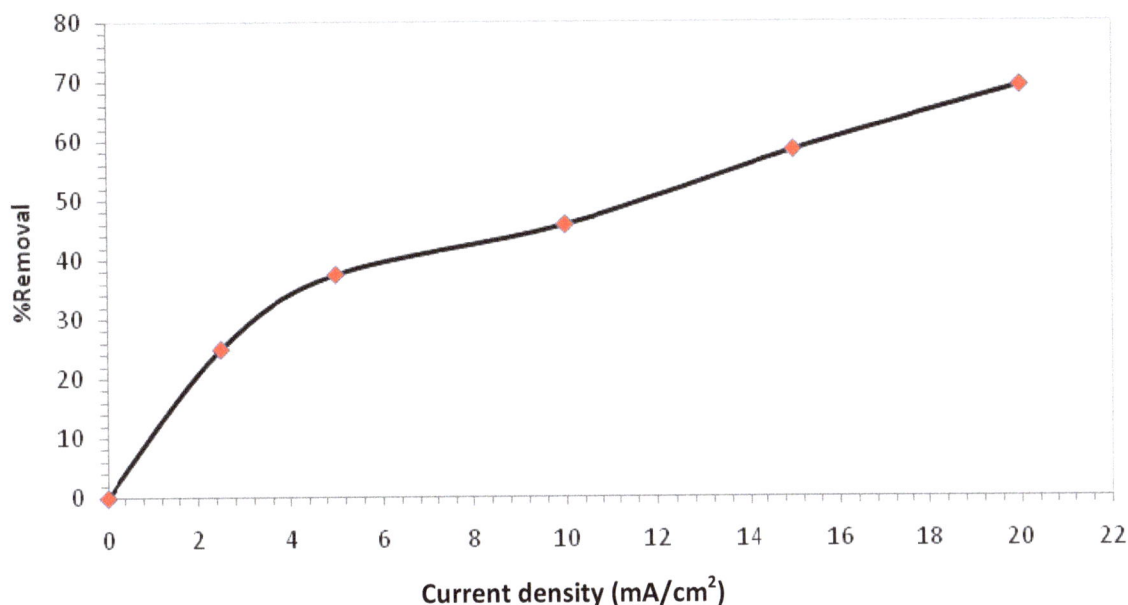

Figure 7. Effect of current density on the removal efficiency (Initial pH7, initial concentration 40 ppm, NaCl concentration 1 g/l, temperature 25°C, time 180 min, cell No. 1).

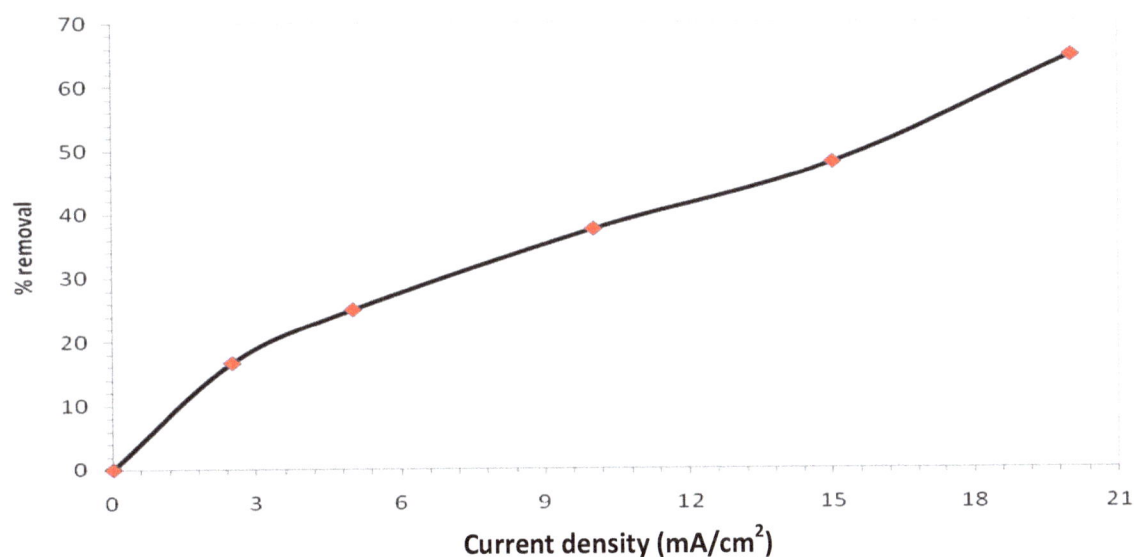

Figure 8. Effect of current density on the removal efficiency (Initial pH7, Initial concentration 40 ppm, NaCl concentration 1 g/l, temperature 25°C, time 180 min, cell No. 2).

processes. Therefore, pH (in the range of 1 to 11) was examined as one of the main variables affecting electrocoagulation removal of β-naphthols from waste water. The results of all experiments of the three cells showed the same trend according to the change in pH as was shown in Figure (10). The trend was that, when the pH was increased from 1 to 7, the percentage removal increased and then decreased for a pH between 9 and 11. This finding supports that electrocoagulation efficiency is a function of pH and these results are in accordance with other researchers who have reported the maximum performance for the ECP at pH between 7 and 8 when using Fe or St.St as the sacrificial electrode (Pulkka et al., 2014; Behbahani et al., 2011). The lower percentage of removal at higer acidic and alkaline pH can be explained by amphoteric behavior of $AL(OH)_3$ which

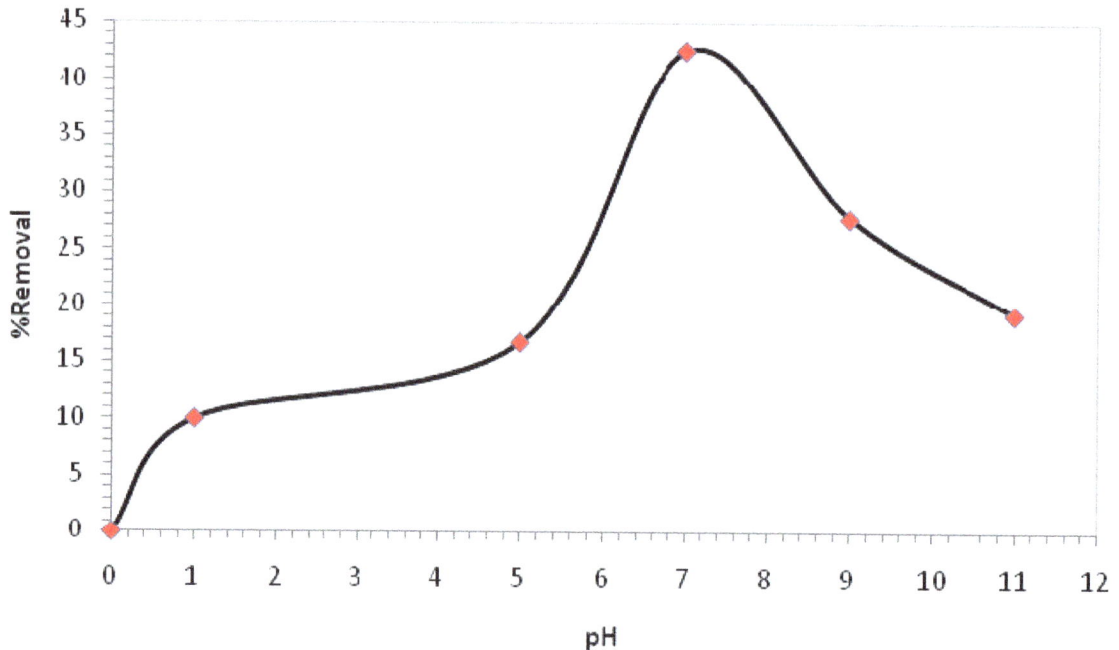

Figure 10. Effect of pH on the removal efficiency (Current density 20 mA/cm^2, initial concentration 40 ppm, NaCl concentration 1 g/L, temperature 25°C, time 180 min, Cell No. 2).

does not precipitate at very low pH. The figure show the effluent pH after EC treatment would increase for acidic influent but decrease for alkaline influent.

Effect of sodium chloride concentration

Table salt is usually employed to increase the conductivity of the water or wastewater to be treated. Besides its ionic contribution in carrying the electric charge, it was found that chloride ions could significantly reduce the adverse effect of other anions such as HCO_3, SO_4^{2-}. The existence of the carbonate or sulfate ions would lead to the precipitation of Ca^{2+} or Mg^{2+} ions that form an insulating layer on the surface of the electrodes. This insulating layer would sharply increase the potential between electrodes and result in a significant decrease in the current efficiency. It is therefore recommended that among the anions present, there should be 20% Cl$^-$ to ensure a normal operation of electrocoagulation in water treatment. The addition of NaCl would also lead to the decrease in power consumption because of the increase in conductivity (Linares-Hernandez et al., 2009). The same trend was also observed for all cells as shown for one of them in Figure 11). From this figure, it was observed that with increasing initial concentration of Nacl from 0.5 to 5 g, the percentage removal increased. But it was also observed that; there is a slight change in results beyond 1 g/L of NaCl; so, the optimum amount of the electrolyte was taken to be 1 g/L. Previous works shows a similar results (El-Naas et al., 2009; Mouedhen et al.,

2008) variety of electrolytes like NaCl, KCl, $NaNO_3$, $NaNO_2$, $NaSO_4$, etc., are available. But, due to low cost and easy availability, NaCl has been selected as the best electrolyte (Zaroual et al., 2009).

Comparison between the anode geometry of three types of cells

Figures 12 to 14 shows the comparison of the three anodes geometry of cells was studied in case of the effect of electrolysis time, pH, and NaCl concentration on the percentage removal of the β-naphthols. As indicated from the previous Figures, the anode geometry of Cell No. 2 which has a cylindrical perforated aluminum anode produced the lowest percentage removal of 55%. While the anode geometry of Cell No. 1 which has a perforated rectangular aluminum sheet as an anode with larger area produced the highest percentage removal of 88% at similar experimental conditions .The geometry of the reactor affects operational parameters including bubble path, flotation effectiveness, floc formation, fluid flow regime and mixing/settling characteristics. From the literature, the most common approach involves plate electrodes (aluminum or iron) and continuous operation (El-Naas et al., 2009). From previous literature (Mouedhen et al., 2008; Zaroual et al., 2009; Lacasa et al., 2011; Kurt et al., 2008), it was found that the perforated anode sheet with larger surface area achieved the maximum percentage removal than the non-perforated anode; this is the same as results obtained

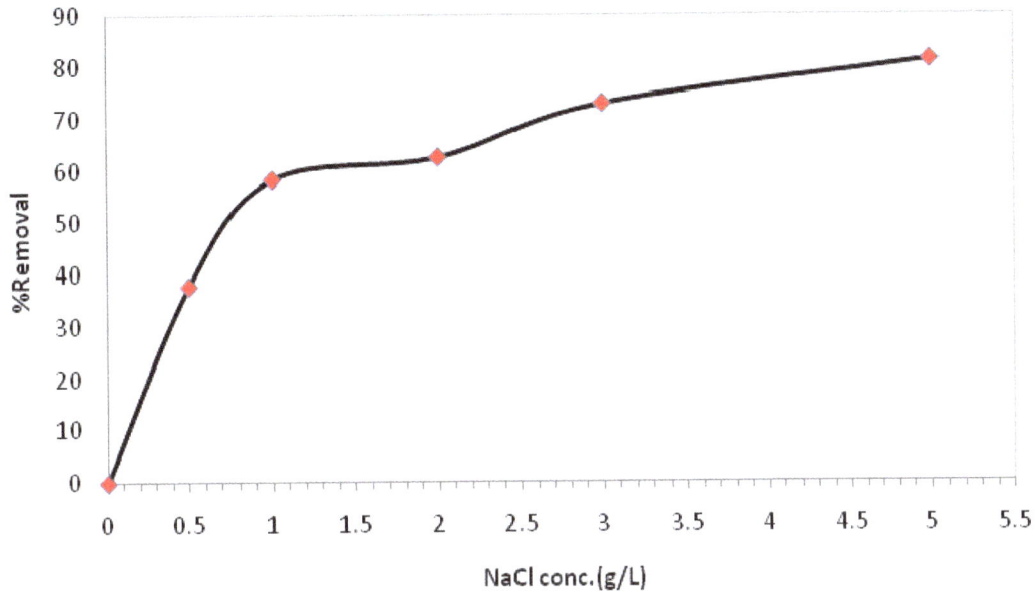

Figure 11. Effect of NaCl Conc. on the removal efficiency (Current density 20 mA/cm^2, initial pH7, initial concentration 40 ppm, temperature 25°C, time 180 min., Cell No. 1).

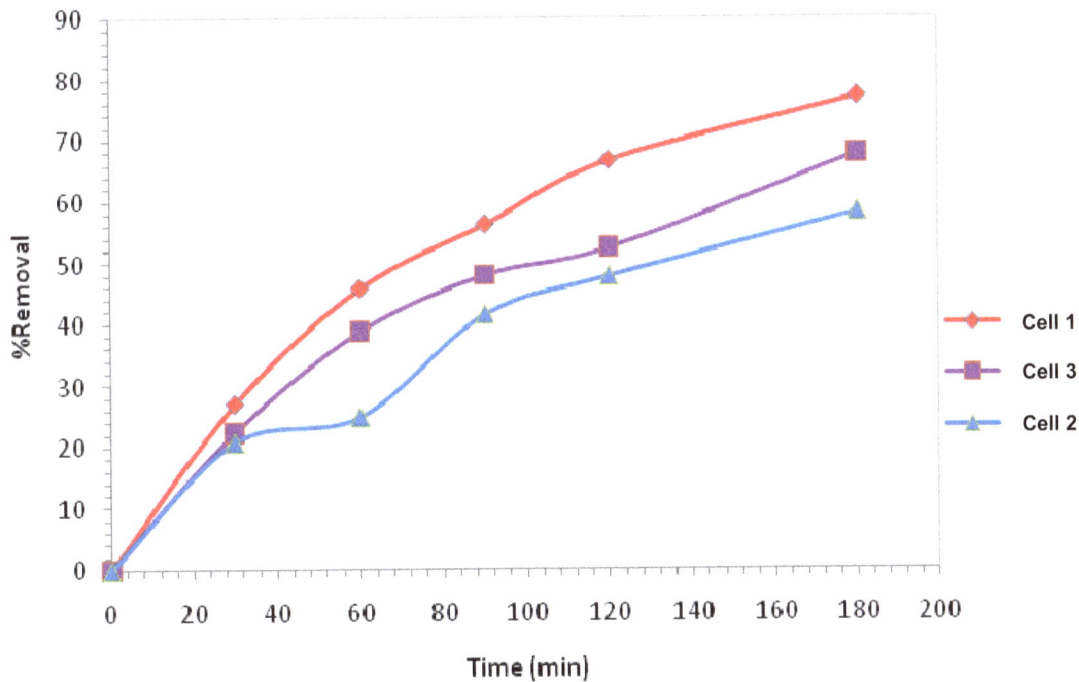

Figure 12. Comparison between the 3 cells in electrolysis time (Current 4A, initial pH7, initial concentration 40 ppm, NaCl concentration 1 g/l, temperature 25°C).

from this study. The shape of the electrodes affects the pollutant removal efficiency in the electrocoagulation process. It is expected that the punched holes type electrodes will result in higher removal efficiency compared to the plane electrodesn (Khandegar et al., 2013) as indicated in this thesis. They have reported higher discharge current for the electrode with punched holes than for plane electrode resulting in higher collection efficiency with punched electrode compared with plane electrode. The electric field intensity at the

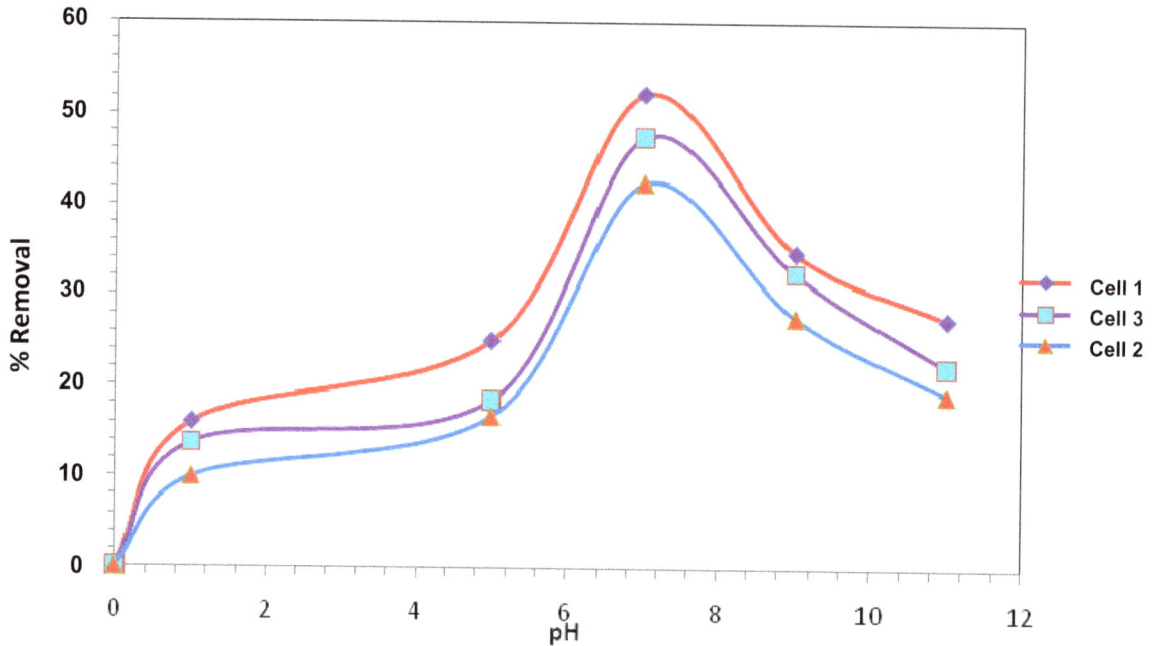

Figure 13. Comparison between the three cells (pH) (Current 4A, initial concentration 40 ppm NaCl concentration 1 g/L, temperature 25°C, time 180 min).

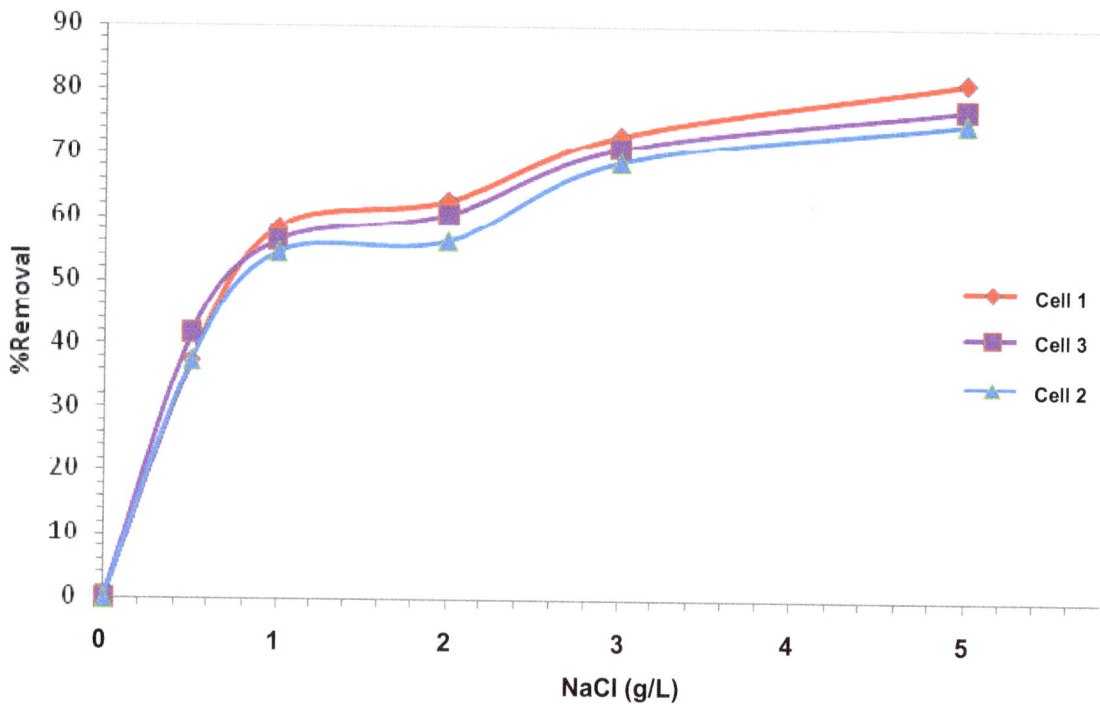

Figure 14. Comparison between the three cells (NaCl) (Current 4A, initial pH7, initial concentration 40 ppm, temperature 25°C, time 180 min).

edge of punched holes type electrodes is higher (1.2 times) than at plane type electrode resulting in an increase in the discharge current at punched type electrode.

Conclusions

Electrocoagulation has successfully treated a wide range of waste streams. Previous research has focused on the application of electrocoagulation to a particular situation (plant and waste stream). The results of this study indicate that EC is a promising treatment for the removal of polyaromatic hydrocarbons (β-naphthols) from waste water. The following conclusions could be drawn from this experimental study:

i. Experimental parameters such as current density, electrolysis time, pH, and electrolyte concentration were investigated for β-naphthols compound removal in batch mode apparatus with three different anode geometries.

ii. The maximum removal of (β-naphthols) was attained at current density 20 mA/cm^2, initial pH7, treated volume 1.5 L, NaCl concentration 1 g/L and temperature of 25°C in Cell 1.

iii. Increasing the current density and NaCl concentration resulted in an increase in the β-naphthols removal.

iv. It was found that Cell No. 1 has a higher efficiency of removal than Cell No. 2 and 3 which has the perforated shape anode.

Conflict of Interest

The authors have not declared any conflict of interest.

REFERENCES

Aakinson R, Arey J, Zielinska B, Aschmann SM (1990). Kinetics and nitro-products of gas-phase OH and NO3 radical-initiated reactions of naphthalene, fluoranthene and pyrene. Int. J. Chem. Kinetics 22:999-1014.
http://dx.doi.org/10.1002/kin.550220910
ATSDR (1995). Agency for Toxic Substances and Disease Registry. US Department of Health andHumanServices,Atlanta, GA.
Arlette M, Cardenas P, Jorge G, Ruben VM (2012). Determination of the point of zero charge for electrocoagulation precipitates from an Iron Anode: Hazard. J. Mater. 7:6142-6153.
Baek M, Goldstone P, Kirk J, Lester R, Perry P (1991). Phase distribution and particle size dependency of polycyclic aromatic hydrocarbons in the urban atmosphere. Chemosphere. J. 22:503-520.
http://dx.doi.org/10.1016/0045-6535(91)90062-I
Behbahani M, Moghaddam AMR, Arami M (2011). "A Comparison Between Aluminum and Iron Electrodes on Removal of Phosphate from Aqueous Solutions by Electrocoagulation Process". Int. J. Environ. 5(2):403-412.
Biswas N, Lazarescu G (1991). Removal of oil from emulsions using electrocoagulation, Int. J. Environ. Stud. 38:65-72.
http://dx.doi.org/10.1080/00207239108710650
Busetti FM, Heitz A, Cuomo M, Badoer S, Traverso P (2006).Determination of sixteen polycyclic aromatic hydrocarbons in aqueous and solid samples from an Italian wastewater treatment plant. A. Chromatogr. J. 1102(1-2):104-115
http://dx.doi.org/10.1016/j.chroma.2005.10.013
PMid:16256127
Chen S, Su B, Chang JE, Lee WJ, Huang KL, Hsieh LT, Huang JC, Lin WJ, Lin CC (2007). Emissions of polycyclic aromatic hydrocarbons (PAHs) from the pyrolysis of scrap tires. Atmosphere. J. Environ. 41:1209-1220.
http://dx.doi.org/10.1016/j.atmosenv.2006.09.041
Daneshvar N, Ashassi-Sorkhabi H, Tizpar A (2003). Decolorization of orange II by electrocoagulation method, Sep. Purif. J. Technol. 31:153-162.

http://dx.doi.org/10.1016/S1383-5866(02)00178-8
Daneshvar N, Oladegaragoze A, Djafarzadeh N (2006). Decolorization of basic dye solutions by electrocoagulation: an investigation of the effect of operational parameters, Hazard J. Mater. B. 129:116-122.
http://dx.doi.org/10.1016/j.jhazmat.2005.08.033
PMid:16203084
Do JS, Chen ML (1994). Decolorization of Dye Containing Solutions by Electrocoagulation. Appl. J. Electrochem. 24:785-790.
http://dx.doi.org/10.1007/BF00578095
Drogui P, Asselin M, Benmoussa SK, Blais JF (2007). Electrochemical removal of pollutants from agro-industry wastewaters. Sep. J. Purif. Technol. 61(3):301-310.
http://dx.doi.org/10.1016/j.seppur.2007.10.013
Ebubekir Y, Murat E, Ercan G (2013). Electrochemical treatment of colour index reactive orange 84 and textile wastewater by using stainless steel and iron electrodes, Envron. J. Process sustainable Energy. 32(1):60-68.
El-Ashtoukhy ESZ, El-Taweel YA, Abdelwahab O, Nassef EM (2013). Treatment of Petrochemical Wastewater Containing Phenolic Compounds by Electrocoagulation Using a Fixed Bed Electrochemical Reactor. Int. J. Electrochem. Sci. 8:1534.
El-Naas MH, Al-Zuhair S, Al-Lobaney A, Makhlouf S (2009). "Assessment of electrocoagulation for the treatment of petroleum refinery wastewater", J. Environ. Manage. 91(1):180-185.
http://dx.doi.org/10.1016/j.jenvman.2009.08.003
PMid:19717218
Emamjomeh MM, Sivakumar M (2009). Review of pollutants removed by electrocoagulation and electrocoagulation/flotation processes. Environ J. Manage. 90(5):1663.
http://dx.doi.org/10.1016/j.jenvman.2008.12.011
PMid:19181438
Eriksson M, Sodersten EY, Mohn WM, Wheatley AD, Sadhra S (2003). Degradation of Polycyclic Aromatic Hydrocarbons at Low Temperature under Aerobic and Nitrate-Reducing Conditions in Enrichment Cultures from Northern Soils., Appl. Environ. J. Microbiol. 69:275-279.
http://dx.doi.org/10.1128/AEM.69.1.275-284.2003
PMCid:PMC152444
Fabbri D, Vassura I (2006). Evaluating emission levels of polycyclic aromatic hydrocarbons from organic materials by analytical pyrolysis. Anal. J. Appl. Pyrolysis. 75:150-158.
http://dx.doi.org/10.1016/j.jaap.2005.05.003
Fan Z, Chen D, Birla P, Kamens RM (1995). Modeling of nitro-polycyclic aromatic hydrocarbon formation and decay in the atmosphere, Atmospher. J. Environ. 29:1171-1181
http://dx.doi.org/10.1016/1352-2310(94)00347-N
Fang GC, Chang KF, Bai H (2002). Toxic equivalency factors study of polycyclic aromatic hydrocarbons (PAHs) in Taichung city, Taiwan J. Toxicol. Industr. Health. 18:279-288.
http://dx.doi.org/10.1191/0748233702th151oa
Gerhard L, Aissa M, Tami C, Bond D, Johann F, Hartmut G (2009). Gas/particle partitioning and global distribution of polycyclic aromatichydrocarbons – A modelling approach, Chemosphere J. in press.
Golder AK, Hridaya N, Samanta AN, Ray S (1994). Electrocoagulation of methylene blue and eosin yellowish using mild steel electrodes, Hazard J. Mater. B. 127:134-1
http://dx.doi.org/10.1016/j.jhazmat.2005.06.032
PMid:16102898
Golder AK, Samantha AN, Ray S (2007). Removal of Cr3+ by electrocoagulation with multiple electrodes: Bipolar mono polar configurations. Hazard. J. Mater. 141:653.
http://dx.doi.org/10.1016/j.jhazmat.2006.07.025
PMid:16938395
Guo H, Lee SC, Ho KF, Wang XM, Zou SC (2003). Particle-associated polycyclic aromatic hydrocarbons in urban air of Hong Kong. Atmospheric J. Environ.37:5307-5317.
http://dx.doi.org/10.1016/j.atmosenv.2003.09.011
Guo ZR, Zhang G, Fang J, Dou X (2006). Enhanced chromium recovery from tanning wastewater. Clean J. Prod. 14(1):75.
http://dx.doi.org/10.1016/j.jclepro.2005.01.005
Khandegar V, Anil K (2013). Electrocoagulation for the treatment of

textile industry effluente A-review. Environ. J. Manage.128:949-963.
http://dx.doi.org/10.1016/j.jenvman.2013.06.043
PMid:23892280

Khandegar V, Anil K, Saroha V (2013). Electrocoagulation for the treatment of textile industry effluent e A. review, Environ. J. Manage.128:949-963.
http://dx.doi.org/10.1016/j.jenvman.2013.06.043
PMid:23892280

Kliaugaitė D, Yasadi K, Euverink G, Martijn FM, Racys V (2013). Electrochemical removal and recovery of humic-like substances from waste water. Separation J. Purif. Technol. 37:108.

Kobya M, Delipinar S (2008).Treatment of the baker's yeast wastewater by electrocoagulation. Hazard. J. Mater. 154:1133.
http://dx.doi.org/10.1016/j.jhazmat.2007.11.019
PMid:18082942

Kurt U, Gonullu MT, Ilhan F, Varinca K (2008). Treatment of domestic wastewater by electrocoagulation in a cell with Fe-Fe electrodes. Environ. Engr. J. Sci. 25(2):153-61.
http://dx.doi.org/10.1089/ees.2006.0132

Lacasa E, Canizares P, Saez C, Fernandez FJ, Rodrigo MA (2011). Electrochemical phosphates removal using iron and aluminium electrodes. Chem. J. Engineering. 172:137-143.
http://dx.doi.org/10.1016/j.cej.2011.05.080

Lin SH, Peng CF (1994). Treatment of textile wastewater by electrochemical method, Water J. Res. 28:277-282.
http://dx.doi.org/10.1016/0043-1354(94)90264-X

Linares-Hernandez I, Barrera-adiaz C, Roa-Morales G, Bilyeu B, Urena-Nunez F (2009). "Influence of the anodic material on electrocoagulation performance", Chem. Eng. J. 148-97-105.

Linares-Hernández I, Barrera-Díaz C, Roa-Morales G, Bilyeu B, Ure-a-Nú-ez F (2009). Influence of the anodic material on.electrocoagulation performance. Chem. Eng. J. 148:97.
http://dx.doi.org/10.1016/j.cej.2008.08.007

Lui K, Han W, Pan WP, Riley JT (2001). Polycyclic aromatic hydrocarbon (PAH) emissions from a coal-fired pilot FBC system, Hazard. J. Mater. 84:171-179.

Marian N (2010). Environmental Remediation: Removal ofpolycyclic aromatic hydrocarbons. PHD Thesis.

Modirshahla N, Behnajady MA, Mohammadi S (2008). Investigation of the effect of different electrodes and their connections on the removal efficiency of 4-nitrophenol from aqueous solution by electrocoagulation. Hazardous J. materials. 24:785-790.

Modirshahla N, Behnajadya MA, Mohammadi S (2008). Investigation of the effect of different electrodes and their connections on the removal efficiency of 4 – nitrophenol from aqueous solutions by electrocoagulation. Hazard J. Mater. 154:778-786.
http://dx.doi.org/10.1016/j.jhazmat.2007.10.120
PMid:18162293

Mohd Salleh MA, Mahmoud DK, Wan Abdul Karim WA, Idris A (2011).Cationic and anionic dye adsorption by agricultural solid wastes: A comprehensive review. Desalination J. 11:280.

Mouedhen G, Feki M, Ayedi HF (2008). Behavior of aluminum electrodes in electrocoagulation process", Hazard. J. Mater. 150:124-135.
http://dx.doi.org/10.1016/j.jhazmat.2007.04.090
PMid:17537574

Nazimek D, Ćwikła W (2004). Influence of the precursors kind of catalysts on the course of a denox reaction. Catalysis J. Today. 90(1-2):39-42.
http://dx.doi.org/10.1016/j.cattod.2004.04.006

Pakpahan EN, Kutty SRM, Malakahmad A (2009). Proceedings of the International Conference on Emerging Science and Engineering, Aligarh, India, P. 569.

Pons MN, Alinsafi A, Khemis M, Leclerc JP, Yaacoubi A, Benhammou A, Nejmeddine A (2005). Electrocoagulation of reactive textiles dyes and textiles wastewater, Chem. J. Eng. Proc. 44:461-470.
http://dx.doi.org/10.1016/S0255-2701(04)00153-9
http://dx.doi.org/10.1016/j.cep.2004.06.010

Pulkka S, Martikainen M, Bhatnagar A, Sillanpaa M (2014). Electrochemical methods for the removal of anionic contaminants from water – A review, Separation Purif. J. Technol. 132:252-271.
http://dx.doi.org/10.1016/j.seppur.2014.05.021

Rajeshwar K, Ibanez JG, Swain GM (1994). Electrochemistry and the environment, Appl. J. Electrochem. 24:1077.
http://dx.doi.org/10.1007/BF00241305

Ravindra K, Sokh R, Van R (2008). Atmospheric polycyclic aromatic hydrocarbons: Source attribution, emission factors and regulation. Atmospheric. J. Environ. 42:2895-2921.

19

Development of simple correlations to evaluate the petroleum potential of Quseir - Safaga oil shale, Egypt

Mina R. Shaker, Shouhdi E. Shalaby and Tarek I. Elkewidy

Faculty of Petroleum and Mining Engineering, Suez University, Egypt.

The specific gravity of oil shale can be used as a practical tool for estimating total organic carbon (TOC) and the oil yield of shales from a given source. Sixteen random oil shale samples, mostly from the Upper Cretaceous Duwi Formation in Quseir - Safaga district in the Eastern Desert of Egypt, were collected. Experiments were conducted to determine the density, TOC and oil yield of such samples. The study has developed empirical correlations that show good relationships between the oil shale density (solid organic and mineral density), its TOC and oil yield. Validation of these correlations has been verified by examining two more samples and tested mathematically. The obtained results of TOC and oil yield from the correlations were not as reliable as values obtained from the experimental work especially for the lower TOC samples. The average error has been evaluated to be 15.9%. However, they are accurate enough for certain processes like oil shale processing, where time and equipment restrictions do not allow experimental determination of the TOC and oil yield, and rapid results are desired. These correlations provide a convenient, nondestructive and rapid means of estimating the oil yield and TOC of shales from the same formation.

Key words: Quseir - Safaga oil shale, correlations, total organic carbon, oil yield.

INTRODUCTION

The evaluation of total organic carbon (TOC) of oil shale source rocks is very important for source rock studies as it gives a direct indication about the petroleum generation potential. The oil yield of the oil shale is also an important parameter that needs to be evaluated to give an indication about the potentiality of the oil shale. However, measuring TOC and oil yield are time consuming processes and could not be accessible at the well sites. Fortunately, such parameters are readily correlated with an easily measurable property of the rock which is the density of the solid part of the rock (mineral + organic).

Measuring the solid density is a rapid and practical methodology that preserves the samples and is widely used for evaluating the oil shale petroleum potential (Braun, 1976; Frost and Stanfield, 1950).

The specific gravity of mineral material is around 2.7 and that of organic matter is near 1. Due to the large difference between these two values and because the specific gravity of mineral material does not differ much from 2.7, the specific gravity of the solid part of the black shale is strongly affected by its organic content. The higher the TOC of the oil shale, the lower is its specific gravity. And since the oil yield of black shales is directly related to the organic content, then it is also in some way related to the solid specific gravity. Thus, if the mineral and organic portions of the rock do not vary much across the deposit of interest, a relationship between the solid specific gravity of the oil shale and its organic content and oil yield can be developed and generally applied to the formation (Smith, 1956).

The relationships between the TOC, oil yield, and oil shale specific gravity have been studied by many for the Green River Formation in the US (Braun, 1976; Frost and Stanfield, 1950; McLendon, 1985; Smith, 1956; 1958).

However, no work was found regarding the Egyptian oil shale. This study attempts to employ experimental methods in order to evaluate a relationship between the solid specific gravity, its organic content, and oil yield of black shales in the Campanian - Maastrichtian Duwi Formation of the Safaga - Quseir District, Eastern Egypt. This will in turn provide - an easily accessible method for evaluating the organic content and oil yield of oil shales of such a formation without the need to perform the experiments under concern.

SAMPLES AND METHODS

Sixteen oil shale samples were collected from shale stocks stored in phosphate mining areas, formerly produced as tailings, whilst phosphate mining in Duwi Formation was taking place in Qusier-Safaga area in the Eastern Desert of Egypt. The only exception is B3 sample which is a weathered outcrop sample from the Dakhla Formation. It was intended to collect large oil shale blocks as possible to perform the experiments on the core of such large blocks which are supposed to be less weathered than their surfaces. Two more oil shale samples, namely B4 and N2, have been obtained from Beida and Nekheil areas respectively to undergo the same experiments. These samples were used for the purpose of verifying the validity of the resulting correlations. Table 1 shows the samples with their locations and names.

Samples were pulverized for total organic carbon (TOC) determination. The samples were then treated with dilute (10 %) hydrochloric acid (HCl) to remove carbonates. Afterwards, the samples were heated in HCl solution on a hot plate for 2 h to ensure the dolomite was completely dissolved. Then they were filtered and mildly heated to evaporate water and HCl. Finally, TOC was measured on carbonate free samples using a LECO C230 analyzer. This measurement was performed at StratoChem Services Company, Cairo, Egypt. For oil yield measurement, the samples were ground and sieved to mesh size +8 (2.38 mm). A mass of 100 g of each sample was heated in a Vinci retort oven to 500°C at an average rate of 15.3°C/m and left at such a temperature for 1 h. The produced oil and water volumes were later measured. For solid density measurement, the samples were dried in a drying oven to evaporate water so that the weight to be measured would be that of the organic and mineral materials only. Then Helium porosimeter was used to measure the samples' porosity which was then used along with the measured bulk volume to compute the solid volume which – in turn – was used with the measured weightto evaluate the density of the solid (organic plus mineral) partorganic plus mineral matrix densityat 20°C.The magnitude of error resulting from such a temperature is negligible. The bulk volume was measured by immersion of wax

coated samples in water. The oil yield and density measurements were made in the laboratories of Petroleum and Energy Engineering Department, School of sciences, the American University in Cairo.

RESULTS AND DISCUSSION

Results of TOC, solid (organic and inorganic matrix) density, porosity and oil yield are shown in Table 2. The TOC values of Duwi Formation samples range from 1.07 to 15.90 wt. % (averaging 4.15 wt. %). These TOC values are greater than the 0.5 wt. % threshold value required for a potential source rock to generate hydrocarbons (Peters and Cassa, 1994; Tissot and Welte, 1984).

For the oil yielding samples, Table 2 shows that the oil yield ranges from 1.85 to 26.42 gal/t. Such values are less than those estimated by Tröger (1984). This could be attributed to the difference in the used retorting technique and the longer atmospheric exposure, which has more affected the samples used here. The Nekheil samples (N and N2) are the most productive, which is in agreement with El-Kammar (1993).

The study arrived at correlations that relate TOC, oil yield, and the solid specific gravity. Such correlations provide practical means to approximately determine the TOC and the oil yield of the black shales of Duwi Formation by just measuring the specific gravity of the solid (organic plus mineral) part of the rock rather than resorting to the time consuming expensive experimental work of determining TOC and oil yield. Presented below are the developed correlations.

TOC and oil yield

Figure 1 shows that the oil yield is strongly correlated to TOC. This is reasonable because the TOC is the source of the generated oil. Samples with TOC < 2% do not yield oil. Geochemical analysis of such samples has shown that they contain kerogen of type III and IV (Shaker et al., 2014). Such refractory organic materials do not combust readily at 600°C (Crain, 2013). Thus no oil is produced at the used retorting temperature (500°C). It is important to observe that the validation samples, B4 and N2, lie so close to the correlation line indicating the validity of the derived correlation.

TOC and solid density

Figure 2 shows an obvious relationship between TOC and the solid specific gravity. It shows that the relationship is restricted to samples with TOC > 2%. Samples with TOC < 2% are divided into three categories:

(1) Outcrop sample: Sample B3, which is the only highly weathered sample due to its nature as an outcrop sample. It has a low specific gravity despite its low TOC.

Table 1. Samples information.

Location	No. of Samples	Names of Samples
Abu-Shgeila	2	AS1
		AS2
Abu-Tundub	1	AT
Beida	4	B1
		B2
		B3
		B4
Hamrawin	3	HR1
		HR2
		HR3
Farah	2	F1
		F2
Hamadat	1	H
Nekheil	2	N1
		N2
Um-Elhwaitat	3	UH1
		UH2
		UH3

Table 2. Results of retorting process, density and porosity measurements.

Sample	TOC	Solid	Porosity	Oil Yield		Water Yield
	wt.%	sp. gr.	fraction	cc/100g	gal/ton	cc/100g
AS1	1.37	2.13	0.035	0.00	0.00	12.1
AS2	1.09	2.265	0.146	0.00	0.00	11.3
AT	12.9	1.83	0.047	8.10	21.40	6.3
B1	4.04	2.42	0.217	1.40	3.70	4.1
B2	2.39	2.5	0.095	1.15	3.04	5.7
B3	0.37	2.27	0.106	0.00	0.00	9.3
HR1	2.14	2.65	0.186	0.80	2.11	3.8
HR2	1.69	2.51	0.165	0.00	0.00	4.6
HR3	5.04	2.36	0.220	2.20	5.81	6.5
F1	2.88	2.53	0.232	1.40	3.70	5.1
F2	3.13	2.35	0.232	1.40	3.70	8.3
H	1.26	1.98	0.127	0.00	0.00	10.2
N	15.9	1.66	0.014	9.30	24.57	4.6
UH1	1.07	2.19	0.152	0.00	0.00	9.1
UH2	4.98	2.23	0.134	2.70	7.13	8.4
UH3	2.43	2.35	0.059	0.70	1.85	12.4
B4	10.38	1.96	0.052	6.05	15.98	4.3
N2	17.43	1.61	0.014	10.00	26.42	4.8

This could be due to its high water content. Water has a specific gravity near one which is low as compared to the solid matrix specific gravity. Thus the high water content contributes to a noticeable reduction in the specific gravity.

(2) Low specific gravity samples: This category includes

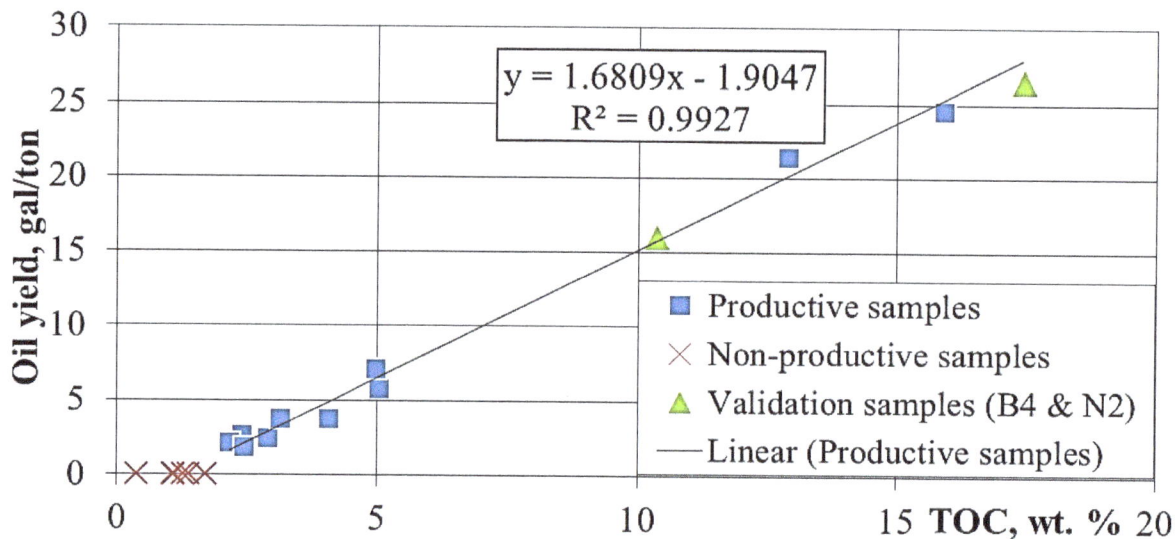

$$y = 1.6809x - 1.9047$$
$$R^2 = 0.9927$$

Figure 1. Oil yield vs. TOC for the 16 original samples plus the two validation samples.

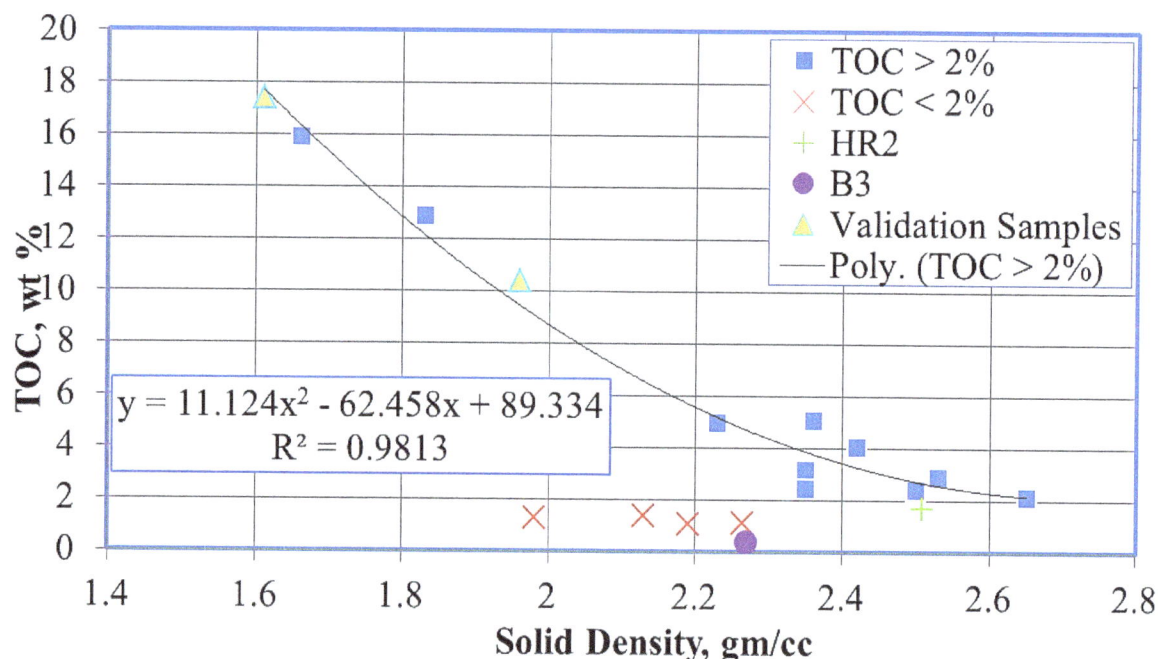

$$y = 11.124x^2 - 62.458x + 89.334$$
$$R^2 = 0.9813$$

Figure 2. TOC vs. solid density.

AS1, AS2, H and UH1. Although such samples have low TOC, they have low specific gravity. This could be attributed to the high water yield of such samples which is among the highest water yields. This could be the factor responsible for the noticeably low specific gravity.

(3)High specific gravity sample: This includes HR2 sample. It is of low TOC and high specific gravity. This is to be expected, especially that it has low water yield unlike those of categories (1) and (2).

Like the oil yield-TOC correlation, samples B4 and N2

have assessed the correlation's validity. This correlation in Figure 2, provides an easily accessible method to evaluate the TOC on condition that the TOC is higher than 2% for samples from Duwi Formation in Quseir - Safaga area.

Oil yield and solid density

Since TOC and solid density are related on one side and

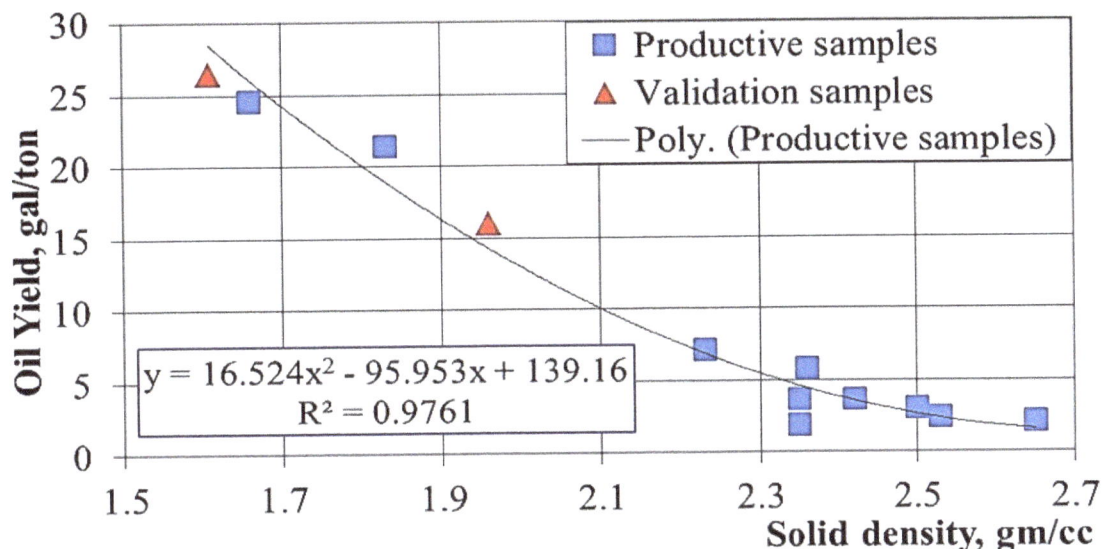

Figure 3. Oil yield vs. solid density.

Table 3. Comparison between true values and correlation based estimates of TOC and oil yield.

Sample	Solid	TOC	Oil Yield	Estimated Yield	Difference		Estimated TOC	Difference	
	sp. gr.	wt.%	gal/ton	gal/ton	gal/ton	%	wt.%	wt.%	%
AT	1.83	12.90	21.40	18.90	2.49	11.65	12.29	0.61	4.74
B1	2.42	4.04	3.70	3.72	0.03	0.72	3.33	0.71	17.52
B2	2.5	2.39	2.64	2.55	0.09	3.37	2.71	0.32	13.56
B4	1.96	10.38	15.98	14.57	1.41	8.83	9.65	0.73	7.03
HR1	2.65	2.14	2.11	0.92	1.19	56.26	1.94	0.20	9.41
HR3	2.36	5.04	5.81	4.74	1.07	18.38	3.89	1.15	22.83
F1	2.53	2.88	2.43	2.17	0.26	10.81	2.52	0.36	12.54
F2	2.35	3.13	3.70	4.92	1.23	33.16	3.99	0.86	27.48
N	1.66	15.90	24.57	25.41	0.85	3.44	16.31	0.41	2.56
N2	1.61	17.43	26.42	27.51	1.09	4.14	17.61	0.18	1.04
UH2	2.23	4.98	7.13	7.36	0.22	3.15	5.37	0.39	7.86
UH3	2.35	2.43	1.85	4.92	3.08	166.31	3.99	1.56	64.20

TOC and oil yield are related on the other side, it is obvious that oil yield and solid density should be related. Figure 3 shows such a relationship. The validity of this correlation has been confirmed as well by the validation samples, B4 and N2. The given correlation is an easy way that can be used to directly estimate the black shaleoil yield without the need to perform the tedious retorting process.

the correlations has been tested mathematically as shown in Table 3. The obtained results of TOC and oil yield from using the correlations are not as reliable as values obtained experimentally, especially for the lower TOC samples. The average error has been evaluated to be 15.9%. The error is roughly inversely related to TOC. However the correlations are accurate enough for certain processes like oil shale processing for which rapid results are desired and accuracy is not highly required.

Correlations validity

Samples B4 and N2 have verified and proved the validity of the strong correlations between solid density, TOC, and oil yield as we have seen. In addition, the validity of

CONCLUSIONS AND RECOMMENDATIONS

The study has arrived at correlations that link between TOC, oil yield and the solid specific gravity. Such correlations provide convenient, nondestructive, and rapid means to approximately determine the TOC and the oil yield of the black shales of Duwi Formation by just measuring the specific gravity of the solid (organic plus mineral) part of the rock instead of the time consuming experimental work of determining TOC and oil yield. The obtained results of TOC and oil yield from the correlations are not as reliable as values obtained from the experimental work especially for the lower TOC samples. The average error has been evaluated to be 15.9%.

However, they are accurate enough for certain processes like oil shale processing where time and equipment restrictions do not allow experimental determination of the TOC and oil yield and rapid results are desired.

It is important to note that the samples used are not fresh samples and the coefficients of the correlations could be modified by further analysis using fresh core samples drilled in Duwi Formation. Also, in determining oil yield, using the standardized Fischer Assay method could be more reliable for developing the correlations which are expected to give more acceptable results.

Conflict of Interest

The authors have not declared any conflict of interest.

ACKNOWLEDGEMENTS

The authors express their gratitude to Dr. Mohammed Said, General Manager of StratoChem Services Company for his assistance in performing the TOC experiment used in this study. They would also like to thank the laboratory technicians of the Petroleum and Energy Engineering Department, School of Sciences and Engineering, the American University in Cairo; Eng. Shadi Azer, Mr. Samer Mina and Mr. Marawan Moussa for their generous help and support during performing the experiments.

REFERENCES

Braun RL (1976). Nondestructive determination of oil yield of oil shale blocks. SPE-5961-MS.
Crain ER (2013). Crain's petrophysical handbook. Retrieved December 2013, from Total organic carbon (TOC) basics: http://spec2000.net/11-vshtoc.htm
El-Kammar M (1993). Organic and inorganic composition of the Upper Cretaceous- Lower Tertiary black shales from Egypt and their hydrocarbon potentialities. Cairo: Ph.D. Thesis, Faculty of Sciences, Cairo University, Egypt.
Frost IC, Stanfield KE (1950). Estimating Oil Yield of Oil Shale from Its Specific Gravity. J. Anal. Chem. 22(3):491-492.
McLendon TR (1985). A correlation for obtaining oil shale grade from water immersion specific gravity measurements. U.S. Department of Energy.
Shaker MR, Shalaby SE, Elkewidy, TI (2014). Egyptian oil shale characterization and exploitation to produce oil and gas. J. Petroleum Mining Eng.17(1):1-10.
Shaker MR (2014). Study on Oil Recovery from Egyptian Oil Shale Formations. Suez: M.Sc. Thesis: Faculty of Petroleum and Mining Engineering, Suez University, Egypt.
Smith JW (1956). Specific gravity-oil yield relationships of two colorado oil-shale cores. Ind. Eng. Chem. 48(3):441-444.
Smith JW (1958). Applicability of a specific gravity-oil yield relationship to green river oil-shale. Ind. Eng. Chem. 3(2):306-310.
Tissot B, Welte D (1984). Petroleum formation and occurrence. Berlin: Springer Verlag.
Tröger U (1984). The oil shale potential of Egypt. Berl. Geowiss. Abh., 50:375-380.

20

Filter cake formation on the vertical well at high temperature and high pressure: Computational fluid dynamics modeling and simulations

Mohd. A. Kabir and Isaac K. Gamwo*

United States Department of Energy, National Energy Technology Laboratory, Pittsburgh, PA 15236-0940, USA.

Oil and gas wells are generally drilled with the intention of forming a filter cake on the wellbore walls to primarily reduce the large losses of drilling fluid into the surrounding formation. Unfortunately, formation conditions are frequently encountered that may result in unacceptable losses of drilling fluid into the surrounding formation despite the type of drilling fluid employed and filter cake created. It is extremely important to optimize filter cake thickness as very thick filter cake can cause stuck pipe and other drilling problems. The focus of this research is to use a computational fluid dynamics (CFD) technique to numerically simulate filter cake formation on the vertical wellbore wall at high-pressure (25,500 psi or 175.8 MPa) and temperature (170°C) conditions. Here, the drilling fluids were treated as a two-phase system of solid particulates suspended in a non-Newtonian fluid. Drilling process simulations were performed for drilling fluid with two particle sizes, 45- and 7- μm, under extreme drilling conditions of high pressure and temperature. The comparison of both scenarios clearly shows that the drilling fluid with larger particles (45 μm) forms thicker filter cake compared to drilling fluids with smaller particles (7 μm). We have further used FLUENT CFD code to successfully simulate filter cake formation on the wellbore wall at moderate pressure (2,000 psi or 13.8 MPa) and temperature (30°C) conditions with drilling fluid of 45 μm particles. The results for axisymmetric and planar wellbore show that the cake formed during extreme drilling processes is thicker than that formed for shallow drilling processes. Filter cake formed on the vertical wellbore wall is nonuniform for both extreme and shallow drilling process.

Key words: Filter cake, two-phase flow, computational fluid dynamics (CFD), deep drilling.

INTRODUCTION

Recently, filter cake formation on the walls of vertical wellbore at extreme pressure and temperature (up to 25,500 psi or 175.8 MPa and 170°C) has attracted the attention of multiphase fluid researchers due to filter cake's crucial role in reducing drilling fluid losses in oil and gas drilling operations (Delhommer, 1987). Filter cake builds up on the wellbore walls in a mechanism similar to soil consolidation during drilling processes, where overbalance pressure forces drilling fluid into the rock formation and leaves solid particles on the walls in the form of a filter cake (Cerasi, 2001).

During the drilling process, solid particulate multiphase drilling fluids are pumped down into the drilling zone through drilling pipe as shown in Figure 1. Modern drilling fluids used in oil production processes are carefully engineered slurries designed to perform several tasks. Among others, the drilling fluids or slurries function to (a) reduce friction and wear on the drilling bit, (b) transport the drilled solids, (c) maintain a favorable pressure difference between the wellbore and the rock formation, (d) cool down the cutters to maintain the temperature below the critical temperature at which cutter properties such as strength and hardness start to change, and (e) generate a filter cake on the wellbore wall to minimize incursion of drilling fluids into the formation (Vaussard, 1986; Maurer, 1997; Spooner, 2004; Ali, 2006; Berry, 2009).

*Corresponding author. E-mail: gamwo@netl.doe.gov.

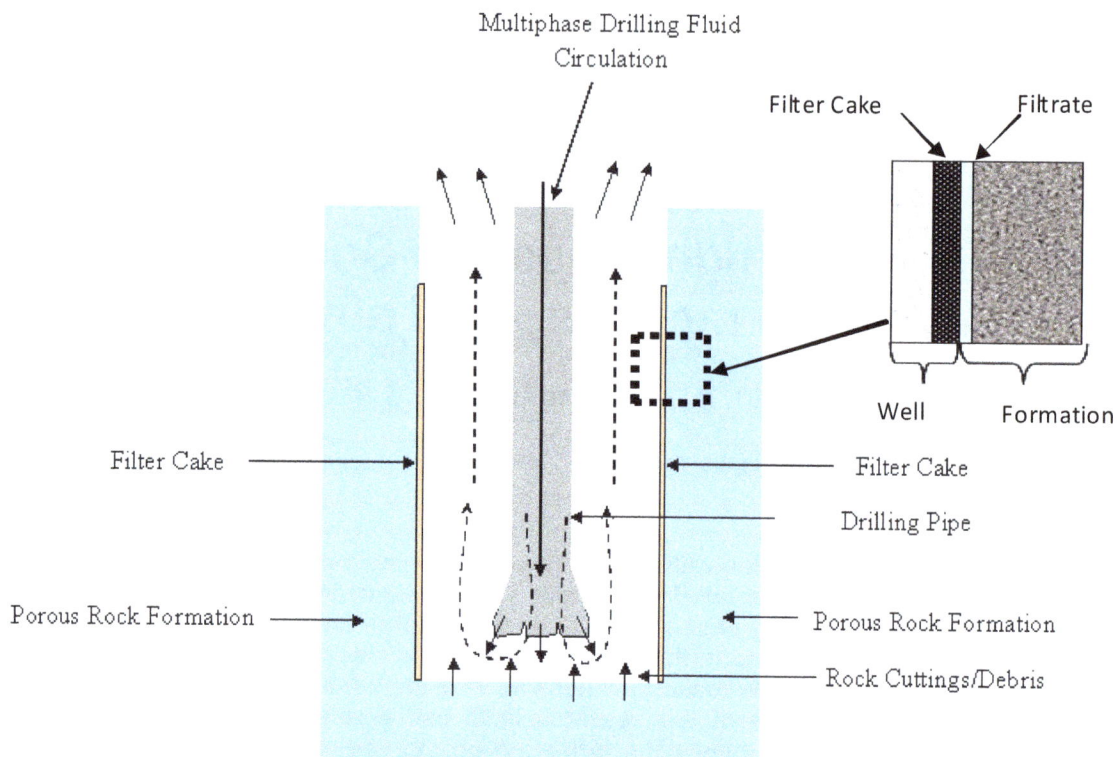

Figure 1. Schematic diagram of drilling fluid circulation in the drilling zone.

The ability to optimize filter cake characteristics is extremely useful (Fisher, 2008). Most wells are drilled with the intention of forming a filter cake of varying thickness on the sides of the borehole. The presence of a filter cake is beneficial since it reduces fluid loss and damage to the formation. However, if the cake is too thick, the effective diameter of the hole is reduced and problems may arise, such as excessive torque when rotating the drill string and excessive drag when pulling it. Thick cakes also contribute to high swab, a decrease in wellbore pressure during the movement of drill strings up the wellbore. Such pressure reduction, if significant, may lead to premature reservoir fluids flowing into the wellbore and towards the surface. Thick cakes may also contribute to sudden increase in pressure (surge pressure) when drill strings or casing is rapidly run into a wellbore, which may be great enough to create lost of drilling fluid circulation.

Earlier research on the filtration (Klotz, 1954; Outmans, 1963; Peden, 1982; Vaussard, 1986; Vaussard, 1986; Delhommer, 1987; Fordham, 1988; Sherwood, 1991) of drilling fluids has suggested that temperature, pressure, hydraulic shear rate, and formation permeability all influence the filtration process. However, the influence of individual factors and their interdependencies remains unclear (Fisher, 2008).

Formation conditions are frequently encountered that may result in unacceptable losses of drilling fluid to the surrounding formation despite the type of drilling fluid employed and filter cake created. The filter cake forms permeable zones in the wellbore wall, which can cause stuck pipe and other drilling problems as well (Delhommer, 1987).

Literature review (Klotz, 1954; Maurer, 1997; Cerasi, 2001; Ali, 2006; Fisher, 2008) shows that very little research has been carried out on filter cake formation during deep drilling. However, limited reports and numerical research do exist on filter cake for shallow-wellbore drilling, although these sources lack detailed information on filter cake formations (Klotz, 1954; Delhommer, 1987; Maurer, 1997; Ali, 2006; Fisher, 2008).

Literature review further reveals that no filter cake formation modeling has yet been performed for deep drilling conditions under high temperature and high pressure. Most of the previous research has been carried out with Newtonian, single phase, and isothermal conditions for shallow drilling process. Hence, we have included the following features in our filter cake formation modeling:

1. CFD simulation of filter cake formation for deep (4 to 5 miles) drilling conditions at high temperature and high pressure.
2. Drilling fluids treated as multiphase non-Newtonian fluid, where solid particulates are suspended in a non-Newtonian fluid phase; the non-Newtonian phase was modeled with power law.

3. Energy equations solved in this numerical modeling and simulation.

In this article, we have used three main interlinked solution aspects to modeling (Fisher, 2008), to simulate the filter cake formation on the vertical wellbore wall in deep (25,000 psi or 172.4 MPa and 170°C) and shallow (2,000 psi or 13.8 MPa and 30°C) drilling processes. The solution aspects used are:

i. Multiphase fluid flow in the pipe and in the annulus between the pipe and the borehole,
ii. Cake formation by deposition of solids from the annulus fluids onto the borehole wall, and
iii. Seepage of mud constituents into the formation during and after cake formation.

The main focus of this research was to investigate the filter cake formation on the vertical well at high temperature and high pressure utilizing computational fluid dynamics (CFD) method tools. The effects of drilling fluid particle sizes on the filter cake thickness were also studied.

MULTIPHASE FLOW AND GOVERNING EQUATIONS

The CFD modeling and simulations involved the solution of Navier-Stokes equations. Hence, detailed equations of multiphase fluids and their relevant theories are also provided here.

The dynamics of solids-in-fluid media have a large effect on various flow phenomena, such as density, viscosity, and pressure. Thus, the hydrodynamics of solids must be modeled correctly (Cornelissena, 2007). The Eulerian approach is preferred over the Lagrangian due to the large volume fraction of solids in the drilling fluid. In the Eulerian approach, fluid and solid phases are treated as interpenetrating continua, and momentum and continuity equations are defined for each phase (FLUENT, 2006). Therefore, the Eulerian-Eulerian multiphase fluid model has been used to simulate fluid flow and filter cake formation in vertical wellbore drilling operations for shallow and deep drilling conditions. The Eulerian model is the most complex of the multiphase models, solving a set of n momentum and continuity equations for each phase. Coupling is achieved through the pressure and inter-phase exchange coefficients. The manner in which this coupling is handled depends upon the type of phases involved. For granular flows, properties are obtained by applying kinetic theory. Mass transfer between the phases is negligible and, therefore, ignored here. The momentum equation for the solid phase differs from the equation used for the fluid phase, since the former contains a solid pressure (Ishii, 1975; Jackson, 1997; FLUENT, 2006; Myöhänen, 2006). Lift and virtual mass forces are assumed to be negligible in the momentum equations.

Modeling fluid flow in the annulus

Multiphase equations for modeling the flow of steady, laminar, non-isothermal, incompressible fluid are given thus (Ishii, 1975; Jackson, 1997; FLUENT, 2006; Myöhänen, 2006; Cornelissena, 2007; Gidaspow, 1994, "Jung and Gamwo, 2008").

Conservation of mass

For liquid, $\nabla.(\alpha_l v_l) = 0$ (1)

For solids, $\nabla.(\alpha_s v_s) = 0$ (2)

where α is the volume fraction and subscripts l and s denote liquid and solid phases, respectively. Moreover, $\alpha_l + \alpha_s = 1$ must be satisfied. v_l and v_s are the velocities of the solid and liquid phases, respectively.

Momentum balance

Liquid phase: The momentum equation for the liquid phase in a solid-liquid system (Ishii, 1975; Jackson, 1997; FLUENT, 2006; Myöhänen, 2006; Cornelissena, 2007) is:

$$\underbrace{\nabla.(\alpha_l \rho_l \vec{v}_l \vec{v}_l)}_{} = \underbrace{-\alpha_l \nabla p}_{} + \underbrace{\nabla.\overline{\tau}_l}_{} + \underbrace{\alpha_l \rho_l \vec{g}}_{} - \left\{ \left(\underbrace{K_{sl}(\vec{v}_l - \vec{v}_s)}_{} \right) \right\}$$

(3)

Convective   Pressure Stress   Body Forces   Momentum exchange

Where ρ_l and ρ_s are the densities of liquid and solid phases, respectively.

To address non-Newtonian behavior of the liquid phase in the multiphase drilling fluid, we have used the power-law model input parameters in the simulation (FLUENT, 2006; Fisher, 2008).

For the fluid, the stress tensor, $\overline{\tau}_l$, is related to the fluid strain rate tensor, $\overline{\dot{\gamma}}_l = \nabla \vec{v}_l + (\nabla \vec{v}_l)^{tr}$, by:

$$\overline{\tau}_l = \alpha_l \tau \overline{\dot{\gamma}}_l + \alpha_l \left(\lambda_l - \frac{2}{3}\tau \right) \nabla.\vec{v}_l \overline{I} \quad (4)$$

where $\tau = k|\overline{\dot{\gamma}}_l|^{n-1}$ and $|\overline{\dot{\gamma}}_l|$ is the magnitude of the strain rate tensor defined as $|\overline{\dot{\gamma}}| = \sqrt{\frac{1}{2}\sum_i \sum_j \dot{\gamma}_{ij}\dot{\gamma}_{ji}}$, and

k and n are consistency factor and power-law exponent, respectively (FLUENT, 2006; Fisher, 2008; Hamed, 2009).

Solid phase: The momentum equation for the solid phase in a solid-liquid system (Ishii, 1975; Jackson, 1997; FLUENT, 2006; Myöhänen, 2006; Cornelissena, 2007) is:

$$\nabla.\left(\alpha_s \rho_s \vec{v}_s \vec{v}_s\right) = -\alpha_s \nabla p - \underbrace{\nabla p_s} + \nabla.\overline{\tau}_s + \alpha_s \rho_s \vec{g} + \left\{\left(K_{ls}\left(\vec{v}_l - \vec{v}_s\right)\right)\right\} \quad (5)$$

Solid pressure

The solids pressure, p_s, stress, $\overline{\tau}_s$ and viscosity, μ_s are determined by particle fluctuations and the kinetic energy associated to these fluctuations, granular temperature Θ. The stress-strain relationship for the solid phase s is:

$$\overline{\tau}_s = \alpha_s \underbrace{\mu_s \overline{\dot{\gamma}}_s} + \alpha_s \left(\underbrace{\lambda_s} - \frac{2}{3}\mu_s\right) \nabla.\vec{v}_s \underbrace{\overline{I}} \quad (6)$$

Shear stress, bulk viscosity and unit tensor

Where solid strain rate tensor $\overline{\dot{\gamma}}_s = \nabla \vec{v}_s + \left(\nabla \vec{v}_s\right)^{tr}$.

Interaction forces are considered here to account for the effects of other phases and are reduced to zero for single phase flow (FLUENT, 2006; Cornelissena, 2007). The momentum exchanges coefficients are indistinguishable ($K_{ls} = K_{sl}$):

$$K_{sl} = \frac{\alpha_s \rho_s f}{T_s^p} \quad (7)$$

This function and coefficients are suitable for drilling process modeling where recirculating multiphase fluids contain high solid fraction.

Here, T_s^p is the particulate relaxation time and f is the model-dependent drag function. The relaxation time is expressed as:

$$T_s^p = \frac{\rho_s d_s^2}{18\mu_l} \quad (8)$$

where d_s is the solid particle diameter.

While the Syamlal-O'Brien drag function f (FLUENT, 2006; Cornelissena, 2007) is used:

$$f = \frac{C_D R_{e_s} \alpha_l}{24 v_{r,s}^2} \quad (9)$$

The relative Reynolds number R_{e_s} can be written as follows (FLUENT, 2006; Cornelissena, 2007):

$$R_{e_s} = \frac{\rho_l d_s \left|\vec{v}_s - \vec{v}_l\right|}{\mu_l} \quad (10)$$

The drag function f includes a drag coefficient C_D and the relative Reynolds number R_{e_s}; however, the drag function differs among the exchange-coefficient models. For the drilling process, multiphase drilling fluid with a high solid fraction continuously cycles through the drill assembly and carry away debris produced by the drilling process.

In the Syamlal-O'Brien model, the drag function of Dalla Valle is used (FLUENT, 2006; Cornelissena, 2007), where $v_{r,s}$ is the terminal velocity correlation:

$$C_D = \left[0.63 + \frac{4.8}{\sqrt{Re_s / v_{r,s}}}\right]^2 \quad (11)$$

The terminal velocity correlation $v_{r,s}$ for solid phase has the following form:

$$v_{r,s} = 0.5\left(A - 0.06 Re_s + \sqrt{\left(0.06 Re_s\right)^2 + 0.12 Re_s \left(2B - A\right) + A^2}\right) \quad (12)$$

where $A = \alpha_l^{4.14}$; $B = 0.8\alpha_l^{1.28}$, $\alpha_l \le 0.85$; $B = \alpha_l^{2.65}$, $\alpha_l > 0.85$

This correlation is based on measurements of terminal velocities of particles in fluidized or settling bed where high solid volume fractions similar to solid volume fractions in drilling fluids are encountered.

The solid pressure P_s is composed of a kinetic term (first term), a particle collisions term (second terms) and a friction term (3rd term) (FLUENT, 2006; Cornelissena, 2007):

$$P_s = \alpha_s \rho_s \Theta_s + 2\rho_s(1 + e_{ss})\alpha_s^2 g_{0,ss}\Theta_s + F_r \frac{(\alpha_s - \alpha_{s,min})^n}{(\alpha_{s,max} - \alpha_s)^p} \quad (13)$$

Both kinetic and collision terms are dependent on the granular temperature Θ. The term e_{ss} is the particle – particle coefficient of restitution (taken here to be e_{ss} = 0.9 - this choice is consistent with literature value under similar simulation conditions) where g_0, is the radial distribution function. This is a correction factor (the non-dimensional distance between spheres) that modifies the probability of collisions between particles when the granular phase becomes dense. The friction is included in this study since the solid volume fraction is relatively high, which may give rise to friction. In this work, the friction pressure is modeled using the semi-empirical model proposed by Johnson et al. (1990). $\alpha_{s,min}$ and $\alpha_{s,max}$ are the minimum and maximum packing respectively. $\alpha_{s,min}$, assumed to be 0.5, is the solid concentration when friction stresses becomes important. The values of empirical materials constants F_r, n, and p are taken to be 0.5, 2.0 and 5.0, respectively, following other investigators (Johnson et al., 1990).

Energy equation

To describe the conservation of energy in Eulerian multiphase applications, a separate steady-state enthalpy equation can be written for each phase q (liquid or solid) (FLUENT, 2006; Cornelissena, 2007) as follows:

$$\nabla \cdot \left(\underbrace{\alpha_q \rho_q \vec{u}_q h_q} \right) = \underbrace{\overline{\tau_q} : \nabla \vec{u}_q} - \underbrace{\nabla \cdot \vec{q}_q} + \sum_{p=1}^{n} \left(\underbrace{\vec{Q}_{pq}} \right) \quad (14)$$

where h_q is the specific phase enthalpy, $\vec{q}_q$ is the heat flux, and $\vec{Q}_{pq}$ is the intensity of heat exchange between phases.

Granular temperature

Particulates' viscosities need the specification of the granular temperature for the solid phase. We used a partial differential equation, which was derived from the transport equation by neglecting convection and diffusion. It takes the following form (FLUENT, 2006; Cornelissena, 2007):

$$0 = \left(-p_s \overline{I} + \overline{\tau}_s \right) : \nabla \vec{v}_s - \gamma_{\Theta_s} + \phi_{ls} \quad (15)$$

where $\left(-p_s \overline{I} + \overline{\tau}_s \right) : \nabla \vec{v}_s$ is the generation of energy by the solid stress tensor, γ_{Θ_s} is the collisional dissipation of energy, and ϕ_{ls} is the energy exchange between the fluid and the solid phase.

The collisional dissipation of energy, γ_{Θ_s}, represents the rate of energy dissipation within the solid phase due to collisions between particles. The term is represented by the following expression derived by Lun (1984):

$$\gamma_{\Theta_s} = \frac{12(1 - e_{ss}^2)g_{0,ss}}{d_s \sqrt{\pi}} \rho_s \alpha_s^2 \Theta_s^{3/2} \quad (16)$$

The transfer of the kinetic energy of random fluctuations in particle velocity from the solid phase to the liquid phase is represented by ϕ_{ls}:

$$\phi_{ls} = -3K_{ls}\Theta_s \quad (17)$$

The radial distribution function, $g_{0,ss}$ is modeled as follows (Ding, 1990; FLUENT, 2006; Cornelissena, 2007):

$$g_{0,ss} = \left[1 - \left(\frac{\alpha_s}{\alpha_{s,max}} \right)^{1/3} \right]^{-1} \quad (18)$$

where $\alpha_{s,max}$ is the maximum packing, assumed here to be 0.63 (the symbols are defined in Table 1).

The viscosity for solids stress tensor is the sum of collisional, kinetic, and frictional viscosity parts:

$$\mu_s = \mu_{s,col} + \mu_{s,kin} + \mu_{s,fr} \quad (19)$$

The collisional part of viscosity is modeled as follows (Ding, 1990; Gidaspow, 1992; FLUENT, 2006; Cornelissena, 2007):

$$\mu_{s,col} = \frac{4}{5}\alpha_s^2 \rho_s d_s g_{0,ss}(1 + e_{ss})\left(\frac{\Theta_s}{\pi} \right)^{1/2} \quad (20)$$

The kinetic part of viscosity is modeled using the equation of Syamlal (FLUENT, 2006):

Table 1. Definition of symbols.

Symbol	Description	Units
Alphabetic		
C_D	Drag coefficient	Dimensionless
d_s	Solid particle diameter	m
e	Coefficient of restitution	Dimensionless
g	Gravitational acceleration	m/s^2
g_0	Radial distribution function	Dimensionless
K	Interphase exchange coefficient,	Dimensionless
K_p	Porous media permeability	m^2
P	Pressure (Fluid)	Pa
R_e	Relative Reynolds number	Dimensionless
t	Time	s
D_p	Porous media mean particle diameter	m
F_r	Materials constant in eq. 14	
p	Materials constant in eq. 14	
n	Materials constant in eq. 14	
Greek letters		
α	Volume fraction (solid or liquid)	Dimensionless
ρ	Density	kg/m^3
Θ	Granular temperature	m^2/s^2
$\bar{I}$	Unit stress tensor	Dimensionless
$\gamma_{\Theta s}$	Collision dissipation of energy	$kg/s^3 m$
$\bar{I}_{2D}$	Second invariant of deviatoric stress tensor	Dimensionless
λ	Bulk viscosity	Pa.s
μ	Shear viscosity	Pa.s
$\overrightarrow{v}_s$	Solid velocity	m/s
$\overrightarrow{v}_l$	Fluid velocity	m/s
v	Seepage velocity	m/s
$\bar{\tau}$	Stress tensor	Pa
ε	Porous media void volume fraction	
Subscripts		
col	collision	
fr	friction	
kin	kinetic	
l	liquid phase	
Max, min	Maximum, minimum value	
q	Either liquid or solid phase	
s	Solid phase	

FLUENT (2006), Cornelissena (2007).

$$\mu_{s,kin} = \frac{\alpha_s d_s \rho_s \sqrt{\Theta_s \pi}}{6(3-e_{ss})}\left[1+\frac{2}{5}(1+e_{ss})(3e_{ss}-1)\alpha_s g_{o,ss}\right] \quad (21)$$

Shear stress includes bulk viscosity, λ_s that in granular flows is related to the particles' resistance to compression and expansion. In Lun et al. (1984), bulk viscosity expression was used in this simulation:

$$\lambda_s = \frac{4}{3}\alpha_s \rho_s d_s g_{o,ss}\left(1+e_{ss}\right)\left(\frac{\Theta_s}{\pi}\right)^{1/2} \qquad (22)$$

When the solids volume fraction is near the packing limit, the friction between particles is important. The friction part of the shear viscosity can be defined using Schaeffer's expression:

$$\mu_{s,fr} = \frac{p_{sfr}\sin\theta}{2\sqrt{I_{2D}}} \qquad (23)$$

where θ is the angle of internal friction and I_{2D} is the second invariant of the deviatoric stress tensor (FLUENT, 2006).

Porous rock formation model

The multiphase fluid flow through the porous rock is modeled using an extension of Darcy's law for multiphase flow, also referred to as the Ergun equation for laminar flow or the Blake-Kozeny equation. This equation reads:

$$\nabla P = -\frac{\mu}{K_p}v \qquad 24)$$

where v is the seepage fluid velocity in the formation and μ the fluid dynamic viscosity. The porous media permeability, K_p, is given below in terms of formation porosity (ε) and the porous media mean pore size (D_p). Here, we set a formation void fraction of 0.2 following Parn-anurak (2003):

$$K_p = \frac{D_p^2}{150}\frac{\varepsilon^3}{\left(1-\varepsilon\right)^2} \qquad (25)$$

The differential pressure in between the porous media formation and annulus was maintained at 500 psi (3.4 MPa).

Drilling process and filter cake formation

Figure 1 illustrates the drilling fluid circulation process during drilling. Here, particulate multiphase fluid is pumped down into the drilling zone through a drilling pipe where drilling fluid interacts with rock debris. As particulate-laden drilling fluid flows upward to the surface through the annulus in between the walls of the well and

the drill string, differential pressure causes filter cake to build up on the porous rock surface as shown in Figure 1.

Since overbalance exists in the annulus, differential pressure forces drilling fluid through the porous rock into the formation and separates the particles on the porous rock surface in the form of filter cake. Fluid that permeates in the porous rock surface is related to the rock resistance, fluid viscosity, and differential pressure. This relationship can be described by Darcy's Law (Fu, 1998; Cerasi, 2001; Parn-anurak, 2003; Fisher, 2008). With time, filter cake will grow on the rock surface; therefore, filter cake itself will also resist fluid permeation into porous rock formations and, hence, fluid permeation will decrease. The resistance from the filter cake can be related to the concentration of mass loading per unit area (kg/m^2) and specific resistance (m/kg). The filter cake builds up to a maximum thickness, which is determined by particle characteristics and fluid shear (Fu, 1998; Fisher, 2008).

Two-dimensional vertical wellbore model

A two-dimensional (2-D) wellbore model was created and meshed with FLUENT- Gambit as shown in Figure 2 (FLUENT, 2006). A symmetry along the central axis was assumed. In this simulation, we zoomed in the drilling zone of a vertical well to capture detailed phenomena occurring in the drilling processes. Hence, the well model is limited to the drilling zone, and the dimensions are 0.24 m wide and 1 m long. To simulate the drilling process, multiphase particulate (α_s = 0.2) drilling fluid was pumped into the main model inlet, and multiphase particulate (α_s = 0.8) rock debris was pushed from the bottom inlet. The main inlet represents drilling fluid pumping in, and the bottom inlet represents rock debris coming from drilling bottom. The solid wall represents the drill string surface. A porous medium with a solid volume fraction of 0.8 next to the drill string represents vertical rock formations on which the filter cake builds up. The pressure and temperature for both inlets are, respectively, 25,500 psi (175.8 MPa) and 170°C for deep drilling conditions, and 2,000 psi (13.8 MPa) and 30°C for shallow drilling conditions. The formation pressure and temperature were maintained at 25,000 psi (172.4 MPa) and 170°C, while the pressure and temperature for shallow drilling conditions were maintained at 1,500 psi (10.3 MPa) and 30°C, respectively, to mimic real-world drilling scenarios. Multiphase particulate non-Newtonian drilling fluids were pumped into the drilling zone where the drilling fluids mingled with rock particles. The particle-laden drilling fluid then flowed upwardly, back to the surface, through the annulus between the walls or sides of the wellbore and the drill string. A variety of drilling fluid types exist, and, as mentioned earlier, the circulation of such fluid functions to, among others things, lubricate the drill bit, remove cuttings from the wellbore as they are produced,

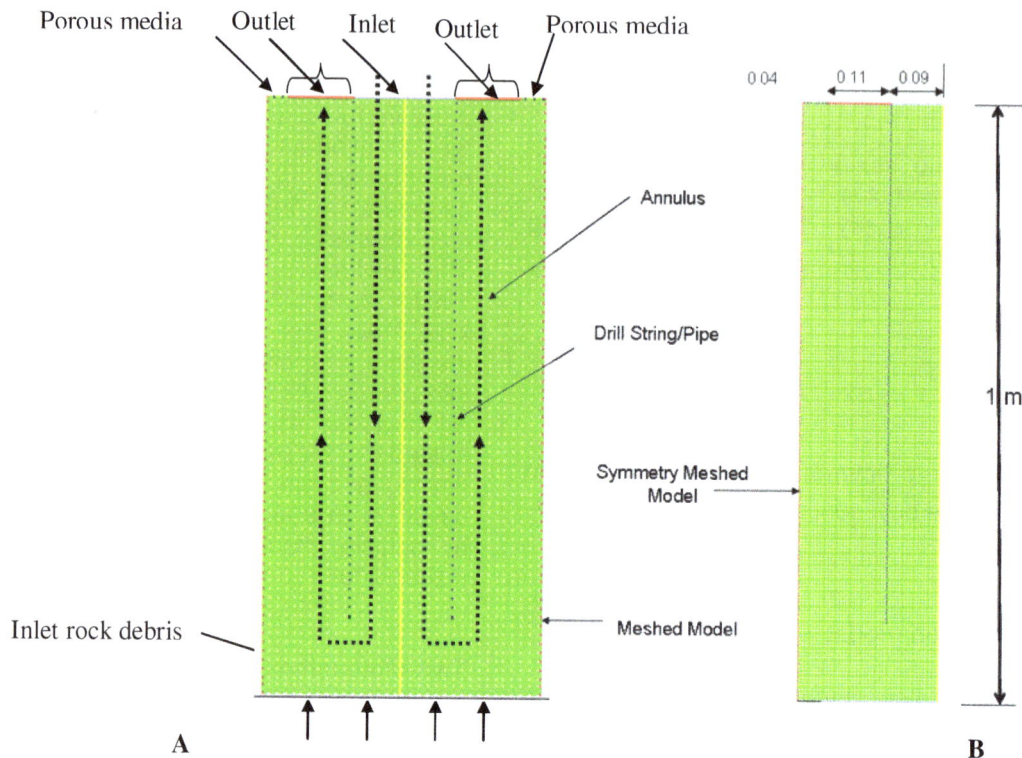

Figure 2. Meshed vertical wellbore model; (a) full-wellbore model; (b) half-wellbore model.

exert hydrostatic pressure on pressurized fluid contained in formations, and seal off the walls of the wellbore so that the fluid is not lost in the permeable subterranean zones (Rogers, 1996).

Initial and boundary conditions

The wellbore was initially filled with multiphase particulate drilling fluid or mud and the bottom portion of the drilling zone was filled with rock debris, as shown in Figure 3. In the model, non-Newtonian power-law fluid properties were given for the liquid phase, and granular properties were given for solid particles. The density of the liquid phase was 999 kg/m^3 with consistency (k) and power-law (n) index of 0.1238 Pa.s^n and 0.67, respectively (Fisher, 2008; Hamed, 2009). The solid phase density was set at 2,350 kg/m^3. Two particle sizes in the fluid were studied: 45 and 7 μm. The domain was discretized with a grid where the flow domain was divided into finite surfaces. As mentioned earlier, axi-symmetry was assumed for modeling the drilling process. There were several trials made (from 5,000 to 11,000 meshes) to eliminate the dependency of the grid size. The half-wellbore model consists of 9,600 numbers of quadrilateral mesh cells with a uniform size of 0.5 x 0.5 cm. The dimension of the porous media formation in the model was 4 x 100 cm, and the porous media formation pressure and temperature were maintained at 25,000 psi (172.4 MPa) and

170°C for deep drilling conditions, whereas the pressure and temperature for shallow drilling conditions were maintained at 1,500 psi (10.3 MPa) and 30°C, respectively.

Model validation for filter cake formation

Our extensive literature review to validate our CFD modeling results revealed very little data available on experimental and numerical filter cake formation on a vertical wall for deep and shallow drilling processes (Sherwood, 1991; Sherwood, 1991; Rogers, 1996; Cerasi, 2001; Usher, 2001; Parn-anurak, 2003; Fisher, 2008; Hamed, 2009). Hence, a single pressure linear filtration process (Figure 4) was chosen to validate our model. Figure 3 shows the initial solid distribution in subsurface vertical wellbore where filter cake forms on the vertical wall, while Figures 4 to 6 show cake formation at the bottom of a laboratory-scale pressure filtration cell. The filter cake formation data were extracted from single pressure filtration CFD simulations and compared with analytically calculated filter cake height to verify the agreement prior to running drilling process simulations.

The simulation was performed for pressure filtration where inlet pressure was kept at 100 kPa. The filtration cell was initially filled with multiphase particulate drilling fluid, and pressure was applied at the top (inlet) with porous media at the bottom (outlet). The applied pressure

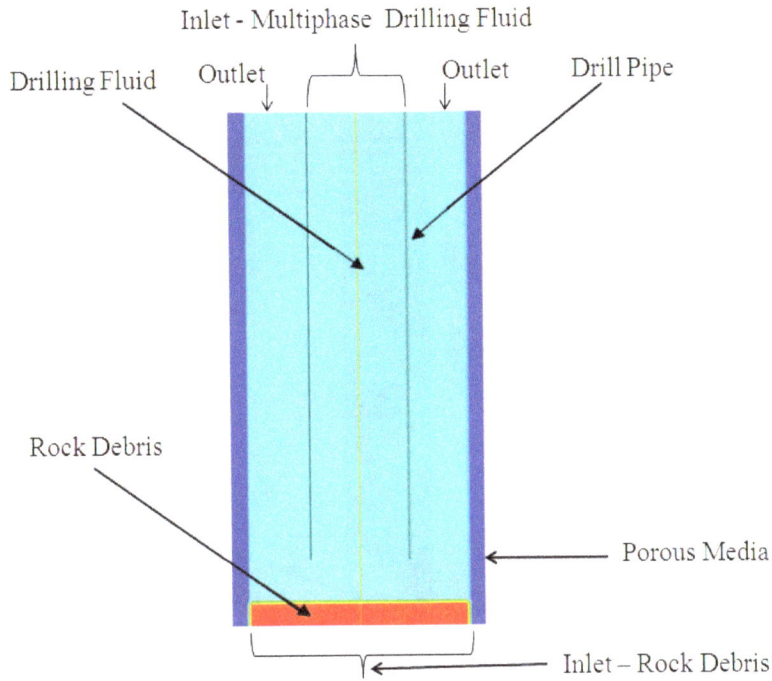

Figure 3. Initial solid volume fraction distribution in the well.

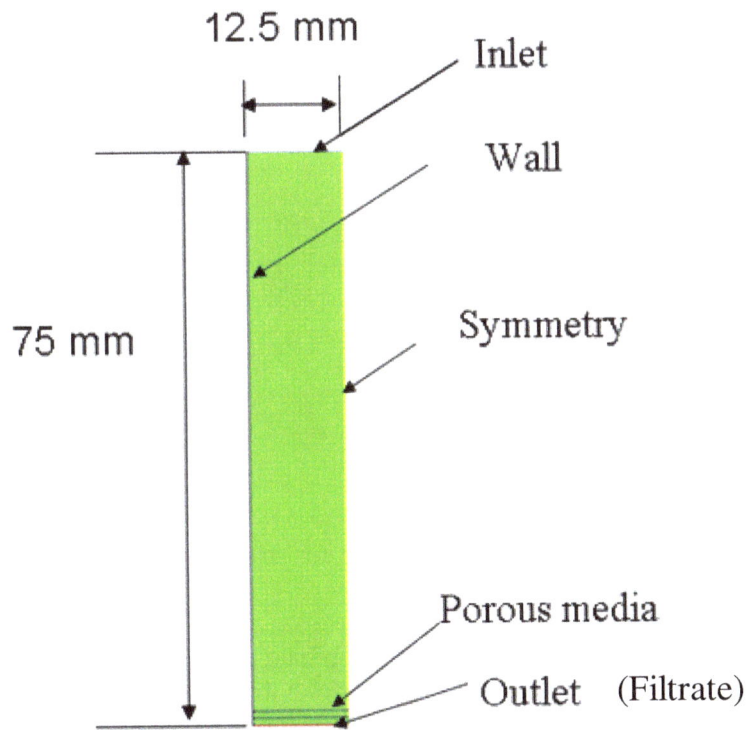

Figure 4. Filtration meshed model created with Gambit (half-filtration cell).

forced fluid through the porous media and separated solid particles in the form of filter cake on the porous media. During filtration, the filter cake that formed reached

equilibrium with the applied pressure at the top. Figure 4 shows the symmetry model of the filtration cell. Multiphase fluid with a solid volume fraction of 0.185 was

Solid Volume Fraction

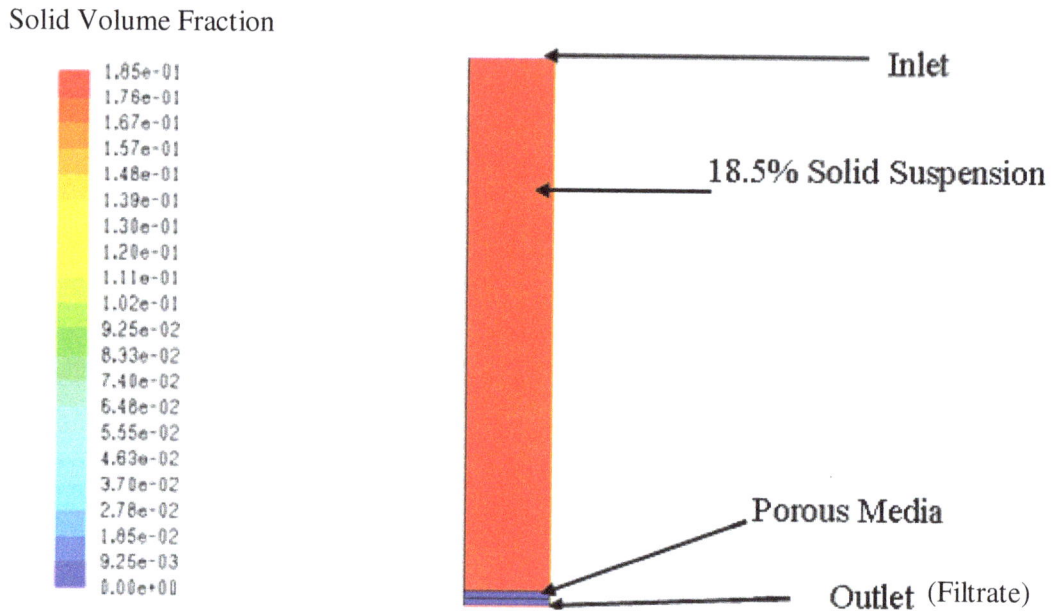

Figure 5. Uniform solid volume fraction of suspension in the filtration model (half-filtration cell).

Solid volume fraction

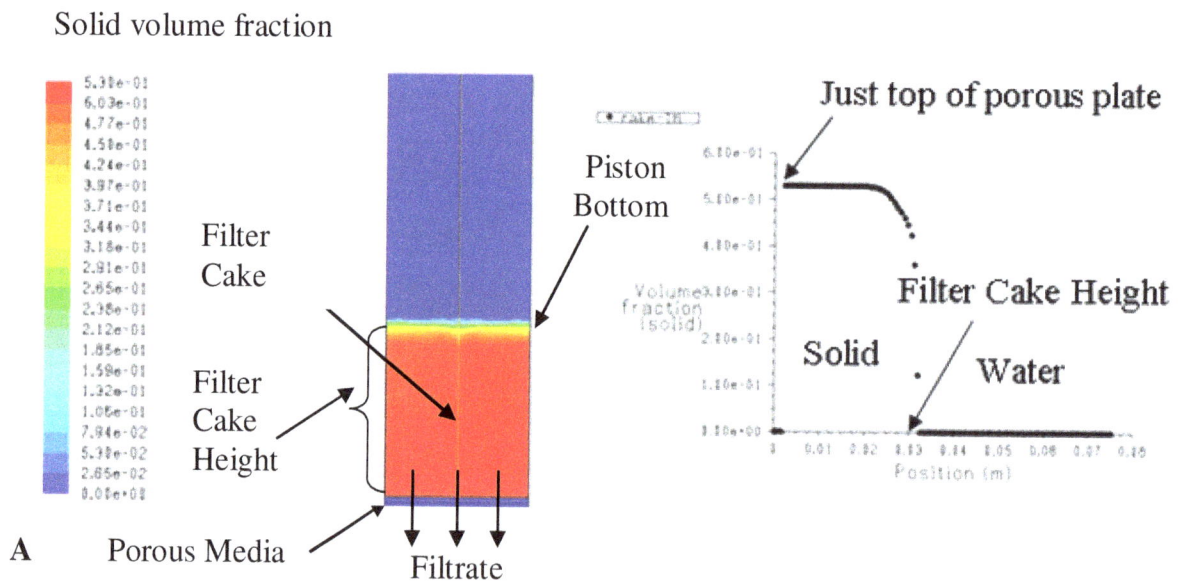

A   B

Figure 6. Filter cake formation and profile; a) solid volume fraction distribution; b) solid volume fraction profile at x = 7 mm.

used as an initial condition for simulation, as can be seen in Figure 5. The simulated fluid had the same properties as that of drilling fluid.

Figure 6a shows the simulated filter cake (red) above the porous media. The solid concentration at the bottom increased to about 0.53 due to the filter cake's formation. The interface between the cake and the fluid has a lower solid concentration of around 0.48. This qualitative simulation results are consistent with experimental observations. Figure 6b shows a quantitative solid concentration

profile along the vertical axis at radial location x = 7 mm from the left edge. It shows no solid present 30 cm above the porous media.

Similarly, we have extracted the filter cake height from the simulated filtration cell. In order to compare the analytically calculated filter cake height, a theoretical equation was used involving the initial solid volume fraction, α_{so}, equilibrium suspension solid fraction, α_{se}, the initial suspension height, h_o, and equilibrium suspension height, h_e (Usher, 2001). In this equation, the initial solid

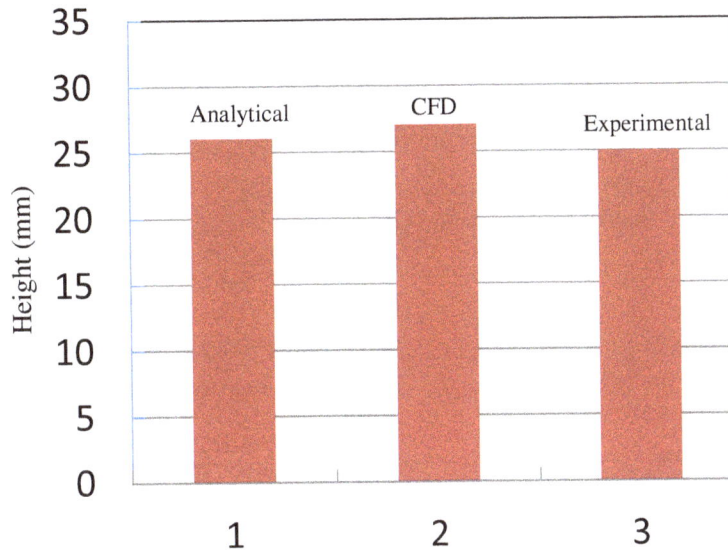

Figure 7. Comparison of filter cake heights.

volume fraction and suspension height of the filtration were α_{so} and $h_{o,}$ respectively. As pressure (P) was applied, filtrate was discharged through the outlet, with the final equilibrium solid volume fraction and the suspension height of α_{se} and h_e, respectively. The conservation of particles mass leads to equation (26) (Usher 2001):

$$h_e = \left(\frac{h_o}{\alpha_{se}} \right) \alpha_{so} \qquad (26)$$

In order to experimentally validate the filter cake thickness, we compared experimental pressure filtration filter cake data of iron ore suspension (Saha, 2009) with CFD simulated and analytically calculated filter cake heights. The details of the iron ore suspension pressure filtration can be found elsewhere (Saha, 2009).

Analytical, experimental, and numerical results of filter cake heights compare reasonably well as shown in Figure 7. The numerical CFD model shows a slightly higher filter cake height compared to analytical results (Figure 7). This is due to the fact that the CFD method accounts for the non-uniform distribution of particles with loosely packed particles near the piston and tightly packed particles near the vicinity of the porous media. The analytical approach assumes uniform distribution of particles throughout the filtration cell.

RESULTS AND DISCUSSION

CFD simulations were performed to simulate filter cake formation in the radial direction on vertical well walls during deep and shallow drilling operations. The initial conditions for both deep and shallow wells are shown in

Figure 3 and described earlier. Here, the drilling pipe is filled with drilling fluid and the bottom portion of the well with rock debris/cuttings (Figure 3). Both initial and boundary conditions were similar to conditions found in field drilling operations (Vaussard, 1986; Delhommer, 1987; Sherwood, 1991; Sherwood, 1991; Rogers, 1996; Maurer, 1997; Cerasi, 2001; Usher, 2001; Parn-anurak, 2003; Spooner, 2004; Ali, 2006; Fisher, 2008; Berry, 2009; Hamed, 2009; Wikipedia, 2010). The details of these conditions for shallow and deep drilling simulations are provided in Table 2.

Deep drilling simulation with 45 -μm particles

The deep drilling process was simulated by setting high-pressure (25,500 psi or 175.8 MPa) and high-temperature (170 °C) conditions at the inlet and bottom portion of the model. The bottom portion was maintained at the same pressure and temperature as that of inlet. It was assumed that the pressure and temperature variations over a 1- m long model are negligible. Following other researchers (Parn-anurak, 2003), formation porosity was assumed to be 0.2. Drilling fluid was pumped down into the drilling zone through the drilling pipe where rock debris mixed with the drilling fluid and was carried away through the annulus. The pressure in the wellbore was maintained higher than the surrounding porous media formation to mimic actual drilling conditions. The differential pressure in the annulus forced the fluid phase through the porous media formation and deposited solid particles in the form of filter cake on the rock surface, as shown in Figure 8; filter cake is defined here as a solid volume fraction above 0.4 at the wall. According to Cerasi and Soga (2001), filter cake grows on the wall in a process similar to soil consolidation, where overbalanced

Table 2. Initial conditions and fluid/formation properties.

Parameter	Shallow well	Deep well
Inlet pressure (drilling fluid/top), psi or MPa	2,000 or 13.8	25,500 or 175.8
Pressure (bottom), psi or MPa	2,000 or 13.8	25,500 or 175.8
Outlet pressure, psi or MPa	1,500 or 10.3	25,000 or 172.4
Formation pressure (porous media), psi or MPa	1,500 or 10.3	25,000 or 172.4
Particle size, μm	45	45 and 7
Formation porosity	0.2	0.2
Temperature, °C	30	170
Solid fraction (drilling fluid/top)	0.2	0.2
Solid fraction (rock/bottom)	0.8	0.8
Particle density, kg/m^3	2,350	2,350
Fluid density, kg/m^3	999	999

Solid volume fraction

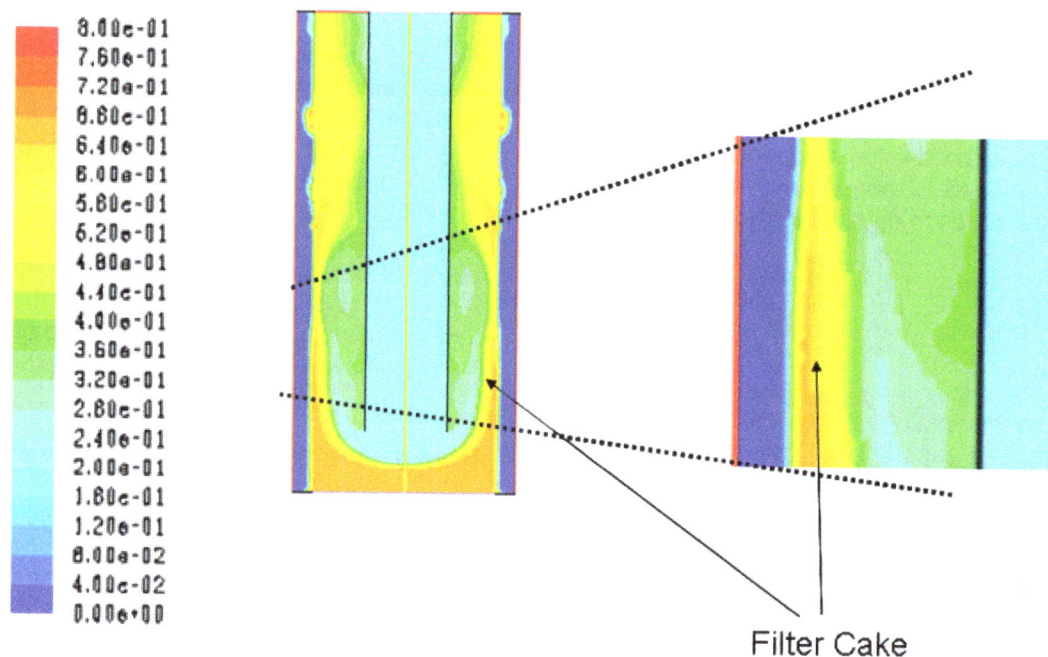

Figure 8. Deep drilling: Filter cake formation on the wellbore wall (solid volume fraction).

pressure will initially force some drilling fluid into the formation, and the solids present in the drilling fluid will clog the pores of the formation and accumulate against the wall under appropriate conditions. As the pressure difference between the wellbore and the formation forces the filter cake to consolidate, the fluid phase (filtrate) invades the formation. The solid particles become more tightly packed, reducing the permeability of the growing cake and, hence, the fluid invasion (Cerasi and Soga, 2001).

The simulated filter cake, as presented in Figure 8, shows that cake forms in non-uniform shapes. This is qualitatively in good agreement with the literature, which reports that non-uniform filter cake forms on the vertical porous rock surface (Sherwood, 1991a, b). Figure 9 shows the solids velocity vector distribution at high-pressure and high-temperature drilling process. Here, the solids maintain nonuniform solid velocities in the annulus, and vortices at the wellbore bottom portion. At the well bottom, the fluid from the pipe encounters drilled solid particles and is forced to drastically change its trajectory. This explains the vortices predicted at the bottom hole.

Solids Vector

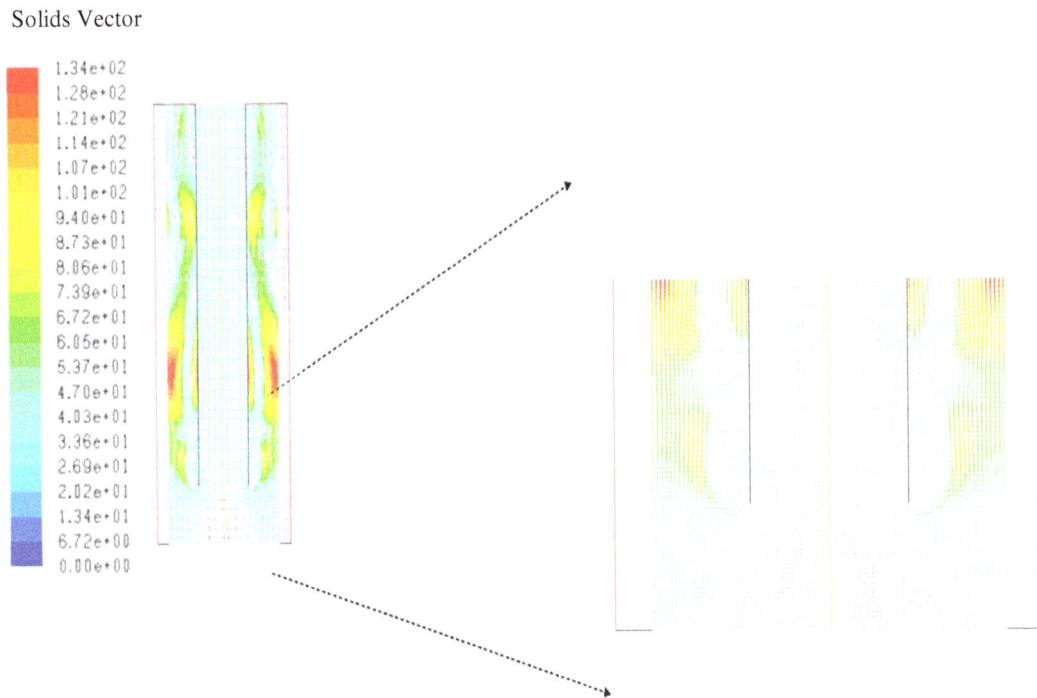

Figure 9. Solids vector distribution.

The vortices also induce nonuniformity in the flow. The ascending fluid loaded with particles rises to the top due to the imposed differential pressure between the hole bottom and the top.

Figure 10a to c show the simulated filter cake in the deep wellbore wall for 45 µm particle in drilling fluid. Figure 10a qualitatively shows filter cake thickness with thinner cake in the lower bottom of the well followed by thicker cake at the upper portion of the well. Figure 10b displays solid volume fractions at different well heights from 0.05 to 0.9 m. Figure 10c exhibits the filter cake thickness extracted from the solid volume fraction graph (Figure 10b). It shows the filter cake thickness versus well heights. The average filter cake thickness varies from 0.023 m near the bottom well to 0.05 m near the top portion. This clearly implies that the simulated filter cake formed on the wellbore wall was non-uniform. This is consistent with experimental observations.

Comparison of deep and shallow drilling simulations

Additionally, nonuniform filter cake was observed in the shallow wellbore, as presented in Figure 11. Figures 11a to c show the simulated filter cake thickness in the shallow wellbore wall with 45 µm particles in the drilling fluid. Figure 11a qualitatively shows the filter cake thickness as thinner in the lower bottom of the well followed by a thicker cake at the upper portion of the well, while Figure 11b shows the solid volume fractions for well heights from 0.05 to 0.9 m.

The filter cake thicknesses were extracted from the solid fraction graph for shallow drilling conditions (Figure 11b) and presented in Figure 11c. It shows the cake thickness over well heights. The average cake thickness next to the bottom was 0.011 m, whereas the average cake thickness near the top was 0.023 m (Figure 11c). The filter cake pattern for shallow drilling conditions was observed to be similar to that of deep drilling conditions.

The non-uniformity of the filter cake thickness in both deep and shallow wells correlates with the magnitude of the vortices in the annulus. It appears large vortices have adverse effects on the filter cake formation with thin filter cake thickness in the bottom region where the vortices intensities are larger and thicker filter cake at the top portion where the vortices intensities are significantly smaller. Hence, vortices' intensities play an important role in the formation of non-uniform filter cake in the well annulus.

Figure 12 compares the simulated filter cake thickness for both deep and shallow drilling conditions. It clearly shows thicker cake for deep drilling conditions. The average cake thickness is 0.04 m for deep drilling processes and 0.0125 m for shallow drilling. Hence, the higher pressure and temperature environment favors thicker cake formation.

Figure 13 compares the solid volume fraction near the well wall for deep and shallow drilling conditions at different heights. It shows higher solid volume fractions for deep drilling. Here, the particles are more consolidated

Solid fraction

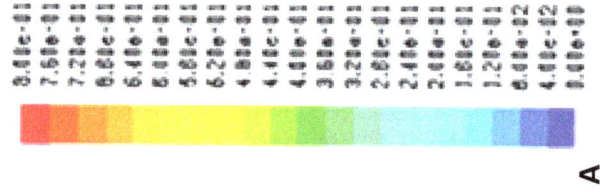

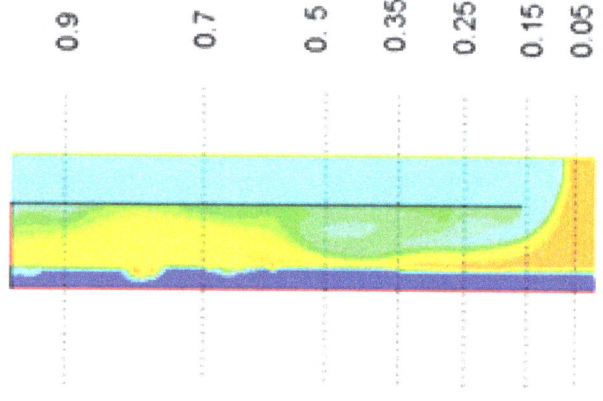

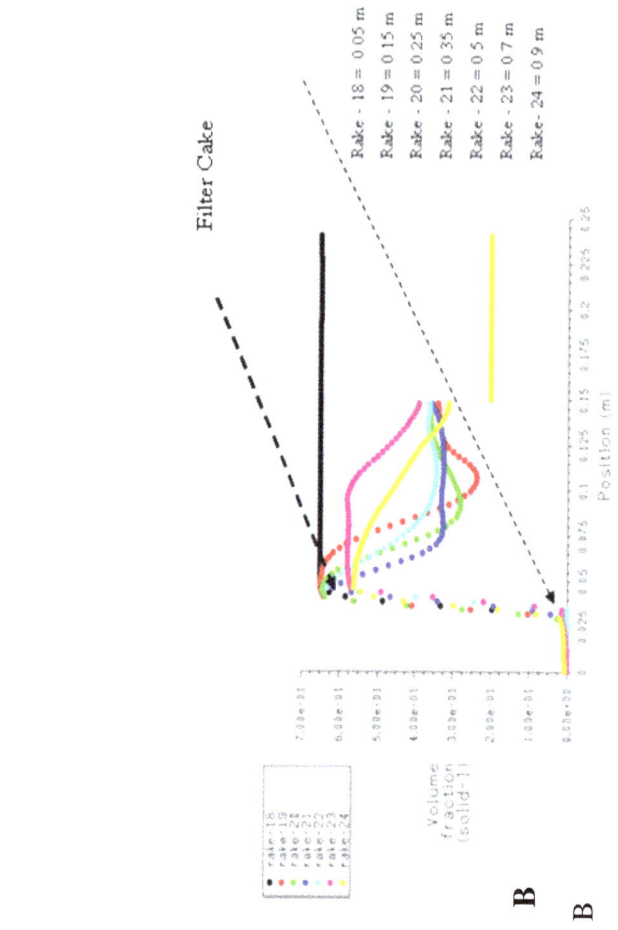

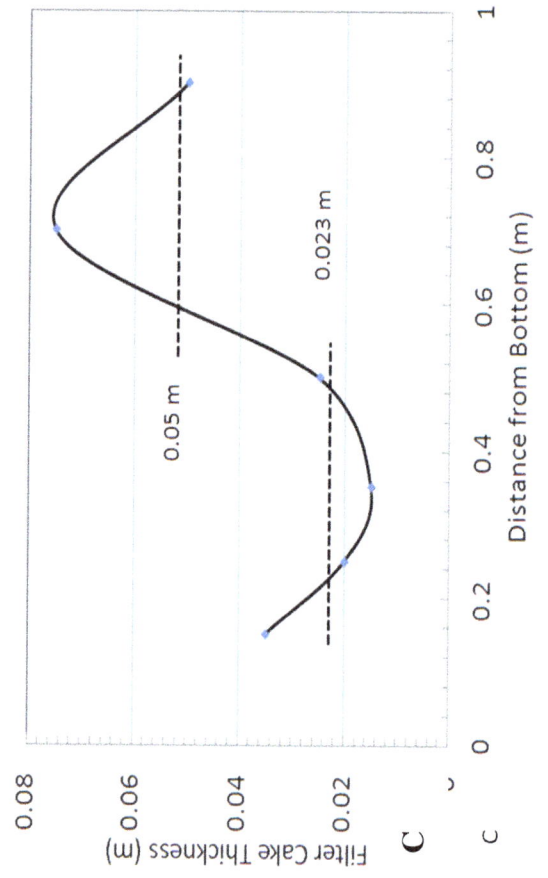

Filter Cake

Rake - 18 = 0 05 m
Rake - 19 = 0 15 m
Rake - 20 = 0 25 m
Rake - 21 = 0 35 m
Rake - 22 = 0 5 m
Rake - 23 = 0 7 m
Rake - 24 = 0 9 m

0.9

0.7

0.5

0.35

0.25

0.15

0.05

Figure 10c. Deep drilling: Filter cake thickness at different heights of wellbore from bottom; a) qualitative solid volume fraction of well; b) solid volume fraction at different well heights; c) filter cake thickness over well heights.

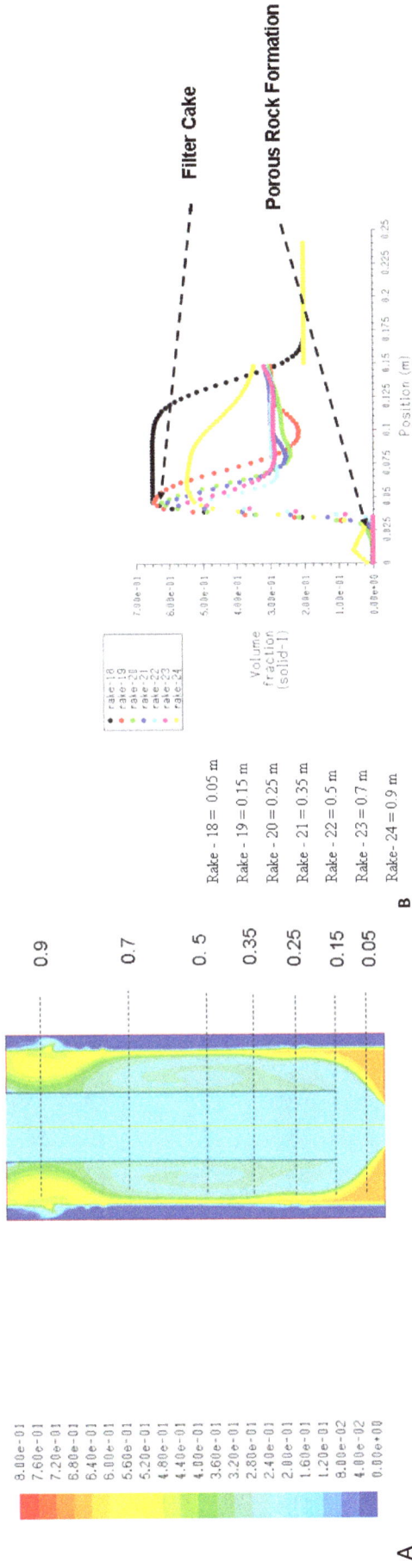

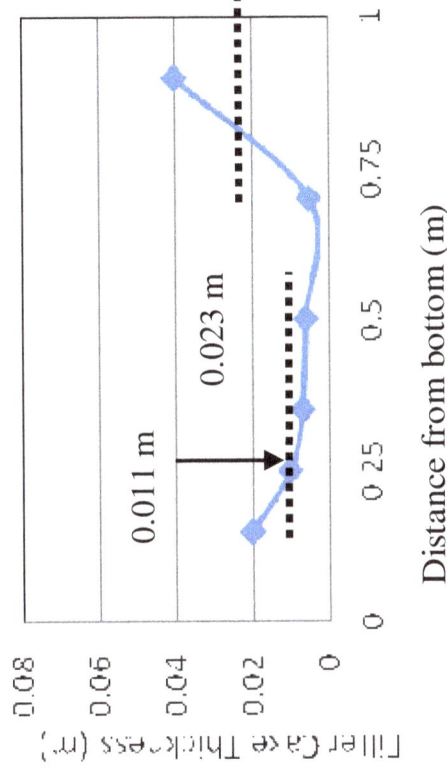

Figure 11a. Shallow drilling: Filter cake thickness at different heights of wellbore from bottom; a) qualitative solid volume fraction of well; b) solid volume fraction at different well heights; c) filter cake thickness over well heights.

due to higher pressure, compared to shallow drilling conditions where the particles remain loosely packed with a lower solid volume fraction. Hence, the higher pressure primarily explains the simulated observations.

Comparison of drilling simulations for 7- and 45- μm particles

Figure 14 shows the simulated filter cake in the deep drilling process with 7-μm particles drilling

fluid. Figure 15 compares the filter cake thicknesses for deep drilling with two drilling fluids that differ solely by the particle sizes: 45- and 7- μm in deep drilling conditions. Clearly, the 45 μm drilling fluid leads to a much thicker filter cake, with a

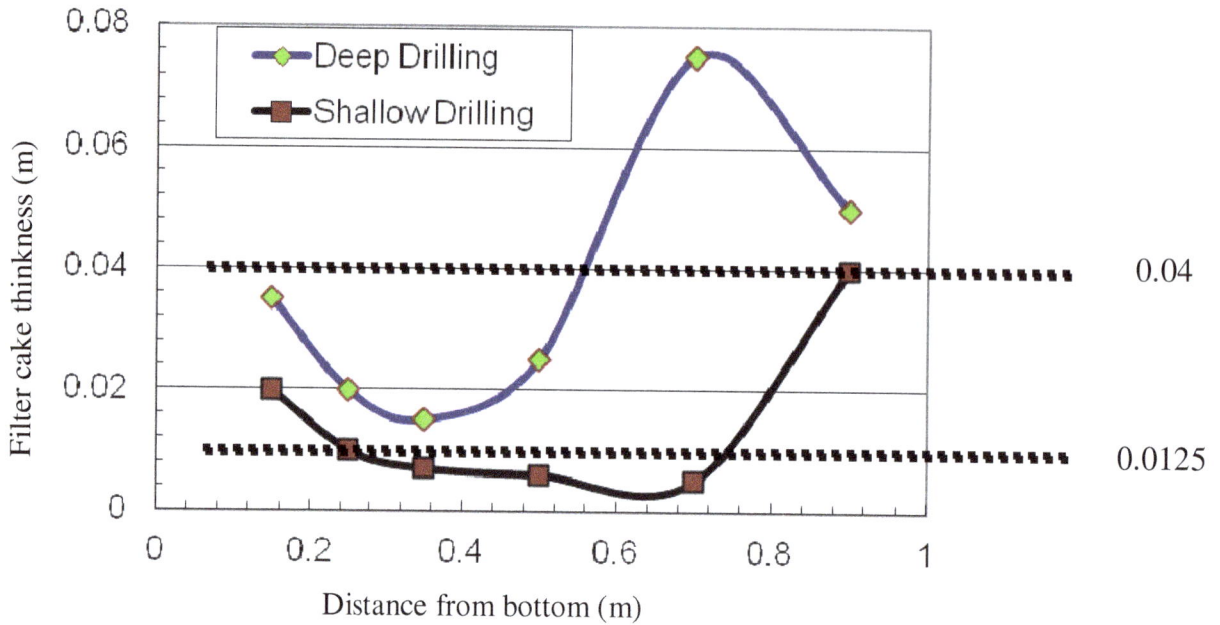

Figure 12. Cake thickness at different heights for deep and shallow drilling.

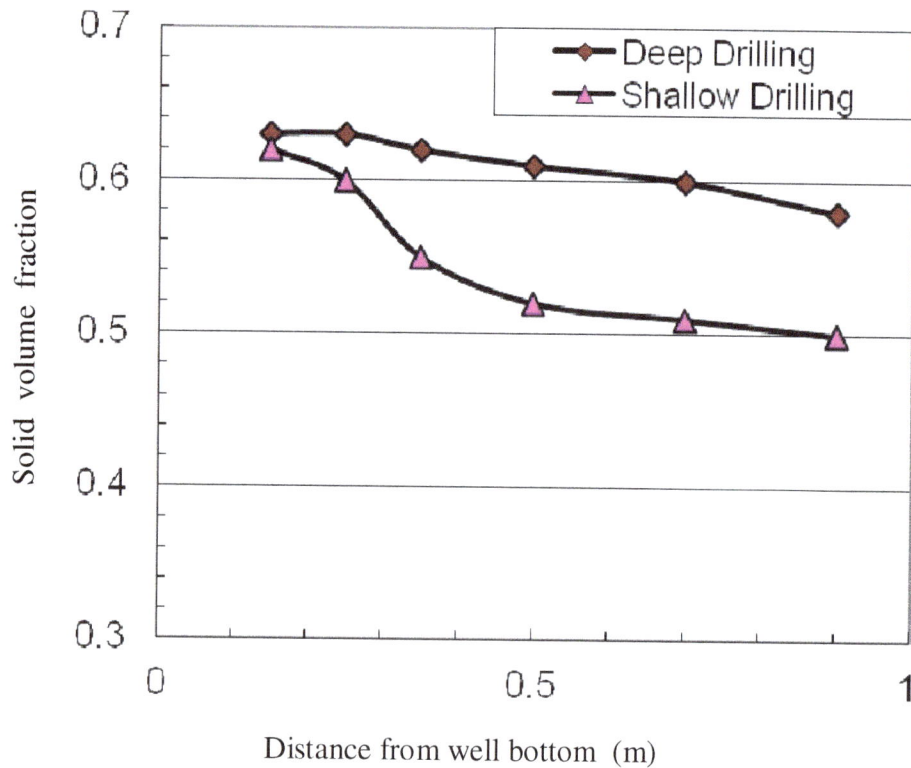

Figure 13. Cake solid volume fraction at different heights for deep and shallow drilling.

average value of 0.04 m compared to 0.008 m for 7- μm drilling fluid. In fact, larger particles formed filter cake five times thicker than smaller particles. Larger particles tend to clog the pores easily, followed by the accumulation of particles on the wall, and this leads to thicker cake formation. Smaller particles tend to travel through the pores easily due to their size; hence, initially they are retained in only a very little amount on the wall to form the filter

Solid Fraction

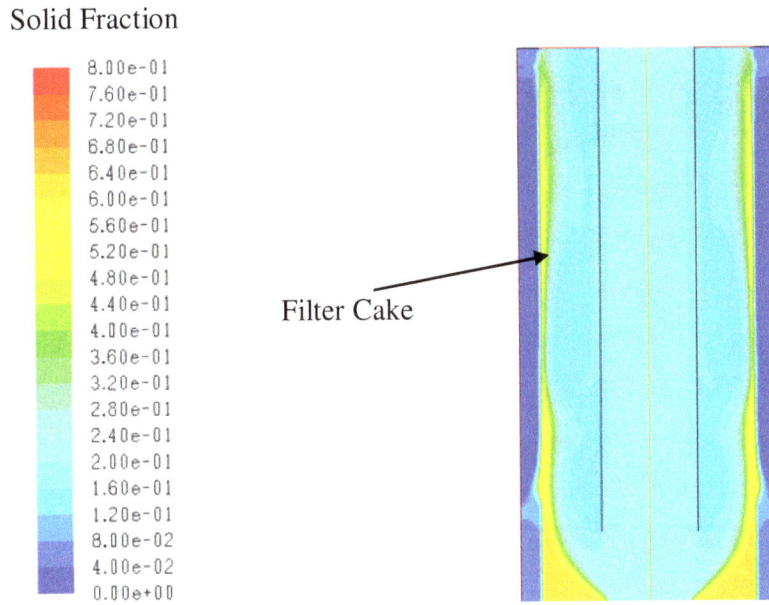

Figure 14. Deep drilling: Qualitative filter cake at different well heights with particle size of 7 μm

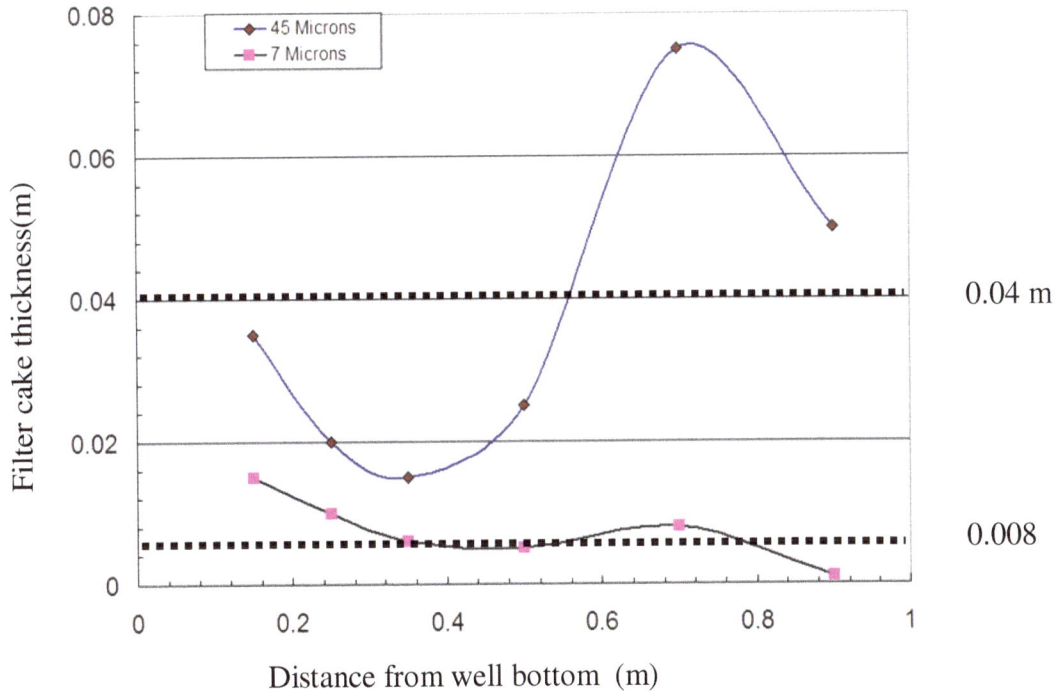

Figure 15. Filter cake thickness at different well heights (high-pressure, high-temperature) for particle sizes of 45 and 7 μm.

cake.

Figure 16 compares the solid volume fraction at x = 0.005 m from the wall for two drilling fluids of different particle sizes of 45- and 7- μm in deep drilling conditions. As expected from our previous simulated results, the solid volume fraction is higher for larger particles. Filter cake thickness optimization based on particle sizes in drilling fluid is important to address in drilling issues arising from very thick cake formation on the wellbore wall.

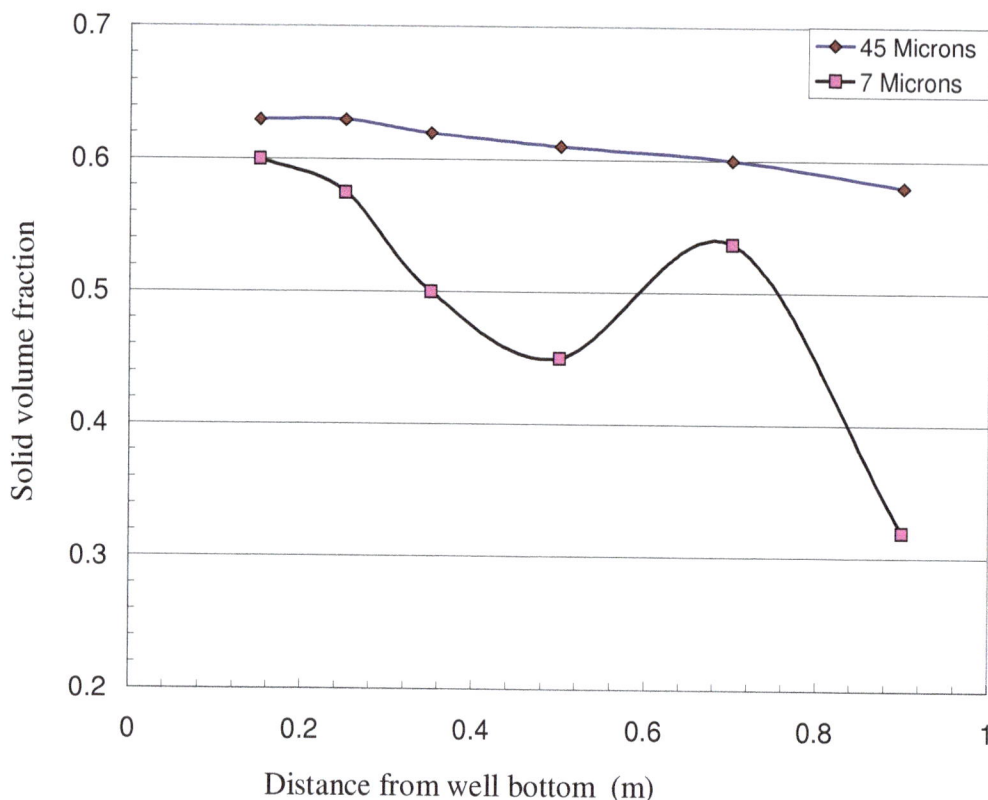

Figure 16. Solid volume fraction of filter cake at different well heights (high-pressure, high-temperature) for particle sizes of 45 and 7 μm.

Conclusion

We have successfully simulated the filter cake formation on the porous rock formation of a vertical wellbore for both deep (5 miles beneath the earth surface) and shallow (0.4 mile) drilling processes using a CFD code FLUENT. The computer-generated filter cakes on porous rock formations are deposited in irregular shapes. This is in agreement with both experimental and analytical claims. The intensity of the vortices observed on the drilling pipe appears to explain the formation of non-uniform filter cake on the well wall. Filter cake thickness and solid volume fractions are higher for extreme drilling processes compared to shallow drilling processes. We believe the higher pressure and temperature surrounding is responsible for thicker cake in the deep drilling process. A parametric study on the effects of drilling fluid particle size clearly shows that larger particles form thicker filter cake compared to smaller particles. Larger particles tend to clog the pores of the porous rock formation while small particles penetrate through the porous rock formation. Hence, it is recommended to use larger particle size in drilling fluids to promote the formation of filter cake, which leads to the prevention of drilling fluid loss through the formation. The model described here may be used to optimize the filter cake thickness during

deep and shallow drilling processes for the production of oil and gas.

ACKNOWLEDGEMENTS

This research was supported in part by an appointment to the National Energy Technology Laboratory Research Participation Program, sponsored by the U.S. Department of Energy and administered by the Oak Ridge Institute for Science and Education. We thank our recent Division Directors J. Brown and J. Thornton as well as our focus area leader G. Guthrie for facilitating this research project.

REFERENCES

Ali S (2006). "Reversible drilling-fluid emulsions for improved well performance." Oilfield Review.
Berry JH (2009). "Drilling Fluid Properties & Function, CETCO Drilling Products." from http://www.getco.com.
Cerasi P, Soga K (2001). "Failure modes of drilling fluid filter cake." Geotechnique, 51(9): 777-785.
Cornelissena JT, Toghipour F, Escudiéa R, Ellisa N, Gracea JR (2007). "CFD modeling of a liquid–solid fluidized bed." Chem. Eng. Sci., 62: 6334–6348.
Delhommer HJ, Walker CO (1987). "Method for controlling lost circulation of drilling fluids with hydrocarbon absorbent polymer" US

Patent Number, 4: 633-950.

Ding J, Gidaspow D (1990). "A bubbling fluidization model using kinetic theory of granular flow." J. A.I. Chem. Eng., 36: 523–538.

Ferguson CK, Klotz JA (1954). "Filtration from mud during drilling." Trans AIME, 201: 29-42.

Fisher KA, Wakeman RJ, Chiu TW, Meuric OFJ (2008). "Numerical modeling of cake formation and fluid loss from non-Newtonian mud's during drilling using eccentric/concentric drill strings with/without rotation." Trans I. Chem. Eng., 78(Part A): 707-714.

FLUENT Inc., (2006) FLUENT 6.3. Lebanon, New Hampshire.

Fordham EJ, Ladva HKJ, Hall C, Baret JF, Sherwood JD (1988). Dynamic filtration of bentonite muds under different flow conditions. 63rd Annual SPE Conference. Houston, Texas SPEFu LF, Dempsey BA (1998). "Modeling the effect of particle size and charge on the structure of the filter cake in ultra-filtration" J. Membrane Sci., 149: 221-240.

Gidaspow D (1994). "Multiphase flow and fluidization: continuum and kinetic theory descriptions" Academic Press, Boston.

Gidaspow D, Bezburuah R, Ding J (1992). Hydrodynamics of circulating fluidized beds, kinetic theory approach In Fluidization VII, Proceedings of the 7th Engineering Foundation Conference on Fluidization.

Hamed SB, Belhadri M (2009). "Rheological properties of biopolymers drilling fluids" SPE, 67: 84-90.

Ishii M (1975). Thermo-fluid dynamic theory of two-phase flow. Collection de la Direction des Etudes et Recherches d'Electricite de France 22. . Eyrolles. Paris. Jackson R (1997). "Locally averaged equations of motion for a mixture of identical spherical particles and a Newtonian fluid." Chem. Eng. Sci., 52(15): 2457-2469.

Johnson PC, Jackson R (1987). "Frictional-collisional constitutive relations for granular materials, with application to plane shearing". J. Fluid Mech., 176: 67-93

Johnson PC, Nott P, Jackson R (1990)."Frictional–collisional equations of motion for participate flows and their application to chutes." J. Fluid Mech., 210:501-535

Jung J, Gamwo IK (2008). "Multiphase CFD-based models for chemical looping combustion process: Fuel reactor modeling". Powder Technol., 183: 401-409.

Lun CKK, Savage SB, Jerey DJ, Chepurniy N (1984). "Kinetic Theories for Granular Flow: Inelastic Particles in Couette Flow and Slightly Inelastic Particles in a General Flow Field." J. Fluid Mech., 140: 223-256.

Maurer Engg. Inc. (1997). "Wellbore thermal simulation model: heory and User's Manual." MAURER ENGINEERING INC. Myöhänen K, Hyppänen T, Kyrki-Rajamäki R (2006). "CFD modeling of fluidized bed systems." SIMS Finland.

Outmans HD (1963). "Mechanics of static and dynamic filtration in the borehole." SPE, 228(236).

Parn-anurak S (2003). "Modeling of fluid filtration and near-wellbore damage along a horizontal well." New Mexico Institute of Mining and Technology. PhDThesis,USA.

Peden JM, Avalos MR, Arthur KG (1982). "The analysis of the dynamic filtration and permeability impairment characteristics of inhibited water based muds." SPE Formation Damage Control Symp. Lafayette.

Rogers HE, Murray DA, Webb ED (1996). "Apparatus and method for removing gelled drilling fluid and filter cake from the side of a wellbore." USA, pp. 5564-500.

Saha H (2009). "Pratical application of filtration theorey to the minerals industry." The University of Melbourne, Australia, PhD Thesis,

Sherwood JD, Meeton GH, Farrow CA, Alderman NJ (1991). "Concentration profile within non-uniform mudcakes." J. Chem. Soc. Far. Trans, 84(4): 611(b).

Sherwood JD, Meeten GH, Farrow CA, Alderman NJ (1991). "Squeeze-film rheometry of non-uniform mudcak." J. Non-Newtonian Fluid Mech., 39: 311-334(a).

Spooner KM, Bilbo D, McNeil B (2004). "The application of high temperature polymer drilling fluid on Smackover operations in Mississippi." AADE-2004 Drilling Fluids Conference. Houston, Texas.

Usher SP, Kretser RG, Scales PJ (2001). "Validation of a new filtration technique for de-waterability characterization." AIChE J., 47(7): 1561-1570.

Vaussard A, Martin M, Konirsch O (1986). "An experimental study of drilling fluids dynamic filtration." SPE, 15412P.

Vaussard A, Martin M, Konirsch O andPatroni JM (1986). "An experimental study of drilling fluids dynamic filtration." 61st Annual Technical Conf. New Orleans, SPE.

Wikipedia (2010). "Oil well," http://en.wikipedia.org/wiki/oil_well.well

21

Chemical method for monitoring demulsification in oil industry

Hikmat S. Al-Salim[1]*, Ahmmed Saadi Ibrahim[1] and Mohammed S. Saleem[2]

[1]Department of Chemical and Petroleum Engineering, UCSI University, Kuala Lumpur, Malaysia.
[2]Chemical Department, University of Mosul, Baghdad, Iraq.

A special and fast chemical method to detect and determine water separation from emulsified crude with accuracy (>97%) was developed. The method was applied successfully to determine the total amount of water in crude and its derivatives. Most methods available for the detection and estimation of low concentrations of water require either sophisticated equipment and/or take a long time. Chemical method depends on a chemical reaction between calcium hydride and water (both soluble as well as suspended droplets). The hygroscopic nature of some petroleum derivatives was followed with this method.

Key words: Diesel fuel, gasoline, kerosene, reaction, demulsification, and water in crude.

INTRODUCTION

Presence of water even in trace amounts in crude or its derivatives such as ATK, kerosene, gasoline diesel fuel, lubricating or hydraulic oil can cause many undesirable effects such as corrosion, raise conductivity, leaching of additives etc. Most methods available for detection and estimation of low concentrations of water require either sophisticated equipment and/or take a long time (Hobson, 1975). There are many different methods for the determination of water in petroleum or its derivatives. These methods can be divided into two groups: Those which depend on measuring a change in a physical property, and those which depend on chemical reaction. Figure 1 shows both main groups (Nelson, 1958; Evans, 1977). This method is capable of detecting and measuring the high accuracy of the total water in crude and its derivatives.

METHODOLOGY

The method is based on a reaction which takes place between calcium hydride and water, be it suspended or dissolved in the medium. The general chemical and physical properties from (New report for, "Acetylene production technology", 2007; Kirkpatrick, 1976).

The generated volume of hydrogen is measured and the quantity

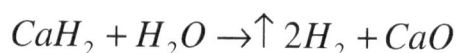

$$CaH_2 + H_2O \rightarrow \uparrow 2H_2 + CaO$$

of water present can be found from the chemical reaction. Trial methods showed that concentrations as low as ppm can be measured accurately. The method is rather simple and fast. Figure 2 shows a diagram for the setup used. One main application of the method was to monitor demulsification process in crude. Depends on an expermental samples specify for expermental work.

This method was used also to monitor the hygroscopic nature of some petroleum derivatives such as kerosene, ATK and gas oil. The results are outlined in Table 1.

Demulsifier[1] was added to a known volume of crude containing some water. The demulcifier was added in trace quantaties (ppm). Not only various concentrations of the additive were examined but also the temperature was varied between 40 to 60°C for each additive concentration. The settling time varied between 1 to 3 h at each temperature. In all about 45 tests were carried out and the results are outlined in Table 3. In each case 150 g of the crude oil was placed in along glass tube 8 cm in diameter and 30 cm in length. Additive was added then stirred mechanically for one hour, during which the glass tube was immersed in a water bath maintained at the required temperature. Settling was carried out at this temperature. Then three equal samples were drawn from the crude oil one from the top of the tube, the second from the middle, and finally the third from the bottom of the tube. These three samples were then analyzed with the hydride method for water content. From the monitor can be measured the amounts of gas hydrogen then depends on chemical reaction can be measured

---

*Corresponding author. E-mail: hikmatsaid@ucsi.edu.my.

---

[1]The demulcification additive used was a commercial one and of unknown structure.

Table 1. Measure the accuracy of chemical method.

Quantity of water in mol/ 50 g (10^{-4})			Mean (10^{-4})	Standard deviation (10^{-4})	Percentage error
Added	**Found**	**Difference**			
2.777	2.7730	0.004	2.7404	0.025	0.144
	2.7600	0.017			0.612
	2.7320	0.045			1.620
	2.7150	0.062			2.230
	2.7220	0.055			1.980
5.555	5.5580	0.003	5.5157	0.029	-0.054
	5.5140	0.041			0.738
	5.4890	0.066			1.188
	5.5020	0.053			0.954
8.333	8.3440	-0.011	8.3166	0.070	-0.132
	8.3200	0.013			0.156
	8.2800	0.053			0.636
	8.4130	-0.080			-0.960
	8.2260	0.107			1.284
16.666	16.6630	0.003	16.618	0.050	0.018
	16.5620	0.104			0.624
	16.6600	0.006			0.036
	16.5890	0.077			0.462

Table 2. Water absorption by three different petroleum products.

Sample	Time (h)	Water content in mole/kg x10^{-4}
Gas oil	3	6.48
	4	13.69
	5	15.51
	6	17.42
	8	19.35
Kerosene	1	7.09
	1.5	7.75
	3	14.26
	4	17.66
	5	23.28
A.T.K.	1	2.53
	4	7.83
	6	12.63
	7	15.47

amounts of moisture. Table 1 gives a very good identifcation about the accuracy of this method.

RESULTS AND CALCULATIONS

From the reaction that has been previously described, it is seen that the volume of hydrogen used to calculate the mole of hydrogen can be measured. Depending on the chemical reaction, each 1 mole from moisture equal 2 moles from hydrogen. So, results in Table 1 illustrate the accuracy of this method. In each case, a known amount of water was added to (150 g) crude and was analyzed. These results indicate that the method employed was

Table 3. Demulsification of crude please to find the total amount of water in organic crude.

Demulsifier concentration (ppm)	Temp. (°C)	Settling time (h)	Water content (ppm.)		
			Top of the tube	Middle of the tube	Bottom of the tube
30	40	1	1129.3	1381.6	1493.8
	40	2	1334.4	1396.0	1274.3
	40	3	1307.5	1340.4	1356.0
	50	1	1125.3	1253.5	1625.9
	50	2	1205.4	1261.5	1534.6
	50	3	1059.2	1038.8	1905.8
	60	1	1518.9	1126.1	1359.2
	60	2	1276.7	1221.0	1506.6
	60	3	1318.8	1301.9	1383.6
40	40	1	1251.5	1181.4	1571.1
	40	2	1068.5	1226.6	1707.6
	40	3	1390.2	1127.5	1487.3
	50	1	1151.4	1279.5	1573.0
	50	2	1256.3	1381.2	1366.8
	50	3	1080.8	1223.4	1700.4
	60	1	1162.2	1135.3	1706.0
	60	2	1246.6	1125.7	1631.5
	60	3	1249.5	1220.2	1533.8
50	40	1	1239.5	1165.4	1597.9
	40	2	1274.7	1178.6	1551.2
	40	3	1257.1	1277.1	1470.1
	50	1	1302.3	1306.7	1395.2
	50	2	1151.8	1331.9	1520.6
	50	3	1164.9	1323.6	1515.3
	60	1	1116.2	1225.8	1661.0
	60	2	1169.5	1224.2	1610.3
	60	3	1164.6	1294.7	1545.3
60	40	1	1328.4	1139.7	1535.8
	40	2	1142.5	1204.9	1656.7
	40	3	1072.5	1172.5	1759.2
	50	1	881.0	1213.8	1907.9
	50	2	1175.8	1045.2	1782.9
	50	3	1338.3	1020.4	1641.9
	60	1	1426.9	949.1	1627.2
	60	2	1344.9	1077.3	1581.8
	60	3	1265.8	1182.9	1555.5
70	40	2	1142.1	1305.1	1557.0
	40	3	1212.6	1216.2	1575.0
	50	1	1283.5	1689.2	1031.2
	50	2	1382.8	1439.3	1182.1
	50	3	1476.5	1292.3	1235.5
	60	1	1180.6	1420.2	1341.6
	60	2	1257.5	1524.2	1221.4
	60	3	1372.0	1251.1	1381.3

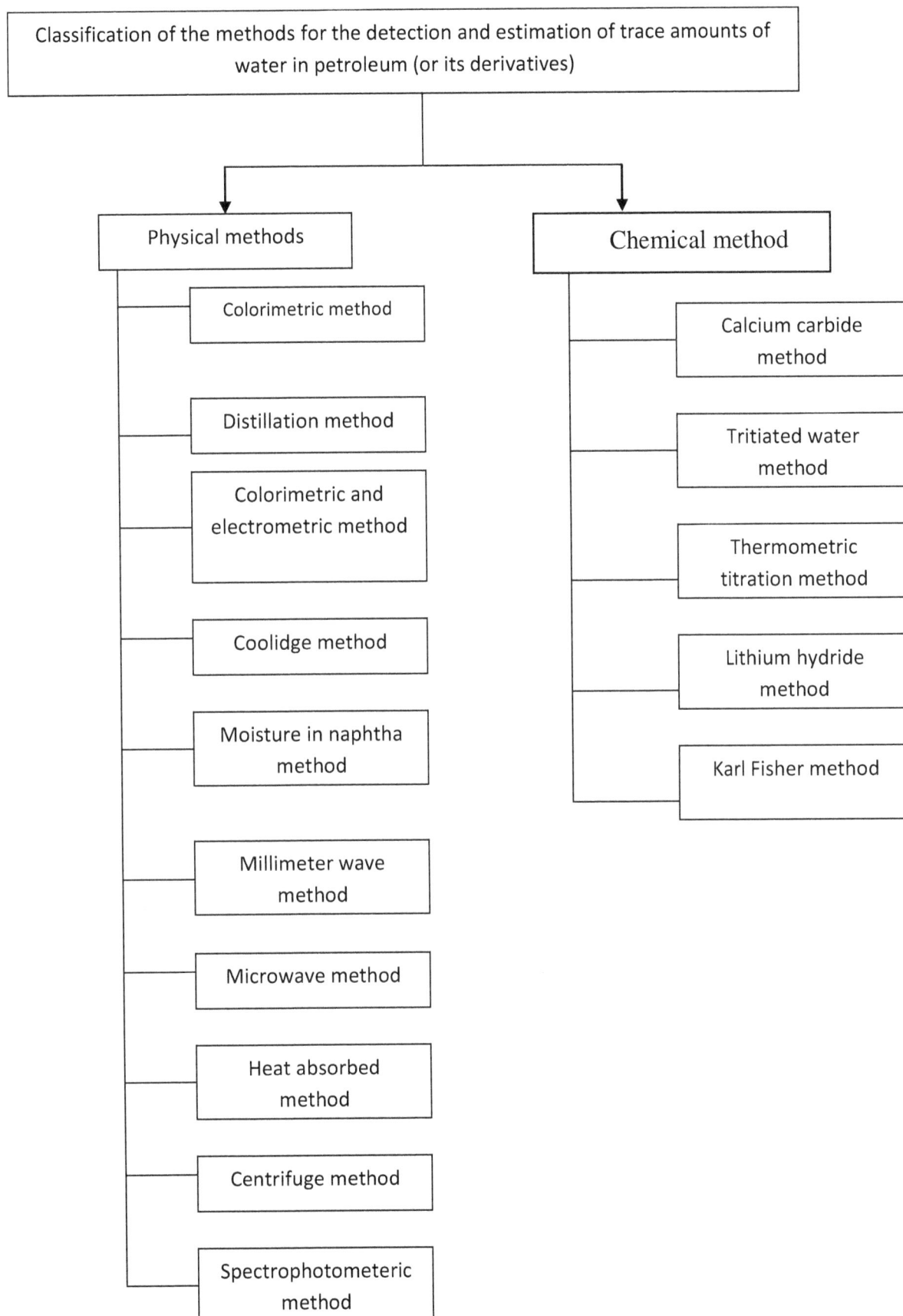

Figure 1. Main methods of detection and estimation of trace amounts of water in petroleum (or its products).

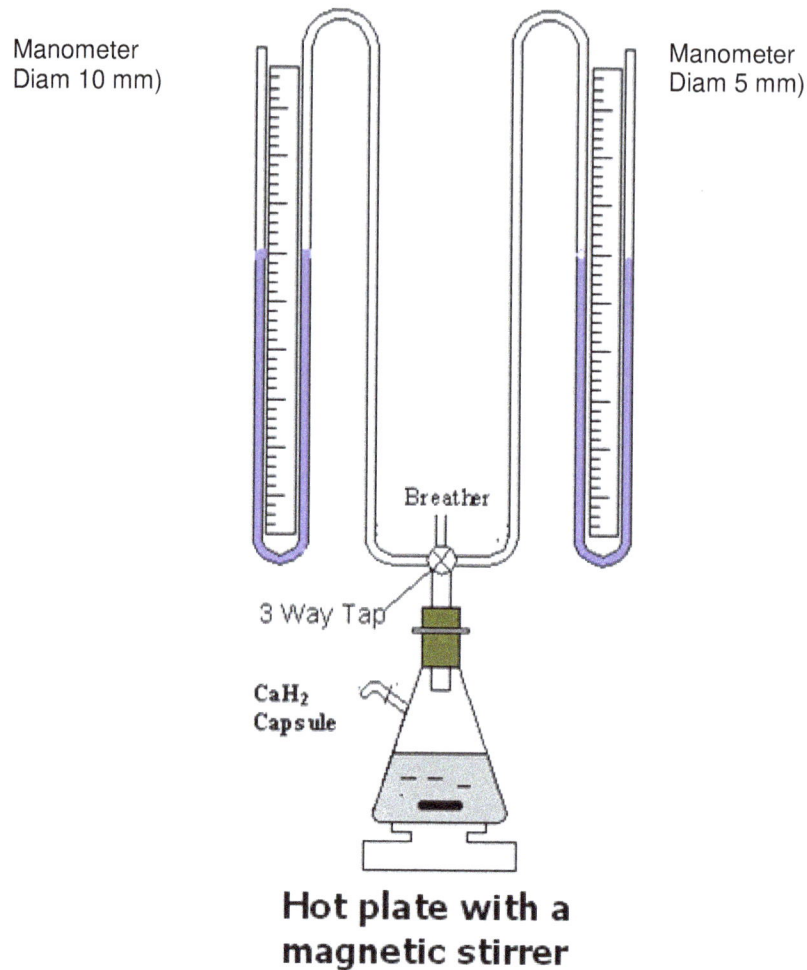

Figure 2. Schematic diagram for the apparatus used.

rather accurate in determining water content with high accuracy. The method was applied in other cases such as:

1. To monitor the hygroscopic nature of some petroleum products (Gas oil, Kerosene and A.T.K) the results are illustrated in Table 2.
2. To follow the settling time during demulcification at three different concentrations and three different temperatures, the results are illustrated in Table 3.

Conclusions

This method for determining trace amounts of water in crude or its derivatives was developed based on the reaction of powdered calcium hydride with free and dissolved water. The accuracy was in most cases more than 97%. The method is rather sensitive and was applied to:

1. Monitor the absorption of water from the atmosphere by kerosene, Gas oil and A.T.K.
2. Monitor settling time when demulcifiers are used with crude.
3. Temperature has no big effects on the system.

REFERENCES

Hobson GD(1975). "Modern petroleum technology". Applied Science Publishers Ltd, Britain, 4th edition, p. 230.
Nelson WL (1958). "Petroleum refinery engineering". McGraw-Hill Book Company, Inc., 4th edition, p. 265.
Evans UR (1977). "The corrosion and oxidation of Metals". Edward Arnold Ltd., p. 296.
New report for "Acetylene production technology", (2007).
Kirkpatrick DM (1976)."Acetylene from Calcium Carbide Is an Alternate Feedstock Route," Oil Gas J., June 7.

22

Evaluating the application of foam injection as an enhanced oil recovery in unconsolidated sand

FALODE Olugbenga Adebanjo and OJUMOOLA Olusegun

Department of Petroleum Engineering, University of Ibadan, Ibadan, Oyo State, Nigeria.

In the Niger Delta, low oil recovery rates less than 30% results mainly due to oil production problems such as water coning, wax deposition and high gas/oil ratios. Meanwhile, the remaining oil becomes a good candidate for EOR methods such as CO_2 injection, polymer and foam injection, in-situ combustion and steam injection. But as it stands, the practice of these known methods of enhanced oil recovery is scarce in the Nigeria's Oil and Gas industry. However, this project is aimed at evaluating the application of foam injection as an enhanced oil recovery method in sandstone reservoir and exploring possible improvement of oil production in the Niger Delta. The project was carried out using two cases. In Case 1, a synthetic model built with static modeling software, later imported to dynamic modeling software and simulated to mimic foam injection process while in Case 2 a real life model with an aquifer fully populated with the necessary reservoir and fluid properties and production history. From the results obtained in Case 1, production indices, field oil recovery, etc. were compared with gas flooding process. For foam flooding, a significant increase in oil recovery as compared to gas flooding and reduction in gas oil ratio and gas produced were observed while in Case 2, field oil recovery, oil production rate, cumulative oil production, gas oil ratio and water cut were compared and significant increase oil recovery using foam flooding, and reduced field water cut was also observed. The economic viability of the project in both cases was also investigated using some economic indicators. Improved displacement efficiency resulting into increased recoverable reserves, and subsequently increased total field oil production has been achieved by foam injection.

Key words: Chemical, foam, enhanced oil recovery, modeling, oil production.

INTRODUCTION

Foam is a colloidal dispersion in which a gas is dispersed in a continuous liquid phase. Surfactants are added to the solution to stabilize foam by reducing interfacial tension. The use of surfactant solutions that increase oil recovery has been deeply studied. In the sixties it was proposed to use foams instead of just aqueous surfactant solutions as displacement agents. The first experiments showed that oil from porous structures unrecoverable by conventional water or gas drives could be displaced by foam. The efficiency of the foam was believed to be the result of the high foam viscosity (apparent viscosity) and its penetration in pores of various sizes. Laboratory research has indicated that the foam-drive process can recover a significant proportion of the oil remaining in unconsolidated sand packs subjected to conventional secondary recovery operations. Exerowa and Kruglyakov (1998) reported that researchers experimented with crude oils and unconsolidated porous media, indicating that total recovery increased from 60% (from water flooding) to 90% after foam injection using 36 foaming agents (23 anionic, 6 non-ionic and 7 amphoteric) to establish the effect of foam quality (gas volume fraction), the surfactant kind and concentration, the mode of the foam injection and the foam bank size on the displacement ability of the foam. The main point of these experiments was that the oil recovery changed with the quality of foam and the permeability.

Yan et al. (2006) investigated different factors' effects on sweep efficiency by foam in smooth heterogeneous fractures and applied their theory to that situation assuming the same gas fractional flow in each portion of the fracture and no cross-flow. Their study was based on the fact that foam can reduce viscous fingering and gravity override caused by the low viscosity and density of the gas.

They consider foam to improve efficiency of a surfactant process for oil recovery in a reservoir consisting of multiple fractures separating matrix blocks where oil is retained by capillarity and/or wettability. Yan et al. (2006) concluded that foam can greatly improve the sweep efficiency in a heterogeneous fracture system.

Sweep efficiencies can be affected by gas fractional flow, aperture ratio and bubble size. The use of foams to improve oil recovery has been used in lab scale and has been tested in real reservoirs, according with Blaker et al. (1999), and predictions based on laboratory experiments and simulations seem to match with results of real processes. Opportunities for research to deeply understand all the phenomena in foam processes as stated by Kovscek and Bertin (2002) are numerous.

The behavior of the foam in porous media is related to the connectivity and geometry of the rock. Porous media have several characteristics that are important to the flow of foam, like the size distribution of the pores and throats.

Foam mechanisms for generation and destruction of lamellae depend strongly on the body to size aspect radio. For large pores occupied mainly by the non-wetting fluid, the wetting fluid resides in the corners and in thin wetting films coating the pore walls. The non-wetting phase resides in the central portion of these large pores. Small pores are filled with wetting fluid. Then the wetting phase remains continuous. During two-phase flow, the non-wetting fluid flows in interconnected large pore channels, while wetting fluid flows in interconnected small channels and in corners of non-wetting phase occupied pores because of pressure gradients in the wetting phase.

Bulk foam is present when the length scale confining the fluids is greater than the length scale of the bubbles,

and can be classified as "kugelschaum" (that is, ball foam) and "polyderchaum" (that is, polyhedral foam). In the first category, spherical bubbles well separated conform the foam, and in the second category the bubbles are separated by thin films or lamellae. When the foam flows in porous media, bubbles and lamellae span completely across the porous space and are called confined foam according to Radke and Gillis (1990).

When the characteristic pore size is comparable to or less than the characteristic size of dispersed gas bubbles, the bubbles and lamellae span pores completely. At low gas fractional flow, the pore spanning bubbles are widely spaced, separated by thick wetting liquid lenses or bridges. At high gas fraction flow, the pore spanning bubbles is in direct, contact, separated by lamellae. Hirasaki and Lawson (1985) denoted this direct contact morphology as the individual lamellae regime.

Although both bulk foam and individual lamellae foam can exist in principle, effluent bubble sizes equal to or larger than pore dimensions are usually reported. It is generally accepted that single bubbles and lamellae span the pore space of most porous media undergoing foam flow in the absence of fractures.

Figure 1 shows the schematic of foam in porous media. The gas can be trapped or flowing as a continuous or discontinuous phase. In discontinuous gas foam, the entire gas phase is made discontinuous by lamellae, and no gas channels are continuous over sample spanning dimensions. Gas is encapsulated in small packets or bubbles by surfactant stabilized aqueous films. In continuous gas foam, the media contain some interconnected gas channels that are interrupted by lamellae over macroscopic distances much greater than pore dimensions.

Discontinuous foam forms under co-injection of gas and surfactant solution, provided that the wetting phase saturation and flow rate are high enough for foam generation. When the wetting phase saturation is low enough, the lamellae generation rate may become lower than the rupture rate, and paths of continuous gas flow may result. Figure 1 is also a summary of the pore level microstructure of foam during flow through porous media. Because of the dominance of capillary forces, wetting surfactant solution flows as a separate phase in the small pore spaces. A minimal amount of wetting liquid transports as lamellae. So the wetting phase relative permeability is unchanged when foam is present. When both flowing and trapped gas exist, flowing foam occurs in large pores because the resistance there is less than in the smaller pores. Then bubble trapping can happen only in intermediate sized pores.

Thus, foam can be classified into "weak" foam and "strong" foam. For "weak foam" with no moving lamellae, the increase in trapped gas saturation is important to the behavior of foam flow as it results in the blockage of gas pathways, which reduces the relative permeability of gas.

The trapped gas reduces mobility, but the rest of gas

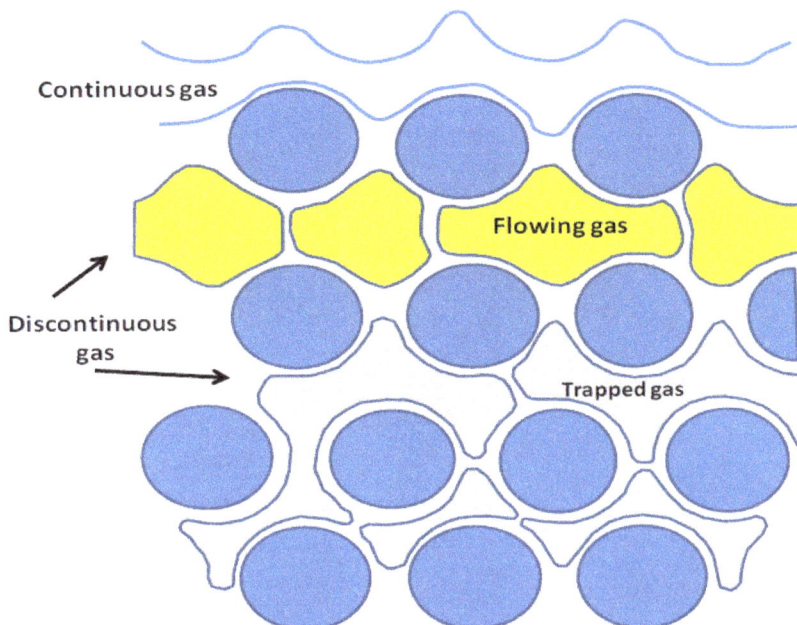

Figure 1. Cartoon of three different forms that gas can take in porous media.

flows as continuous gas.

"Strong" foam flows by a different mechanism. The lamellae make the flowing gas discontinuous. Then the bubbles trains face much higher resistance than in continuous gas flow. The apparent viscosity of the discontinuous foam is much greater than in continuous foam. The combined effect of the reduction of gas relative permeability and the increase of apparent gas viscosity greatly increases the mobility reduction effect of foam. The most important factors that affect foam trapping and mobilization are pressure gradient, gas velocity, pore geometry, bubble size, and bubble train length. Increasing the pressure gradient can open new channels which were occupied by trapped gas.

Statement of problem, objective and limitation of study

The major objective of this study is to evaluate the application of foam injection as an enhanced oil recovery method in sandstone reservoirs and exploring possible improvement of oil production in the Niger Delta (Figure 2), since foam can help reduce gas mobility and affect the oil recovery in three ways:

1. By stabilizing the displacement process as the displacing fluid (gas) viscosity increases;
2. By blocking the high-permeable swept zones and diverting the fluid into the unswept zones; and
3. By reducing the capillary forces via reducing the interfacial tensions due to the presence of surfactant.

The evaluation of the application of foam injection as an enhanced oil recovery method in sandstone reservoir was carried out using two cases and will follow the steps listed as follows:

1. Obtaining a real life reservoir model/building a synthetic reservoir model;
2. Simulation of the foam injection process;
3. Comparison of results (such as production indices, field oil recovery and so on) for foam injection with gas flooding/water flooding.

Generally, water /gas flood efficiency in sandstone reservoirs is relatively low, due to heterogeneity. In the case of gas flooding, major problems that are usually encountered are poor volumetric sweep efficiency and low incremental oil recovery due to channeling or fingering and gravity segregation, which are caused by rock heterogeneity as well as the low density and viscosity of the injected gas. The need for mobility control in gas flooding has led to the use of foam for sweep improvement and profile modification. Foam is employed to improve the efficiency by which the displacing fluid sweeps the reservoir and contacts and recovers oil.
Limitations are:

1. The ECLIPSE Foam model does not attempt to model the details of foam generation and collapse.
2. Detailed laboratory studies (pilot test) need to be done before implementation of foam flooding.
3. There are very few real reservoir models to be used for

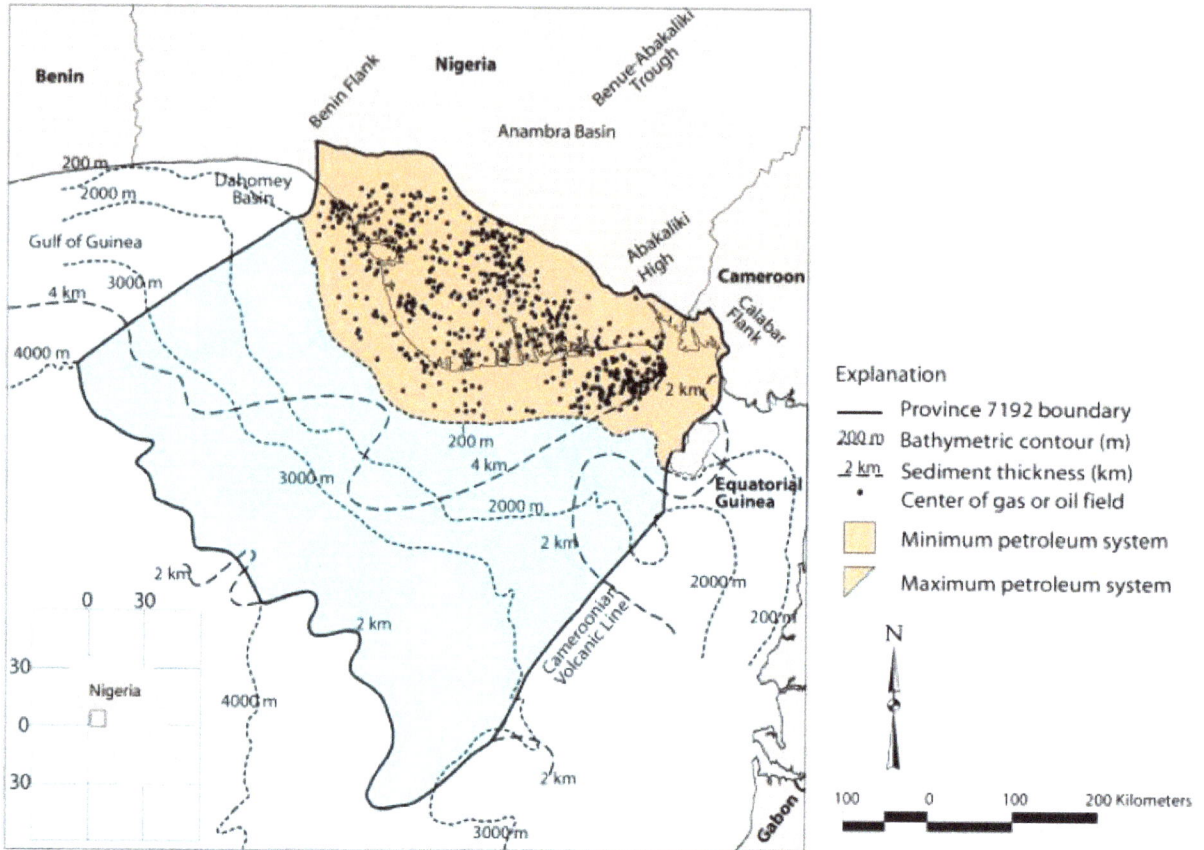

Figure 2. Index map of Nigeria and Cameroon. Map of the Niger Delta showing Province outline. Source: Petrocosultants, 1996a

carrying out the study.
5. Foam is modeled as tracer which may be transported with either the gas or the water phase with account taken of adsorption on to the rock surface and decay over time.

Foams in porous media

Gas injection for enhanced oil recovery can be efficient at mobilizing oil where gas sweeps, but suffers from poor volumetric sweep efficiency because of reservoir heterogeneity, viscous instability and gravity segregation of injection gas to the top of the formation (Lake, 1989). Foam can address all three sources of poor sweep of gas (Schramm, 1994; Rossen, 1996).

   Foam is a dispersion of gas in liquid (Bickerman, 1973). The dispersed phase is sometimes referred as the internal or discontinuous phase, and the liquid phase as the external or continuous phase. In foam, gas bubbles are separated by thin film of fluid called lamella. The lamella surrounding gas bubbles are normally unstable and break very quickly. However, the presence of surface active agents (surfactants) stabilizes the lamellae, thus improving foam stability.

Foam generation mechanisms inside a realistic porous media

Snap off: Roof (1970) showed that when oil emerges from a water-wet constriction into a water filled pore, the interfacial forces are such that a leading portion of the oil may separate into a droplet (snap off). The same mechanism occurs during invasion of gas to pores filled with liquid. It takes place regardless of the presence or absence of surfactant, but if a stabilizing surfactant is not present, snapped off bubbles quickly coalesce (Kovscek and Radke, 1993). The snap-off process is a result of the difference in the capillary pressure between the pore body and pore throat. Thus occurrence of the process is a function of ratio of the body-to-throat equivalent diameters. Kovscek and Radke (1993) and Li et al. (2010) presented details of the snap-off process.

Leave behind: The leave behind mechanism also occurs during invasion of a gas phase to a porous medium saturated with a liquid phase. Foams generated solely by leave-behind give approximately a five-fold reduction in steady-state gas permeability (Ransohoff and Radke, 1988; Kovscek and Radke, 1993), whereas discontinuous-gas foam created by snap-off resulted in a several-hundred fold reduction in gas mobility (Persoff et al., 1991; Ettinger and Radke, 1992; Kovscek and Radke, 1993). This indicates that the strength of foam (that is, number and stability of lamellae) is affected by the dominant mechanism of foam generation.

Lamella division mechanism: Increasing number of lamellae or bubbles by lamella division mechanism can be existed when mobile foam bubbles are pre-existed in the porous medium. When a moving lamella train encounters a branch in the flow path, it may split into two, one in each branch of the path (Tanzil et al., 2002). Lamella division is thought to be the primary foam-generation mechanism in steady gas-liquid flow (Gauglitz et al., 2002; Li, 2006).

Flow characteristics of foam in porous media

In porous media, foam flow is characterized by the location of wetting and non-wetting phases in pores. The non-wetting phase resides in the central portion of the large pores, while wetting phase resides in corner of the gas-occupied pores and in thin wetting films coating the pore walls.

During the flow of foam in porous media, gas can be trapped, or flowing as a continuous or discontinuous phase. In continuous-gas flow, the porous media contains some channels uninterrupted by lamella. In discontinuous-gas flow, the entire gas phase is made discontinuous by lamellae, and no gas channels are continuous over macroscopic distances. Discontinuous foam is usually associated with stronger foam strength. All three types of channels can be present in the porous medium during foam flow.

METHODOLOGY

This study was carried out using two cases with one being a synthetic model and the second one being a real model.

Case 1

This case entails building a synthetic model using a static modelling software and populating the model with average petrophysical properties, pressure, volume, and temperature properties, as well as saturation dependent data unique to unconsolidated sandstone as generated using relevant correlations. The model will then be imported to dynamic modeling software and simulated to mimic foam injection process.

Case 2

As regards this case, the real life model was obtained, it is a case in which the static model had already been built with an aquifer and simulated using dynamic software to mimic water flooding process. This case therefore, entails importing the already built static model

to dynamic modeling software for simulation to mimic foam injection.

An economic analysis was run to check for the viability and possible implementation of the project.

Building a simple synthetic model

A synthetic model is an artificial 3-D reservoir model whose surface and horizons are created using artificial algorithm. This 3-D model incorporates all the geologic attributes of the reservoir to be built. These attributes include the structural shape and thicknesses of the formations within the subsurface volume being modeled, their lithology, and the porosity and permeability distributions.

The static model as shown in Figure 3 (synthetic) was developed using Petrel; static modelling software.

Steps to building a synthetic model are:

1. Creation of grided surfaces: Anticlinal surfaces were created and served as input data required for the creation of a 3-D grided dome-like structure.
2. Making horizons from created surfaces: The created surfaces were converted into horizons as soon as they were made to form a 3-D grid.
3. Making zones from created horizons: Zones are geological portion in the stratigraphic intervals above, in-between and below the horizons. This process defines the sub units of the 3D grid and is carried out in the "make zone" section. It inserts additional horizons and zones into the 3D grid by inserting isochores up or down from the previously input horizons. The model was divided into three zones by four horizons.
4. Layering: This is the final step; it involves making the final vertical resolution of the 3D grid and this is done by dividing each zone into layers. The model was divided into nine layers.

There are nine layers in the model, which are grouped into three zones. The first three red layers makes up the first zone, the next three purple layers makes up the second zone, while the last three blue layers makes up the third zone.

The reservoir model is a 30 by 30 by 9 model and was discretized into 8100 cells.

With the aid of the property calculator the model was populated with some petrophysical properties. Using arithmetic averaging (random distribution), average porosity of 27% was assigned the cells. Arithmetic averaging was used for populating porosity because we assume isotropic nature of the property. While triangular distribution was used in populating the permeability of the reservoir with the minimum, medium and maximum values being 300, 1000 and 1700 md respectively. Triangular distribution was used to populate the permeability because of the anisotropic nature of the property. A net to gross ratio of 0.8 was used to describe proportion of the gross rock volume formed by the reservoir rock.

For this study, investigation was done for an undersaturated reservoir and light oil fluid model wa s used. Since there is no compositional variation with temperature and pressure of the reservoir fluid, a black oil simulator (ECLIPSE 100) was used to simulate the process.

Reservoir simulation basics can be summarized in these steps.

Divide the reservoir into several cells

Case 1: In the case of the synthetic model, this step has been carried out in the static modelling software and can be imported into the dynamic modelling software (ECLIPSE 100) using an INCLUDE file keyword. Figure 4 is the reservoir model showing the location of wells (Case 1)

Case 2: The reservoir model in this case is a real one. This implies

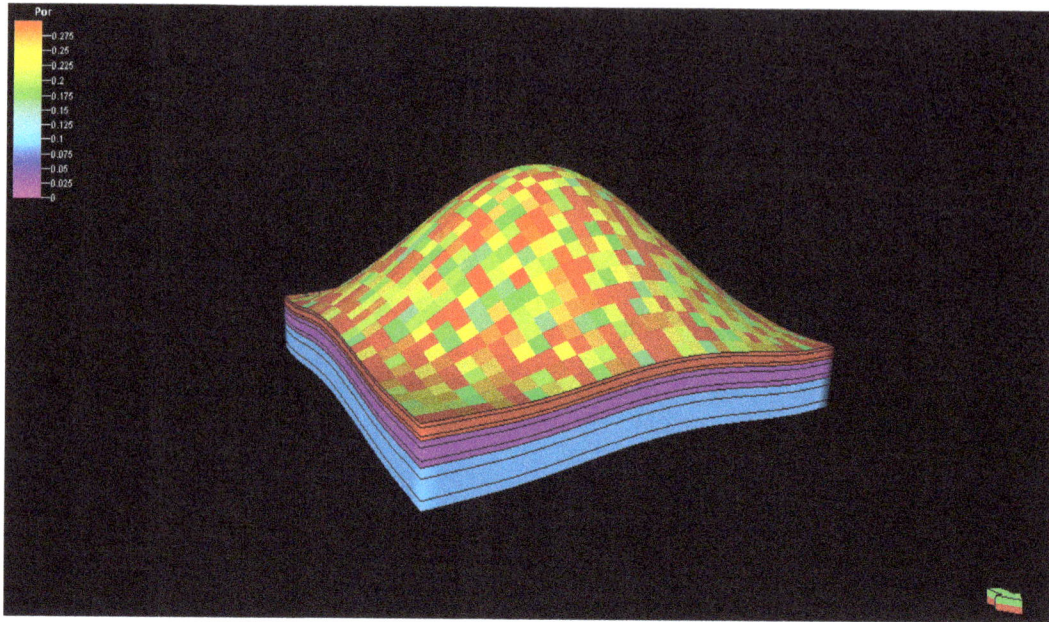

Figure 3. Static geologic model showing zones and layers and distribution of porosity.

that the model has been built already all it needs is for it to be imported to the dynamic modelling software for simulation. Figure 5 is the reservoir model showing the location of wells (Case 2).

Provide basic data for each cell:

1. Fluid and rock properties: The PVT data are used to translate produced volumes to reservoir conditions and to convert these to mass, ready for the simulator's mass balance equations. PVT tables are derived from a combination of laboratory experiments, field tests or correlations.

The PVT properties of the reservoir model in Case 1 such as solution gas oil ratio, formation volume factor and viscosity were defined using correlations. The reservoir fluid data are needed to evaluate phase density at reservoir and stock tank conditions.

Case 1: The reservoir pressure is 3814.7 psi (26 301.4 kpa) at 7100 ft (2164 m) and reservoir thickness of 400 ft (122 m) with gas/oil contact of 7100 ft (2164 m) and oil/water contact of 7450 ft (2271 m).
Case 2: The PVT data for this reservoir is as specified in the already built model. The reservoir pressure is 3035.7 psi (20 930 kpa) at 7000 ft (2134 m) with gas oil contact of 7000 ft (2134 m) and oil water contact of 8200 ft (2499 m).

2. Petrophysical data:

Case 1: The rock compressibility data for unconsolidated sandstone was generated using Newman's correlation and also the saturation dependent data were also generated for sandstone using relevant correlations.
Case 2: Average porosity is 18%, average permeability is 308md, while other data such as rock compressibility data and saturation dependent data are as specified in the already built model.

Positioning/completion of wells within the cells:

Case 1: A total of eleven wells were drilled; consisting of three gas injectors and ten producing wells. The gas injectors were located at the crest of the structure while the producers were located at the

flanks of the reservoir. The producers were perforated at layers 3 to 7, while the gas injectors were perforated at layers 1 to 3 so as to efficiently push the oil towards the perforations. The gas injectors will be used to inject the surfactant into the reservoir.
Case 2: A total of seven wells were drilled; consisting of three gas injectors and ten producing wells. With four producers and one gas injector located on the center fault block of the structure while the last producer and the second injector were located on the west fault block of the reservoir as shown in Figure 5. The producers (1, 2, 3, 4) were perforated at layers 1 to 12, 1 to 10, 2 to 9 and 1 to 3 respectively while the gas injector was perforated at layers 1 to 4 (on the central fault block) so as to efficiently push the oil towards the perforations while the producer and injector on the west fault block were perforated at layers 1 to 10 and 1 to 4 respectively. The gas injectors will be used to inject the surfactant into the reservoir:

Specify well production rates as a function of time

The well production rates are controlled using the required keywords.

Solve equations to yield the pressure and saturation for each block, as well as production of each phase from each well

Each cell is solved simultaneously, so the number of cells in the simulation model is directly related to the time required to solve a time step. In general, short time steps are easier (quicker) to solve than long ones.

The distribution of the injected foam is solved by a conservation equation: as a tracer in the gas phase with decay, or as a tracer in the water phase with decay (Equations 1 and 2 respectively):

$$\frac{d}{dt}\left(\frac{VS_gC_f}{B_rB_g}\right) + \frac{d}{dt}\left(V\rho_r C_f^a \frac{1-\phi}{\phi}\right) = \sum\left[\frac{Tk_{rg}}{B_g\mu_g}M_{rf}(\delta P_g - \rho_g gD_z)\right]C_f + Q_gC_f - \lambda(S_w, S_o)VC_f$$
(1)

$$\frac{d}{dt}\left(\frac{VS_wC_f}{B_rB_w}\right) + \frac{d}{dt}\left(V\rho_r C_f^a \frac{1-\phi}{\phi}\right) = \sum\left[\frac{Tk_{rw}}{B_w\mu_w}(\delta P_w - \rho_w gD_z)\right]C_f + Q_wC_f - \lambda(S_w, S_o)VC_f$$
(2)

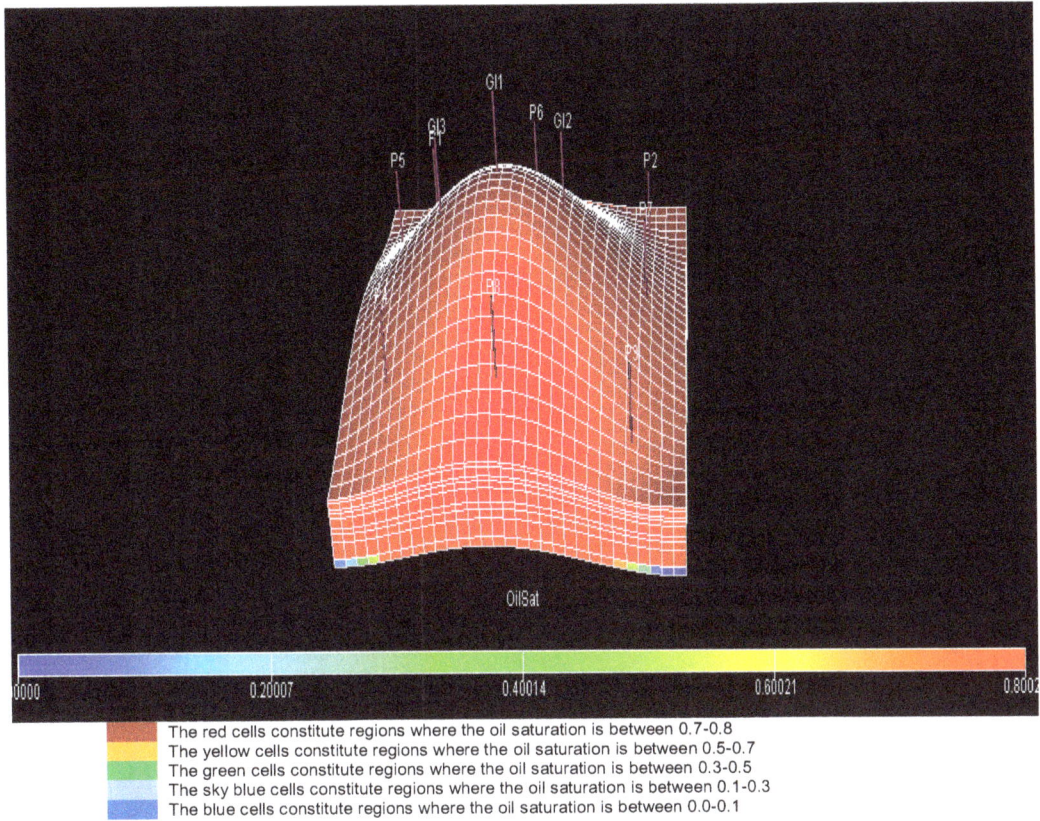

Figure 4. Reservoir model showing the location of wells (Case 1).

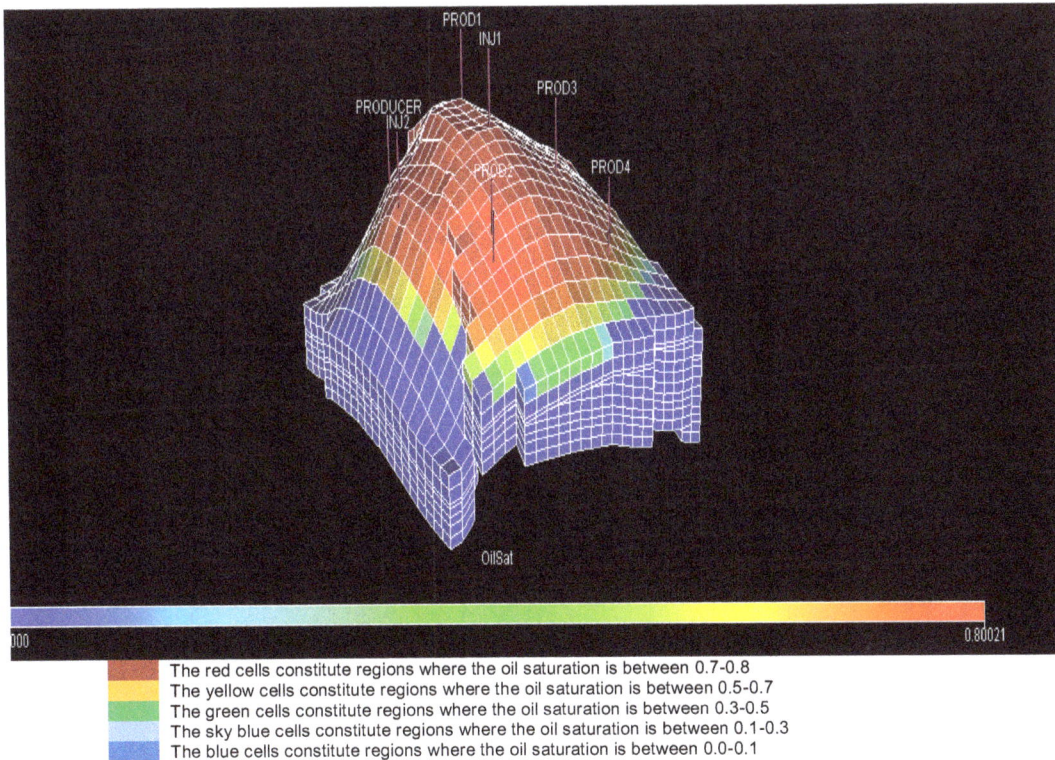

Figure 5. Reservoir model showing the location of wells (Case 2).

Where, C_f denotes the foam concentration; ρ_w, ρ_g denotes the water and gas density respectively; Σ denotes the sum over neighboring cells C; C_f^a denotes the foam concentration; μ_w, μ_g denotes the water and gas viscosity respectively; D_g is the cell center depth; B_r, B_w, B_g is the rock, water and gas formation volume respectively; T is the transmissibility; K_{rw} and K_{rg} is the water and gas relative permeability respectively; S_w, S_g is the water and gas saturation respectively; V is the block pore volume; Q_w, Q_g is the water and gas production rate respectively; P_w, P_g is the water and gas pressure respectively; λ is the rate decay parameter function of oil and water saturation; M_{rf} is the gas mobility reduction factor described below, and G is the gravity acceleration

Modeling foam injection in reservoir using eclipse

Foam flooding through the reservoir can be modeled using the ECLIPSE 100 simulator. The ECLIPSE Foam model does not attempt to model the details of foam generation and collapse. Foam is modeled as tracer which may be transported with either the gas or the water phase with account taken of adsorption on to the rock surface and decay over time.

Case 1

A 30 by 30 by 9 model consisting of 8100 cells was used. The reservoir model was simulated using ECLIPSE black oil simulator E100 due to the constant composition of the reservoir fluid with respect to temperature and pressure. The reservoir is an undersaturated reservoir as the reservoir pressure is greater than the bubble point pressure. Upon initialization, the oil in place was discovered to be 2402MMSTB while the dissolved gas was 1806BSCF.

In exploiting the reserve, a total of 11 wells was proposed consisting of 3 gas injectors and eight producers. The gas injectors would be used for co-injection of surfactant and gas into the formation. The simulation was carried out to mimic foam injection for a period of sixty four years starting September 2013 and the rate of production was maintained at 150,000STB/D.

Case 2

A 24 by 25 by 12 model consisting of 7200 cells was used. The reservoir model was simulated using ECLIPSE black oil simulator E100 due to the constant composition of the reservoir fluid with respect to temperature and pressure. The reservoir is an undersaturated reservoir as the reservoir pressure is greater than the bubble point pressure. Upon initialization, the oil in place was discovered to be 260MMSTB while the dissolved gas was 252BSCF.

In exploiting the reserve, a total of 7 wells was proposed consisting of 2 gas injectors and five producers. The gas injectors would be used for co-injection of surfactant and gas into the formation. The simulation was carried out to mimic foam injection for a period of sixty four years starting January, 1988 and the rate of production was maintained at 15,000STB/D.

Economics

In evaluating the economic viability of this project, economic indicators (yard stick) that will be used are the net present values (NPV) and the discounted profit to investment ratio. In all the cases that are to be considered, some of the economic data input will be assumed and sensitivity analysis will be adopted such that the views of a pessimist, optimist and inbetweenist will be portrayed and how their view affects the economics of the projects.

RESULTS AND INTERPRETATIONS

Case 1

In this case, results obtained for foam injection was compared with that gas injection (that is, without foam). The results include a plot of field oil recovery, field oil production rate, gas/oil ratio and field oil production total. From the plot of the field oil recovery obtained (Figure 6), it can be seen clearly that the recovery factor for foam injection is about 62% while that of gas injection is about 46%, thus, showing an incremental recovery of 16% which is quite significant.

From the plot of the field oil production rate obtained (Figure 7), for foam injection the restricted production rate (150,000 STB/D) was maintained for about twenty years before the decline in production rate while for gas injection the restricted production rate (150,000 STB/D) was maintained for about fourteen years before production started declining. This implies that the foam injected was able to sustain the reservoir pressure above the pressure at which the production target will no longer be met, for a longer period of time as compared to when gas was injected.

From the plot of the field gas oil ratio obtained (Figure 8), a significant reduction in gas oil ratio was noticed with foam injection confirming the mobility reduction effect that foam has on gas. It shows that foam injection successfully delayed gas breakthrough/gas channeling (fingering) for over 40 years. For the gas injection without foam case, there was an increase in the field gas oil ratio once the reservoir crossed its bubble point.

Case 2

In this case, results obtained for foam injection was compared with that water injection. The results include a plot of field oil recovery, field oil production rate, field water cut and field oil production total.

Foam injection gave a total cumulative oil production of 1490 MMSTB while total oil production was 1106 MMSTB for gas injection (Figure 9). It can be inferred that foam injection enabled the optimization of recoverable reserve by squeezing out more oil from the reservoir compared to the gas injection.

From the plot of the field oil recovery obtained (Figure 10), it can be seen clearly that the recovery factor for foam injection is about 51% while that of water injection is about 40%, thus, showing an incremental recovery of 11% which is quite significant.

From the plot of the field oil production rate obtained (Figure 11), the quick commencement of the decline in

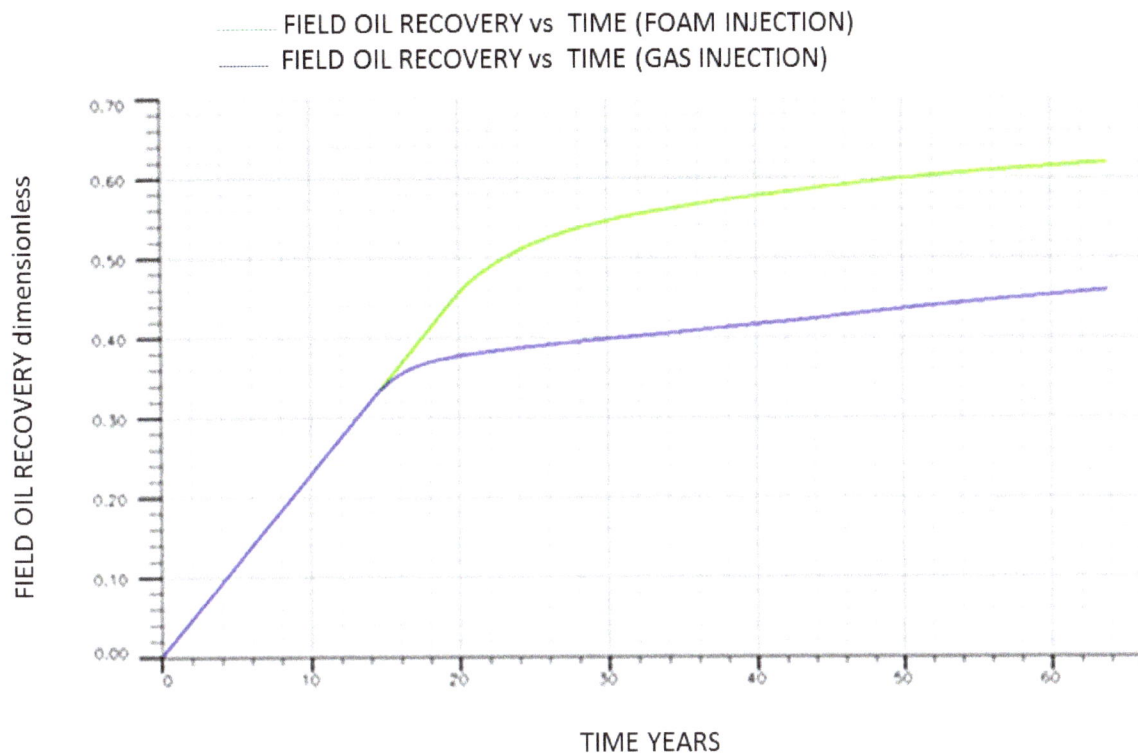

Figure 6. Plot of field oil recovery for foam injection versus gas injection (Case 1).

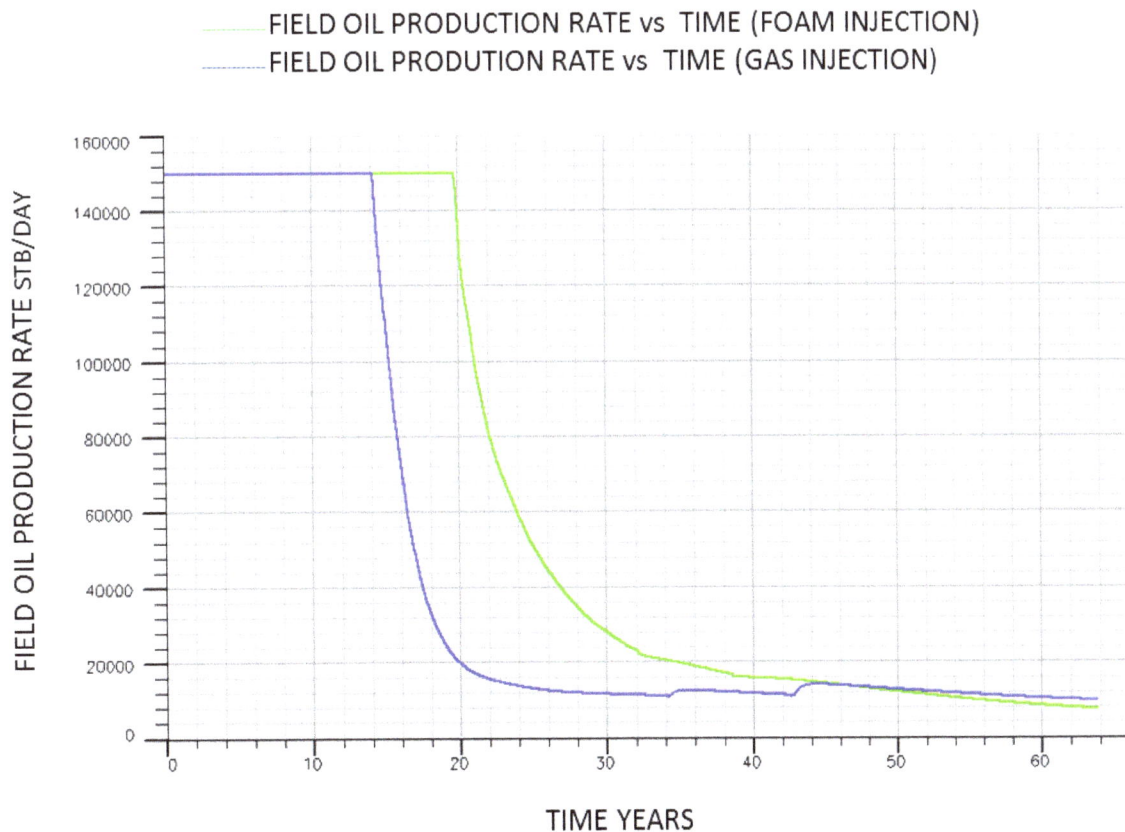

Figure 7. Plot of field oil production rate for foam injection versus gas injection (Case 1).

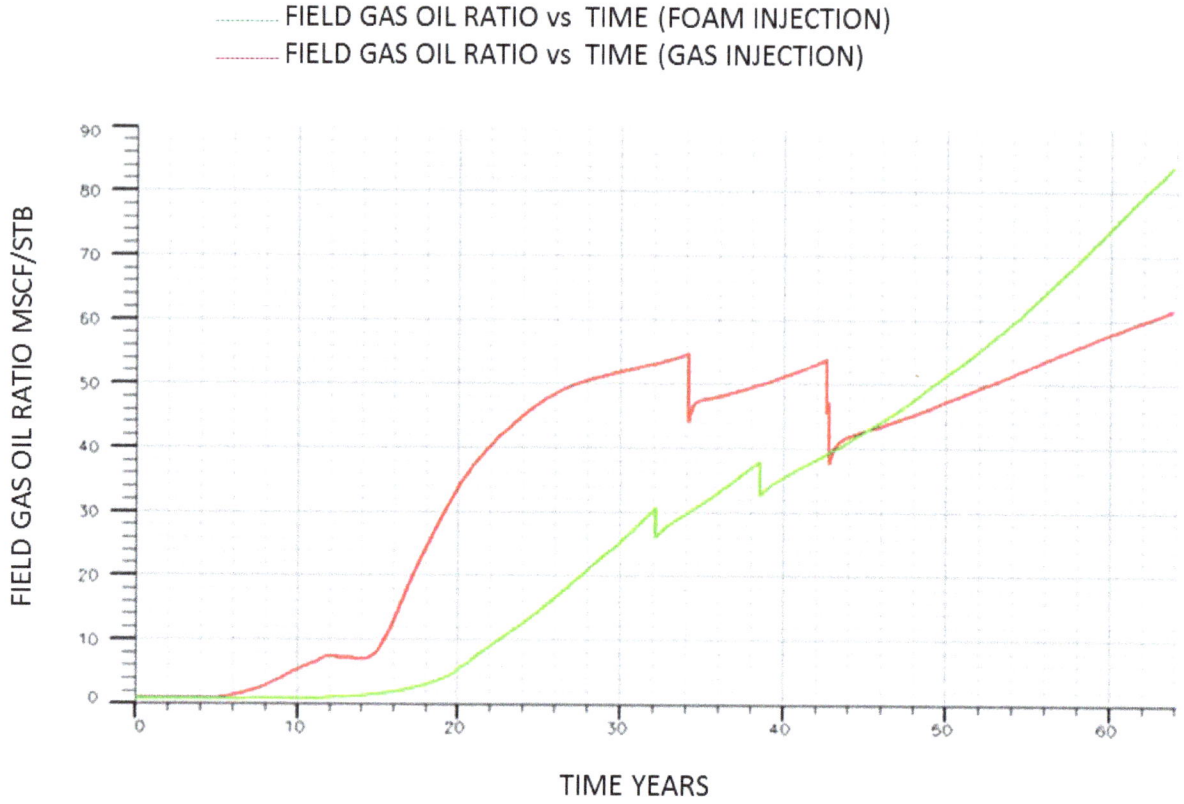

Figure 8. Plot of field gas oil ratio for foam injection versus gas injection (Case 1).

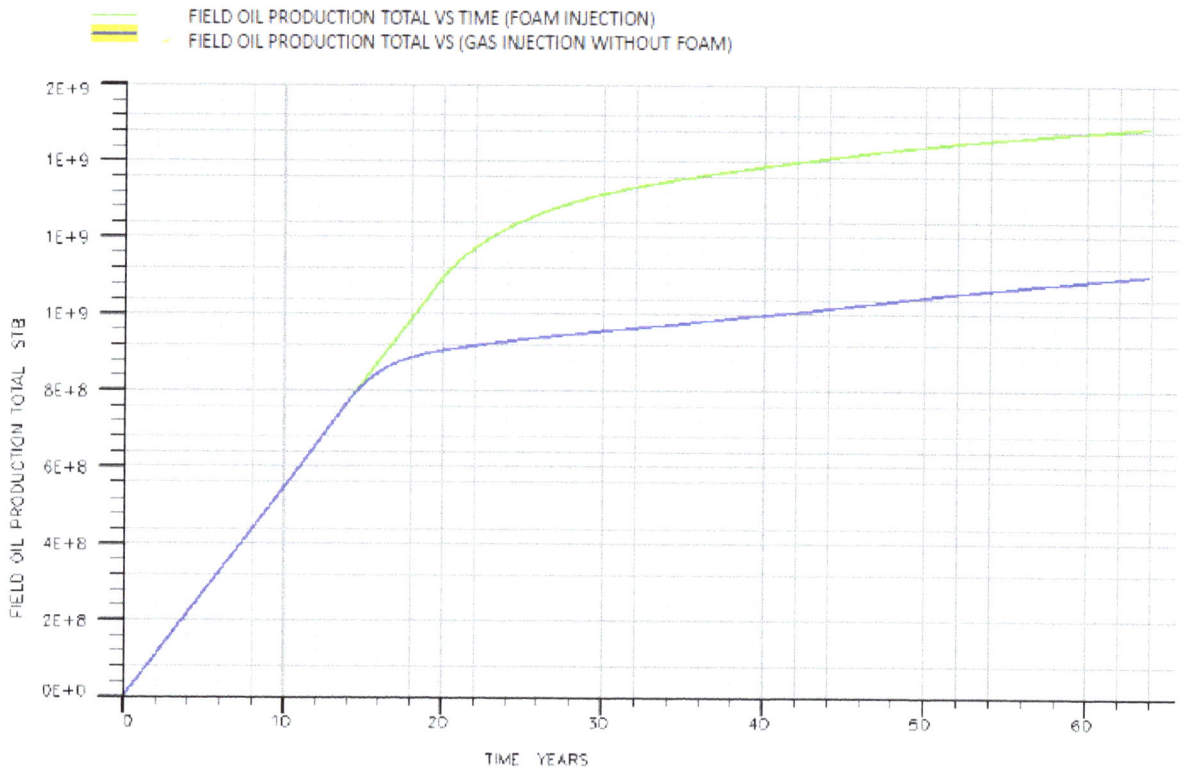

Figure 9. Plot of field oil production total for foam injection versus gas injection.

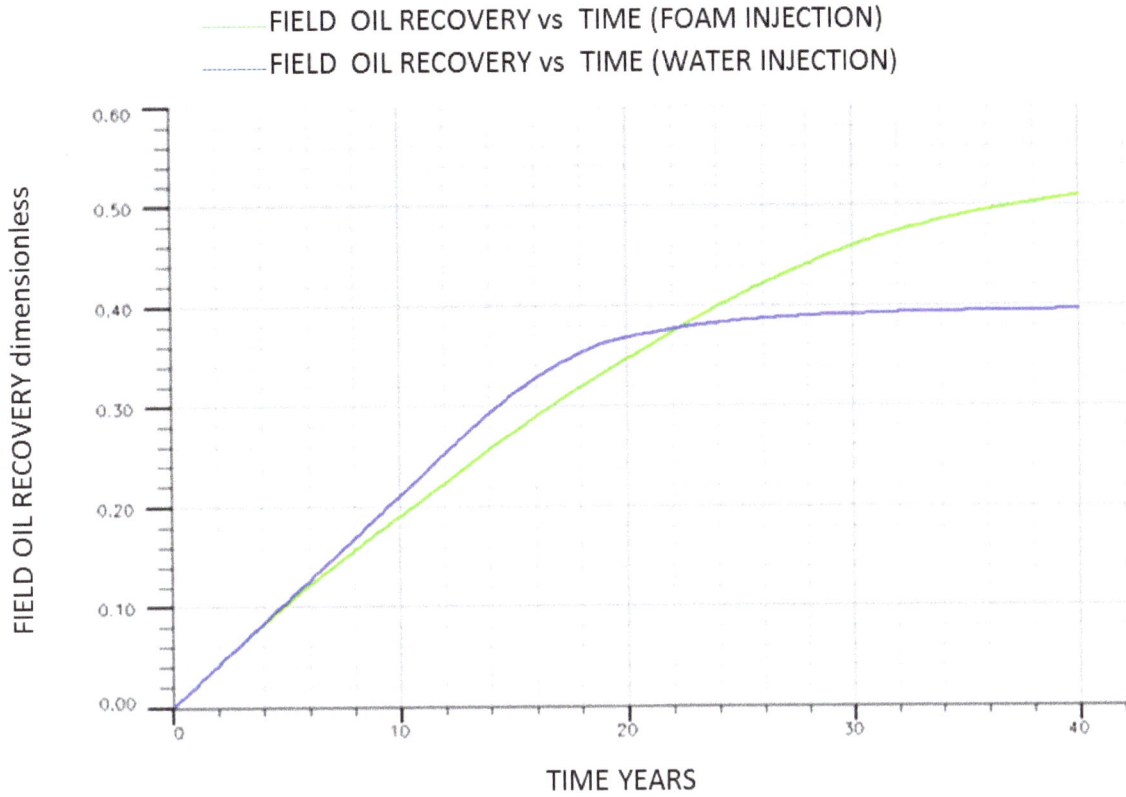

Figure 10. Plot of field oil recovery for foam injection versus water injection (Case 2).

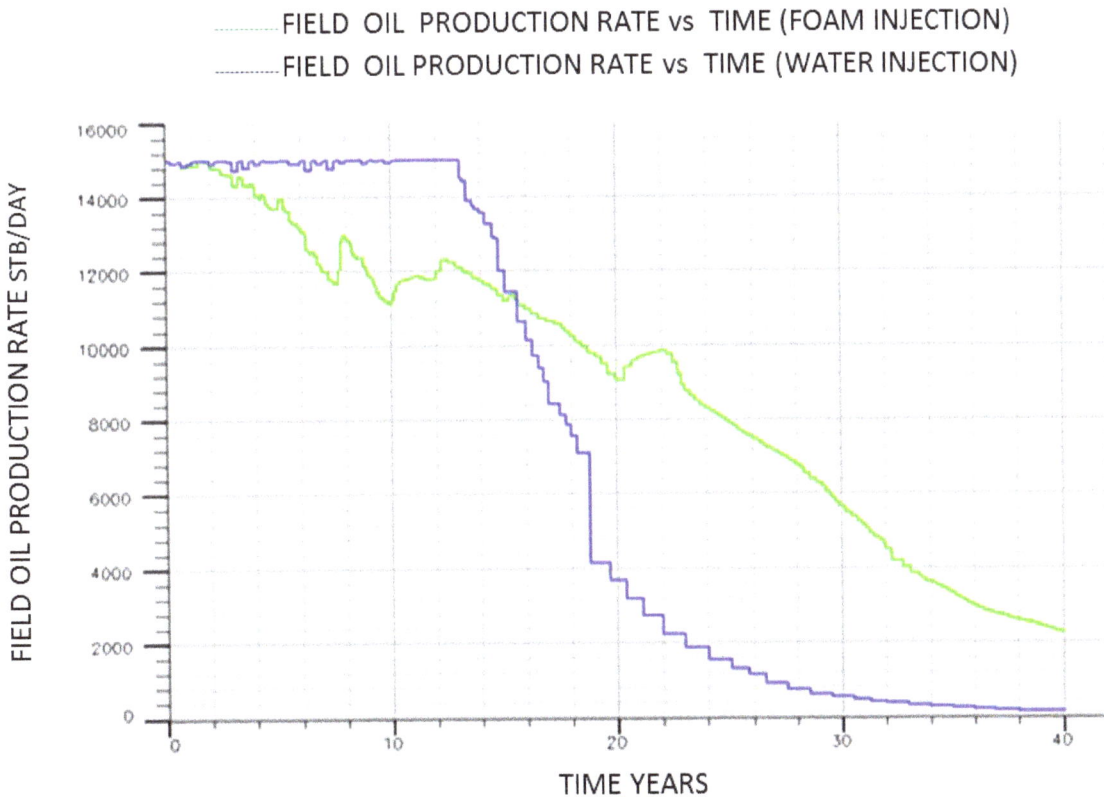

Figure 11. Plot of field oil production rate for foam injection versus water injection (case 2).

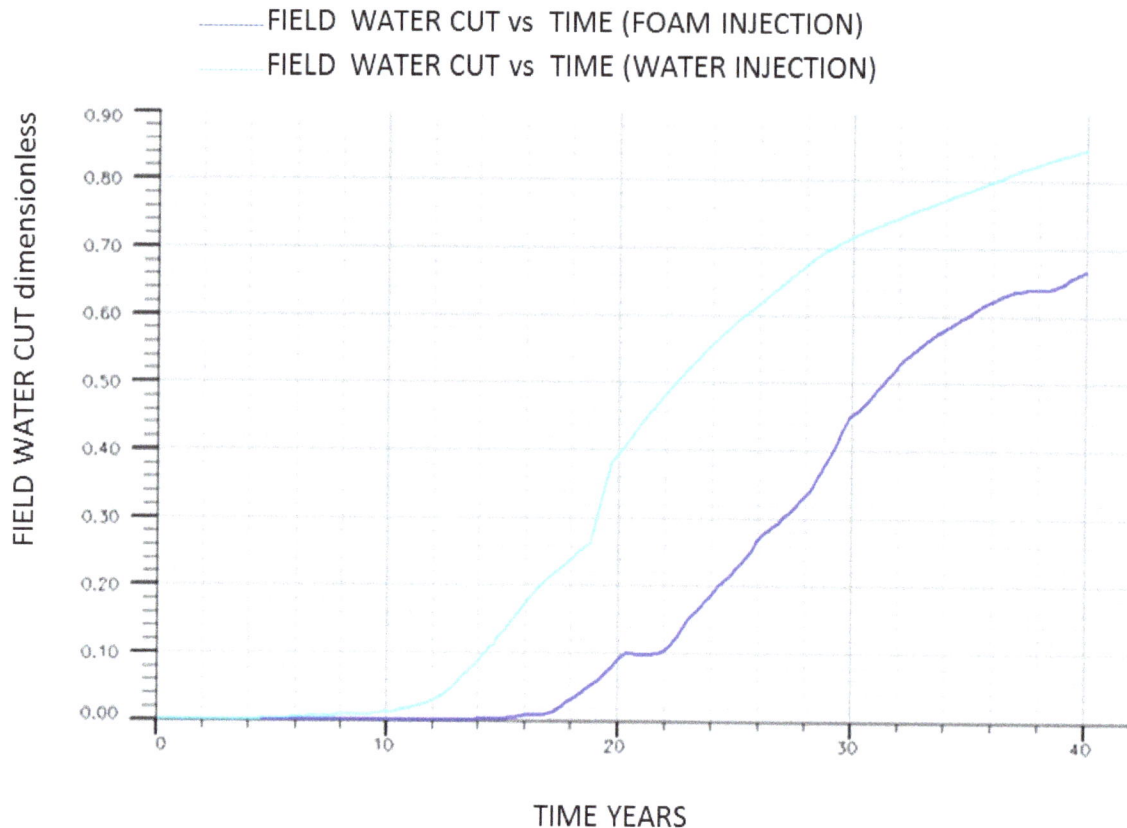

Figure 12. Plot of field water cut for foam injection versus water injection (CASE 2).

production can be attributed to the fact that foam decays faster in the presence of oil and water, thus reducing the strength of the foam which in turn affects the efficiency of the foam injected into the reservoir. It can also be deduced from the production profile that even though the decline for foam injection started earlier as compared to water injection, foam injection still squeezed more oil from the reservoir over time than water injection. A better profile and recovery will be obtained if water is injection is employed for about ten to thirteen years before foam injection is initiated.

From the plot of the field water cut obtained (Figure 12), a significant reduction in water cut was noticed with foam injection confirming the fact that foam has an excellent blocking effect. It tends to block the high permeable zones which could enhance quick water breakthrough, thus significantly reducing water cut. For the water injection case, there was an increase in the field water cut.

Foam injection gave a total cumulative oil production of 133 MMSTB while total oil production was 103 MMSTB for water injection. It can be inferred that foam injection enabled the optimization of recoverable reserve by squeezing out more oil from the reservoir compared to the water injection.

Economics

Reserves can be defined as estimated quantities of crude oil, natural gas, condensates, natural gas liquids and other related substances which are known to be recoverable and marketable from known accumulations using established technologies and operating conditions under approved regulations.

From the definition of reserves giving above, it is worthy of note that the economics of a project is equally as important as the technical aspect of the project, because there is no point embarking on a project that is not profitable.

In the light of the foregoing, economic indicators such net present value (NPV) and discounted profit to investment ratio (DPI) were used to determine the economic viability of the project. The economic assumptions made in carrying out the analysis and the results are shown in Tables 1 to 4 while the plots of the cumulative discounted cash flow against time are shown in Figures 13 to 15.

For case 1 the economics for foam flooding proved (using NPV and DPI as yard stick) to be more viable than that of gas injection. For case 2 the result for foam flooding was not provided because the values obtained

Table 1. Summary of economics for foam flooding project.

Input data				
Foam flooding		**Risk-averse**	**Risk-neutral**	**Risk-tolerant**
Royalty	13%			
Operating cost, $MM/year		50	40	30
injection cost, $/MScf		3	2	1
Petroleum profits tax, PPT		70%	50%	40%
Exploration cost , $MM		200	140	100
Facilities		600	450	400
Development well cost , $MM	600.00			
Price, $/STB	77.50			
Drilling and completion cost, $MM (Injector)	40.00			
Drilling and completion cost, $MM (Producer)	60.00			
Result				
NPV, $MM (i=20%)		3026.670	6757.870	9121.55
DPI		2.162	5.679	8.292

Table 2. Summary of economics for gas flooding project.

Input data				
Gas flooding		**Pessimist**	**Inbetweenist**	**Optimist**
Royalty =	13%			
Operating cost, $MM/year		46	36	26
injection cost, $/MScf		3	2	1
Petroleum profits tax, PPT		70%	50%	40%
Exploration well cost , $MM		200	140	100
Facilities		600	450	400
Development well cost , $MM	600			
Price, $/STB	77.5			
Drilling and completion cost, $MM (Injector)	40.00			
Drilling and completion cost, $MM (Producer)	60.00			
Result				
NPV, $MM (i=20%)		2821.6	6397	8607.85
DPI		2.015	5.376	7.825

Table 3. Summary of economics for water flooding project.

	Input data			
Water flooding		**Pessimist**	**Inbetweenist**	**Optimist**
Royalty	13%			
Operating cost, $MM/ye		17	16	15
injection cost, $/bbl		4	3	1
Exploration well cost, $MM		70	66	60
Facilities		140	120	110
Development well cost, $MM	284.00			
Price	77.50			
Drilling and completion cost, $MM (Injector)	20.00			
Drilling and completion cost, $MM (Producer)	28.00			
Result				
NPV, $MM (i=20%)		1084.654	1140.689	1215.813
DPI		2.196	2.427	2.678

Table 4. Summary of economic assumptions for foam flooding project.

Input data				
Water flooding		**Pessimist**	**Inbetweenist**	**Optimist**
Royalty	13%			
Operating cost, $MM/year		30	28	25
injection cost, $/bbl		4	3	1
Exploration well cost , $MM		70	66	60
Facilities		140	120	110
Development well cost , $MM	284			
Price	77.5			
Drilling and completion cost, $MM (Injector)	20			
Drilling and completion cost, $MM (Producer)	28			

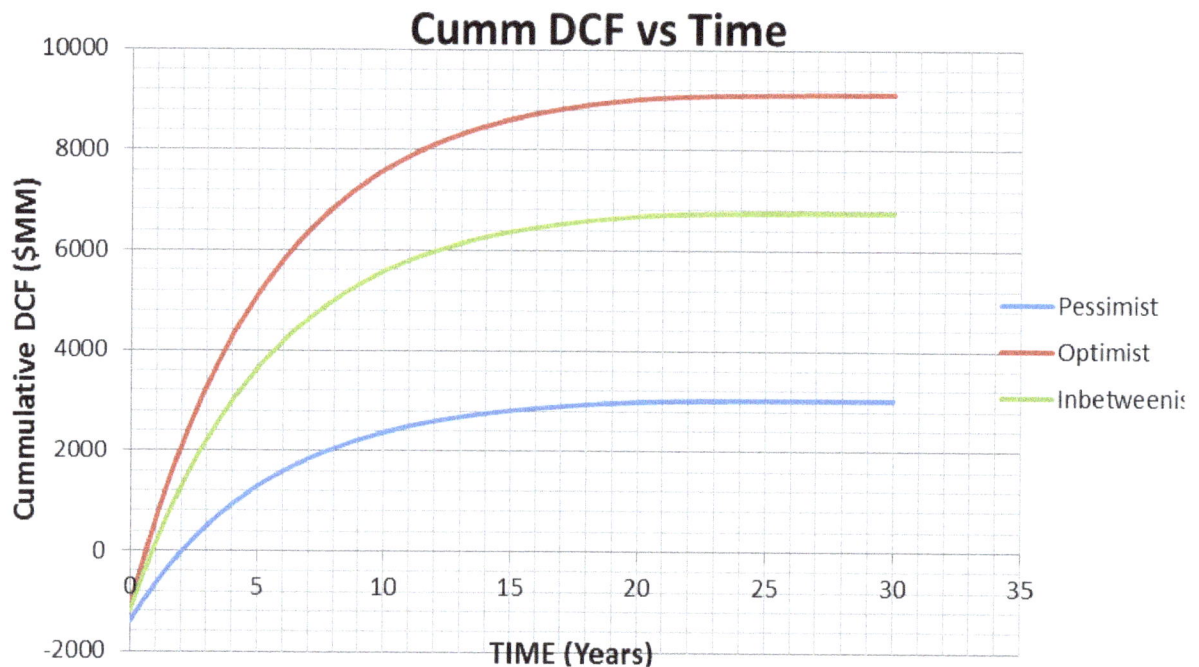

Figure 13. Cumulative discounted cash flow vs time (years) for foam flooding.

for NPV and DPI are negative (implying the project is not viable). This can attributed to the high influence of the Opex and the injection cost on the cash flow. Water flooding on the other hand proved to be more viable because of its higher early cash flow and lesser effect of injection cost.

Conclusion

From the results obtained, foam injection has shown clearly the possibility of increasing recoverable reserves, thus, total field oil production by improving the displacement efficiency and subsequently mobilizing more oil towards the producers. Foam injection therefore stands a great chance of helping to provide a means to increase Nigeria's recoverable reserves and optimize the nation's oil reserves.

Further studies should be carried out on the use of foam injection in Nigeria, owing to the fact that before this EOR method can be used pilot tests have to be carried out and also the need for series of laboratory tests, which will help in determining the foaming agents that will be suitable for use in the Niger Delta reservoirs.

Foam injection however provides a way of reducing Nigeria's gas flaring to near zero thereby preserving the nation's asset. Foam injection reduces gas mobility thereby reducing the production of unwanted gas thus

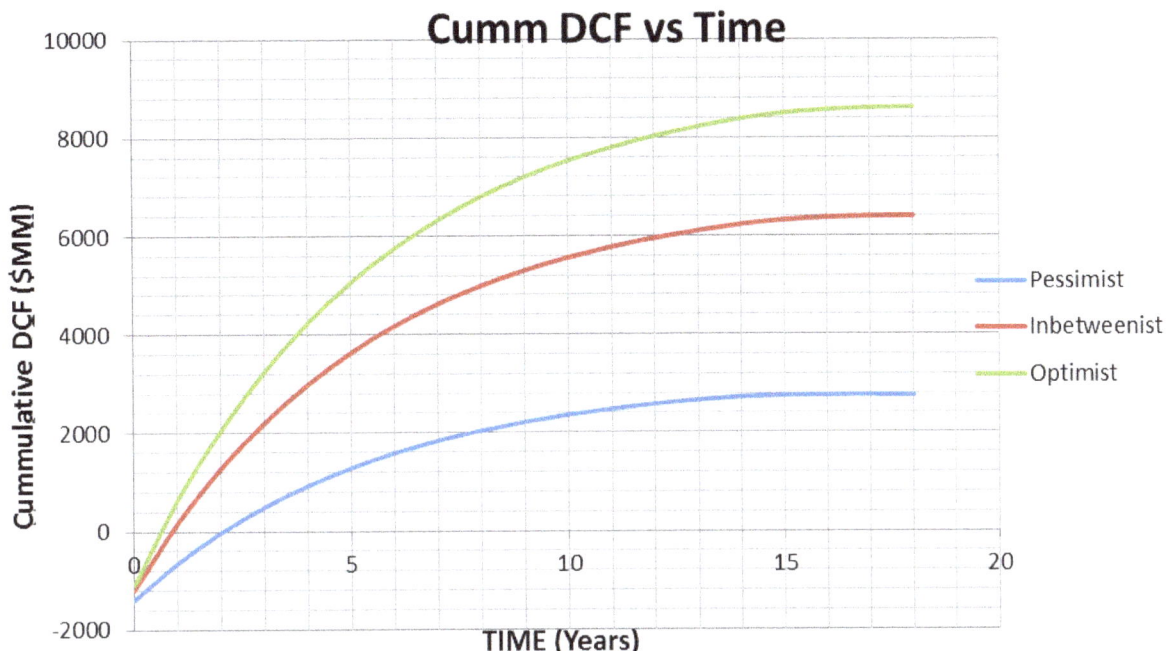

Figure 14. Cumulative discounted cash flow vs time (years) for gas flooding.

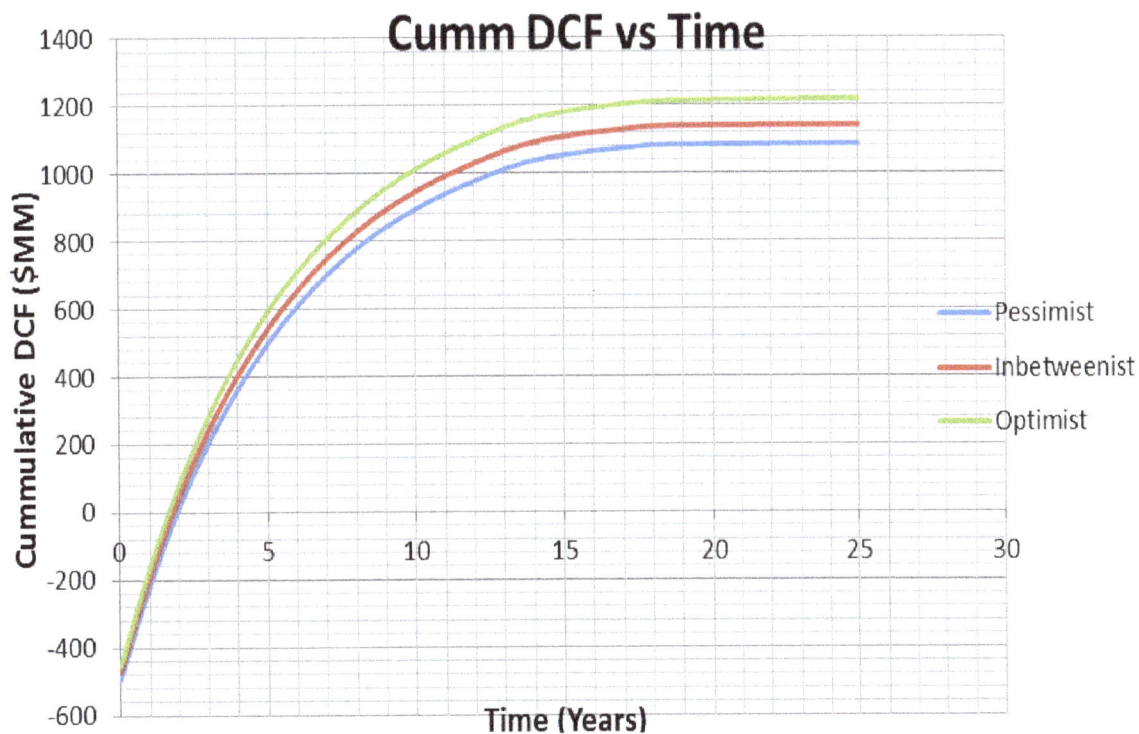

Figure 15. Cumulative discounted cash flow vs time (years) for gas flooding.

preventing its flaring.

Conflict of Interest

The authors have not declared any conflict of interest.

ACKNOWLEDGEMENTS

The authors would like to acknowledge the management of the Schlumberger Learning Centre of the University of Ibadan for allowing us to use ECLIPSE 100 and Petrel E&P suite of Software donated by Schlumberger for this study.

REFERENCES

Bickerman JJ (1973). Foams. New York: Springer-Verlag. http://dx.doi.org/10.1007/978-3-642-86734-7 PMid:4569373.

Blaker T, Celius HK, Lie T, Martinsen HA, Rasmussen L, Vassenden F (1999). Foam for Gas Mobility Control in the Snorre Field: The FAWAG Project. SPE Annual Technical Conference and Exhibition, Houston, Texas. http://dx.doi.org/10.2118/56478-MS

Ettinger RA, Radke CJ (1992). The Influence of Texture on Steady Foam Flow in Berea Sandstone. SPERE 7(1):83-90. http://dx.doi.org/10.2118/19688-PA

Exerowa D, Kruglyakov PM (1998). Foam and Foam Films. Elsevier.

Gauglitz PA, Friedmann F, Kam S I, Rossen WR (2002). Foam Generation in Homogeneous Porous Media. Chem. Eng. Sci. 57(19):4037-4052. http://dx.doi.org/10.1016/S0009-2509(02)00340-8

Hirasaki GJ, Lawson JB (1985). Mechanisms of Foam Flow in Porous Media: Apparent Viscosity in Smooth Capillaries. SPE J. 25(2):176-190.http://dx.doi.org/10.2118/12129-PA

Kovscek AR, Bertin HJ (2002). Estimation of Foam Mobility in Heterogeneous Porous Media. SPE/DOE 75181 Improved Oil Recovery Symposium, Tulsa, Oklahoma 17-20 April.

Kovscek AR, Radke CJ (1993). Fundamentals of Foam Transport in Porous Media, in Foams in the Petroleum Industry. L.L. Schramm, Editor., American Chemical Society, Washington, D.C., pp. 115-163.

Lake LW (1989). Enhanced Oil Recovery, Prentice Hall, Upper Saddle River, NJ.

Li, Q. (2006) Foam Generation and Propagation in Homogeneous and Heterogeneous Porous Media. PhD Dissertation, University of Texas at Austin, USA.

Li RF, Yan W, Liu S, Hirasaki GJ, Miller CA (2010). Foam Mobility Control for Surfactant Enhanced Oil Recovery. SPE J. 15(4) 1-15.

Persoff P, Radke CJ, Pruess K, Benson SM, Witherspoon PA (1991). A Laboratory Investigation of Foam Flow in Sandstone at Elevated Pressure. SPERE (Aug.) pp. 185-192. http://dx.doi.org/10.2118/18781-PA

Radke CJ, Gillis JV (1990). A Dual Gas Tracer Technique for Determining Trapped Gas Saturation During Steady Foam Flow in Porous Media SPE Annual Technical Conference and Exhibition, New Orleans, Louisiana.

Ransohoff TC, Radke CJ (1988). Mechanisms of Foam Generation in Glass-Bead Packs. SPERE (May), pp. 573-585.

Roof JG (1970). Snap-Off of Oil Droplets in Water-wet Pores. SPE J. 10(1970):85-90. http://dx.doi.org/10.2118/2504-PA

Rossen WR (1996). Foams in Enhanced Oil Recovery, In: Foams: Theory Measurement and Applications, R.K. Prud'homme and S. Khan (Eds), Marcel Dekker, New York City.

Schramm LL (1994). Foams: Fundamentals and Applications in the Petroleum Industry. American Chemical Society: Washington, D.C. (1994). http://dx.doi.org/10.1021/ba-1994-0242

Tanzil D, Hirasaki GJ, Miller CA (2002). Conditions for Foam Generation in Homogeneous Porous Media. SPE 75176 presented at the 2002 SPE/DOE Symposium on Improved Oil Recovery, Tulsa, OK, 13-17 April.

Yan W, Miller C A, Hirasaki G J (2006). Foam Sweep in Fractures for Enhanced Oil Recovery. Colloids and Surfaces A: Physicochem. Eng. Aspects (282-283):348-359.

Permissions

List of Contributors

O. Obodeh
Mechanical Engineering Department, Ambrose Alli University, Ekpoma, Edo State, Nigeria

N. C. Akhere
Mechanical Engineering Department, Ambrose Alli University, Ekpoma, Edo State, Nigeria

J. D. Udonne
Department of Chemical and Polymer Engineering, Lagos State University, Lagos, Nigeria

A. Aboulkas
Laboratoire de Recherche sur la Réactivité des Matériaux et l'Optimisation des Procédés «REMATOP», Département de chimie, Faculté des Sciences Semlalia, Université Cadi Ayyad, BP 2390, 40001 Marrakech, Maroc, Morroco
Laboratoire Interdisciplinaire de Recherche en Sciences et Techniques, Faculté polydisciplinaire de Béni-Mellal, Université Sultan Moulay Slimane, BP 592, 23000 Béni-Mellal, Maroc, Morroco

K. El harfi
Laboratoire de Recherche sur la Réactivité des Matériaux et l'Optimisation des Procédés «REMATOP», Département de chimie, Faculté des Sciences Semlalia, Université Cadi Ayyad, BP 2390, 40001 Marrakech, Maroc, Morroco
Laboratoire Interdisciplinaire de Recherche en Sciences et Techniques, Faculté polydisciplinaire de Béni-Mellal, Université Sultan Moulay Slimane, BP 592, 23000 Béni-Mellal, Maroc, Morroco

M. Nadifiyine
Laboratoire de Recherche sur la Réactivité des Matériaux et l'Optimisation des Procédés «REMATOP», Département de chimie, Faculté des Sciences Semlalia, Université Cadi Ayyad, BP 2390, 40001 Marrakech, Maroc, Morroco

M. Benchanaa
Laboratoire de Recherche sur la Réactivité des Matériaux et l'Optimisation des Procédés «REMATOP», Département de chimie, Faculté des Sciences Semlalia, Université Cadi Ayyad, BP 2390, 40001 Marrakech, Maroc, Morroco

M. A. Ekpo
Department of Microbiology, University of Uyo, Uyo, Akwa Ibom State, Nigeria

A. J. Nkanang
Department of Microbiology, University of Uyo, Uyo, Akwa Ibom State, Nigeria

Mehrdad Alemi
Department of Petroleum Engineering, Amirkabir University of Technology, Tehran, Iran

Mansour Kalbasi
Department of Petroleum Engineering, Amirkabir University of Technology, Tehran, Iran
Department of Chemical Engineering, Amirkabir University of Technology, Tehran, Iran

Fariborz Rashidi
Department of Petroleum Engineering, Amirkabir University of Technology, Tehran, Iran
Department of Chemical Engineering, Amirkabir University of Technology, Tehran, Iran

J. D. Silva
University of Pernambuco-UPE, Polytechnic School in Recife, Rua Benfica - 455, Environmental and Energetic Technology Laboratory, Madalena, Cep: 50750-470, Recife - PE, Brazil

P. A Enikanselu
Department of Applied Geophysics, Federal University of Technology, P. M. B. 704, Akure, Ondo state, Nigeria

A. O Ojo
Department of Applied Geophysics, Federal University of Technology, P. M. B. 704, Akure, Ondo state, Nigeria

A. A Adepelumi
Department of Geology, Obafemi Awolowo University, Ile-Ife, Osun State, Nigeria

O. A Alao
Department of Geology, Obafemi Awolowo University, Ile-Ife, Osun State, Nigeria

T. F Kutemi
Department of Geology, Obafemi Awolowo University, Ile-Ife, Osun State, Nigeria

Liu Renqing
Daqing E&P Institute, Daqing, Heilongjiang, China

Akbar Mohammadi Doust
Department of Chemical Engineering, Faculty of Engineering, University of Sistan and Baluchestan, Zahedan, Iran

Farhad Shahraki
Department of Chemical Engineering, Faculty of Engineering, University of Sistan and Baluchestan, Zahedan, Iran

Jafar Sadeghi
Department of Chemical Engineering, Faculty of Engineering, University of Sistan and Baluchestan, Zahedan, Iran

A. D. I. Sulaiman
Petroleum Engineering Programme, Abubakar Tafawa Balewa University, Bauchi, Nigeria

A. J. Ajienka
Petroleum and Gas Engineering Department, University of Port-Harcourt, Nigeria

I. S. Sunday
Petroleum and Gas Engineering Department, University of Port-Harcourt, Nigeria

Mehdi Mohammad Salehi
Chemical Engineering Department, Sahand University of Technology, Tabriz, Iran

Eghbal Sahraei
Chemical Engineering Department, Sahand University of Technology, Tabriz, Iran

Seyyed Alireza Tabatabaei Nejad
Chemical Engineering Department, Sahand University of Technology, Tabriz, Iran

Dong Liu
State Key Laboratory of Offshore Oil Exploitation (CNOOC Research Institute), China

Wenlin Li
E&P Institute, Sinopec, Beijing, China 100101

Hongyan Wang
Geological and Scientific Research Institute, Shengli Oilfield, SINOPEC, Dongying 257015, PR China

Brian Miller
EnProTech, Houston, USA

E. B. Olanisebe
Department of Petroleum Engineering, University of Ibadan, Nigeria

S. O. Isehunwa
Department of Petroleum Engineering, University of Ibadan, Nigeria

C. O. C Oko
Department of Mechanical Engineering, University of Port Harcourt, Rivers State, Nigeria

O. E. Diemuodeke
Department of Mechanical Engineering, University of Port Harcourt, Rivers State, Nigeria

Seema Dhail
Department of Biotechnology, Manipal University, Jaipur-302004, India

Ehssan Mohamed Reda Nassef
Petrochemical Engineering Department, Faculty of Engineering, Pharos University, Canal El Mahmoudia Street, Beside Green Plaza Complex Alexandria, Egypt

Mina R. Shaker
Faculty of Petroleum and Mining Engineering, Suez University, Egypt

Shouhdi E. Shalaby
Faculty of Petroleum and Mining Engineering, Suez University, Egypt

Tarek I. Elkewidy
Faculty of Petroleum and Mining Engineering, Suez University, Egypt

Mohd. A. Kabir
United States Department of Energy, National Energy Technology Laboratory, Pittsburgh, PA 15236-0940, USA

Isaac K. Gamwo
United States Department of Energy, National Energy Technology Laboratory, Pittsburgh, PA 15236-0940, USA

Hikmat S. Al-Salim
Department of Chemical and Petroleum Engineering, UCSI University, Kuala Lumpur, Malaysia

Ahmmed Saadi Ibrahim
Department of Chemical and Petroleum Engineering, UCSI University, Kuala Lumpur, Malaysia

Mohammed S. Saleem
Chemical Department, University of Mosul, Baghdad, Iraq

FALODE Olugbenga Adebanjo
Department of Petroleum Engineering, University of Ibadan, Ibadan, Oyo State, Nigeria

OJUMOOLA Olusegun
Department of Petroleum Engineering, University of Ibadan, Ibadan, Oyo State, Nigeria

www.ingramcontent.com/pod-product-compliance
Lightning Source LLC
Chambersburg PA
CBHW050445200326
41458CB00014B/5072